W9-AQT-639

Sex, Gender, and the Politics of *ERA*

SEX, GENDER, AND THE POLITICS OF *ERA*

A State and the Nation

DONALD G. MATHEWS
JANE SHERRON DE HART

New York Oxford
OXFORD UNIVERSITY PRESS
1990

Oxford University Press

Oxford New York Toronto
Delhi Bombay Calcutta Madras Karachi
Petaling Jaya Singapore Hong Kong Tokyo
Nairobi Dar es Salaam Cape Town
Melbourne Auckland

and associated companies in
Berlin Ibadan

Published by Oxford University Press, Inc.,
200 Madison Avenue, New York, New York 10016

Library of Congress Cataloging-in-Publication Data
Mathews, Donald G.
Sex, gender, and the politics of ERA :
A state and the nation /
Donald G. Mathews, Jane Sherron De Hart.
p. cm. ISBN 0-19-503858-4
1. Women's rights—North Carolina.
2. Equal rights amendments—North Carolina.
3. Feminism—North Carolina.
I. De Hart, Jane Sherron.
II. Title. HQ1236.5.U6M38 1990
305.42'09756—dc20 90-7174

Photographs courtesy of the Raleigh *News and Observer* Collection,
North Carolina State Archives.

2 4 6 8 9 7 5 3 1

Printed in the United States of America
on acid-free paper

To the Memory of Alice Elaine Mathews

To Ruby Sherron De Hart

Preface

1. Equality of rights under the law shall not be denied or abridged by the United States or by any State on account of sex.
2. The Congress shall have the power to enforce, by appropriate legislation, the provisions of this article.
3. This amendment shall take effect two years after the date of ratification.

COMPLETE TEXT OF THE EQUAL RIGHTS AMENDMENT

"E. R. A. WON'T GO AWAY!"

The words were unfurled on banners at countless marches. They were chanted at rallies and translated from public display into private conviction by people who wanted to forbid the government to deny citizens their rights "on account of sex." The chant prefaced angry denunciations of state legislators for betraying the popular will by rejecting the Equal Rights Amendment (ERA). Public opinion polls had, after all, indicated that most Americans favored sexual equality. The chant also dramatized a defiant pledge to renew the struggle after the deadline for adding the amendment to the U.S. Constitution—June 30, 1982—had passed in defeat. To fulfill the pledge, supporters reintroduced the amendment in Congress the following year, and they failed once again. The majority they won on Capitol Hill was not large enough to return the amendment to the states for action. Political realists were thus forced to concede that ERA might indeed "go away"—for awhile. Whether there would be a second coming was a matter of faith; and the faithful remained defiant. Even after eight years of an administration hostile to legislative action on behalf of equality, the nation's largest advocacy group for women continued to call for "*a constitutional guarantee* [*of equality*] *so fundamental, so clear, that no court, no Congress, no President can ever again abridge it.*"[1]

The chant, therefore, may be—as articles of faith frequently are—true. That ERA should remain a priority of activist women on the eve of the twenty-first century after having been introduced by women born in the nineteenth century suggests indeed that it will not go away. To transform it into a viable political option, however, will require perspective as well as persistence. With the following account of intense drama, deep convictions, ordinary politics, and extraordinary personalities, we hope to contribute to that perspective.

This is not the first book to discuss ERA. Writing in 1977–78, after the last state had ratified but before the deadline for ratification had expired, Janet K. Boles devoted *The Politics of the Equal Rights Amendment* to an examination of political decision making under conditions of intense conflict. Observing that lawmakers dislike issues generating such conditions, Boles argues that legislators in unratifying states voted against the conflict over ERA rather than against the amendment itself. In *Constitutional Inequality,* Gilbert Y. Steiner argues that the opportunity for ratification existed only between 1971 and 1973. Thereafter, success was made difficult by three developments: the association of the amendment with abortion, the enhanced saliency of the draft as an issue in the wake of the Soviet invasion of Afghanistan, and the increased prestige of U.S. Senator Samuel James Ervin, Jr., following the Watergate hearings. Steiner argues that decisions by the Supreme Court in 1980 (*Harris* v. *McRae*) and 1981 (*Rostker* v. *Goldberg*) enabled legislators credibly to connect ratification with strengthening abortion rights and the likelihood of drafting women. The opposition of traditionalist women is discounted because they were ignorant. Citing changes in judicial practice regarding alimony and child support, Steiner concludes that "any woman who by the end of the 1970s still believed marriage to mean a permanent guarantee of support would have been so out of touch with reality as to cast doubt on her credibility as a lobbyist."[2]

The legal historian Mary Frances Berry also discusses "why ERA failed" in a book of the same title. Identifying elements required for ratification, she convincingly argues that supporters failed to recognize that the consensus so carefully constructed to achieve passage in Congress should have been replicated on a state-by-state basis. Proponents failed to do so; they began the ratification process without the resources and strategy necessary to make a compelling case that sex-based discrimination in the law was an urgent problem solved *only* through an amendment to the Constitution. As the ratification process was drawn out, the case for amendment became harder to make because some states successfully implemented their own ERAs and the Supreme Court moved toward a more rigorous standard of judicial review. The initial failure of advocates to counter objections by opponents, Berry argues, also enabled the latter to divert the issue from one of legal equality to one of sex role. Jane J. Mansbridge also addresses the question of failure in her prizewinning analysis of *Why We Lost the ERA*. She agrees that the Supreme Court's movement toward stricter scrutiny under the equal protection clause of the Fourteenth Amendment weakened the case for ERA. She also points out that support for ratification had actually diminished by 1982 and provides compelling evidence that public support was weak from the start. Poll data indicating initial support were deceptive because they represented popular endorsement of equality in the abstract rather than real changes in the roles and behavior of both sexes. Believing that ratification would have strengthened women's position, Mansbridge correctly argues that women on both sides of the controversy exaggerated its immediate effects. Hers is also an impressive analysis of the predictable tensions between ideological purity and political realism in social movements.[3]

Our own study complements these books. It "began" in 1973 when a North Carolina legislator asked if opponents had warned that woman suffrage would "destroy the family." It "ended" in Ohio in 1987 with an interview of a former Washington-based activist who had worked in North Carolina during the final battles for ratification. Now a state legislator, she had just returned from the Ohio Governor's Mansion, where she had been honored with a baby shower. Life went on after ratification failed, in both the political and the personal spheres, but the struggle for ERA had politicized women as had no other issue since suffrage. The nature of that politicization was especially evident in those important unratifying states where the length and intensity of conflict revealed a complex mixture of emotions, convictions, and values that made of the ratification process something far more complicated than a referendum on equal rights for women.

Focusing our study on North Carolina provides an excellent opportunity to analyze the ratification process in depth. The state's Division of Archives and History—one of the best historical resources in the nation—has extensive legislative papers on ERA from 1972 through 1982. In addition, the Southern Historical Collection at the University of North Carolina at Chapel Hill contains the papers of the late Samuel James Ervin, Jr., the amendment's chief foe in the U.S. Senate and an architect of antiratificationist strategy. These papers, supplemented by those in the possession of individual activists and soon to be deposited at the University of North Carolina at Greensboro, constitute the kind of sources upon which historians traditionally rely. We supplemented written materials with participant observation at rallies, marches, retreats, prayer vigils, and legislative hearings. Moving beyond North Carolina borders, an associate reported on the 1977 National Women's Conference in Houston that supported ERA; another attended a profamily conference in Washington, D.C.

Focusing on North Carolina also defines a reasonable scope within which to conduct extensive interviews. Subjects included political experts, legislators, activists in both state and nation, and participants in female networks throughout the state. The interviews varied in function. Some helped us assess events in the immediate aftermath before details were lost. Others helped us see various linkages that explained what ERA meant. All helped answer questions about differnces between antagonists and suggested, by implication, the need to analyze nuances of belief in the self-understanding of observers. To be sure, analysis of constituent mail helped explore those "goes-without-saying" assumptions and personally confirmed meanings that help shape collective consciousness, but interviews were our most important sources. They helped us interpret the surprisingly personal statements in letters to public figures as well as public positions that seemed more symbolic than substantive. We knew from studies of ERA in scholarly journals[4] that we should reject familiar stereotypes of "libbers" bent on destroying the family and of "angry housewives" rushing into the political arena to defend it. But we had to learn how to approach partisans on both sides. Ratificationists were amenable to any approach interviewers chose; antiratificationists were not. Initially we opted for relatively structured interviews designed to elicit a broad range of information with opportunity at the end for comments on matters not covered. What worked with advocates failed with opponents. Sometimes reluctant to be interviewed, opponents frequently spoke elusively of *how* they had worked, preferring instead to reiterate standard arguments against ERA. Candor about our own position (we supported ratification), assurance of scholarly intention (we wanted to understand the actions of *both* parties), promise of confidentiality (we would use quotations without identifying sources if requested)—nothing we said generated opponent interviews comparable to those provided by ratificationists. Troubled by the asymmetry of the responses, we settled for a less-structured format with antiratificationists and simply asked their help in understanding why they opposed ERA. They obliged us by telling their own stories—not ours. They provided patterns of meaning and coherence that otherwise might not have been understood. They helped us see that what ratificationists called ignorance or irrationality could actually provide insight into the meaning of action.

These exchanges helped us see that the battle for the Equal Rights Amendment had important psychological ramifications, especially among those who believed that sexual distinctiveness dictated rules and expectations so appropriately fitted to women that attempts to ignore it in the law were dangerous. That feeling had to be not simply reported but respected. Our objective was, after all, to produce an explanation of the ERA conflict that, if not stating the case for or against ratification precisely the way the parties involved would have done so, would nonetheless be consistent with their positions. An approach long congenial to most historians and anthropologists, this interpretive mode has recently gained favor among social scientists. It can, as Bruce Jennings argues, provide a critically important tool to policy analysts.[5] As public policy is formulated and evaluated, it is essential to be able to identify why certain policy initiatives fail, why significant groups may not respond as it is assumed that they will. With respect to ERA, for example, why did women oppose a measure designed to benefit women? The question, however, should not be taken to imply that our work is designed merely as an explanation of why ERA failed; it failed essentially because of timing and the fact that a constitutional amendment requires not only a two-thirds majority in each house of Congress but also a majority vote of over 50 percent (three-fifths in Illinois) in the legislatures of three-fourths of the states. Ratification of the amendment would not have changed the value of our study because the book is not about political calculation, but about meaning.

In following the rubric of empathic listening, we came to the conclusion that the ERA controversy was over exactly what opponent women said it was—sex. And sex meant something much broader and more complicated than physical distinctiveness. The attempt to distinguish between that distinctiveness and the way in which people had inferred from it patterns of behavior was alien to the way in which opponents understood sex. The social and biological sciences through which feminists could lay open the intricacies of power relations, physical distinctiveness, and gender definitions—while conceded, perhaps, an appropriate place in the academy—were rejected by opponent women as the means of understanding human behavior. The modern temper of breaking down everything from society and language to organisms into their constituent parts for analysis was not as "liberating" to opponents as it was to supporters. The distinctively antimodern embrace—on the opponents' part—of the holistic meaning of human distinctiveness made sex the basis for a broad range of personal and social obligations. This freighted "sex" with such tremendous emotional and social implications that it functioned as a religious concept; it linked personal and social identities with a feeling of ultimate obligation and destiny. The perception on the part of anti-ERA women that proponents had repudiated womanhood in order to be men was born of the former's commitment to traditional obligations defined at birth (i.e., sex) that they believed to be part of the essential order of creation. We soon learned that we were dealing with two different ways of understanding the universe. One was comfortable with modern social analysis and challenges to tradition; the other believed such critiques eroded the foundations of self and community.

It is this lesson, and the fact that arguments used in North Carolina were the same as those used throughout the nation, that makes this a national study. The issues were not Southern issues; the same arguments were used in Massachusetts, for example. If murmurs about busing and desegration suggest a distinctly Southern accent, one has only to turn to the Bay State's debate over its own ERA to read the same complaints about the amendment and the Department of Health, Education, and Welfare. The intersection of individual activists, locality, state, and nation was obvious throughout the ratification struggle. Proponents thought of themselves as part of a general social trend; their ideas, values, expectations, and sense of themselves were parts of a broad ideology that knew no sectional or state boundaries. The television that brought them welcome news of the women's movement evoked unwelcome responses from legislative opponents; the airplanes that brought travelers from Japan to do business brought Phyllis Schlafly from Illinois to testify. The "literature" of both sides was printed in California, New York, and Texas. The expectations of outsiders looking to North Carolina while state legislators voted on ratification felt no doubt that North Carolina was the nation. That ratification failed in the state legislature was a function of state politics, to be sure. And to the extent that the narrow defeats in the state mirrored rather than replicated the slim majority throughout the nation, the focus on North Carolina is state defined. But we are ultimately not trying to explain why ERA failed so much as what the debate surrounding the issue tells us about the women's movement and its place in this country during the 1970s. Understanding that debate helps us also to understand the appeal of a presidential candidate in the 1980s who campaigned against feminism and the movement culture out of which it came.

We begin our study with the struggle for woman suffrage because it helps us understand the dynamics of gender politics in the public arena. The intricate and subtle fusion of sex, gender, race, and power that underlies resistance to change in 1990 should be understood as the persistence of themes which were evident in the fight for equal suffrage but are hidden to popular history because women's issues were so seldom a subject of policy debates in the postsuffrage decades. Gender, power, race, and change were at the crux of the ratification struggle of the 1970s, just as they were at the crux of the suffrage struggle in the early twentieth century. If, as Mary Berry contends, fifty years after suffrage ERA supporters were so unaware of their own history that they had no knowledge of the requirements for successful ratification of a controversial amendment, we would argue that they lacked other knowledge as well. As Carroll Smith-Rosenberg and other historians have demonstrated, history recounts social and political conflicts that illustrate the intricate linkages among biological categories, gender definitions, social cohesion, and moral order.[6] Understanding these linkages is critical to understanding what happened to ERA.

Discourse provides the crucial key to understanding. By discourse we mean not simply language—nonverbal as well as verbal—but the meanings, relationships, and connections associated with it for both speakers and listeners. Politicians have always known how to say something without saying it, using "loaded" words, gestures, and pauses symbolic of a range of meanings. The most evocative public speaker or partisan writer knows how to elicit responses from audiences with implicit as well as explicit language. It is the analysis of discourse that allows us to explore contested cultural and political terrain as well as the symbolic landmarks identifying it in ways that add an important new dimension to current perceptions of the ERA conflict. What at first glance appears to have been a legislative battle over ratification of a constitutional amendment emerges as something more. It was more than community conflict in which political decisions were made simply to avoid increasingly rancorous debate. It was more than symbolic conflict in which groups jockeyed for the legitimation conferred by having their own values inscribed in the Constitution—although it was most certainly that, too.

Essentially, the conflict was about gender. The amendment's focus was "on account of sex." And "sex" was the activating word even though it was misleading. As one North Carolina activist recalled, "The debate on the Equal Rights Amendment was about gender; people got excited about it because they thought about it as 'sex.' "[7] "Gender" is the meaning our society and culture attaches to sexual differences. "Sex" is the biological distinction between male and female that—at least before the performance of sex-change operations and chromosome counts in East German female athletes—was thought to be natural and immutable. "Sex" is also an obsession in American life, whether in entertainment, politics, religion, or commerce. While "sex" was the word written into the amendment and the word opponents used to give their position an explicit concreteness, the issue of "gender" suffuses ratification discourse. This issue transformed the contest for ratification from the politics of sex into the politics of gender.

The organization of our text is both topical and chronological. The first chapter tells the story of woman suffrage; the second spans the time between enfranchisement and congressional passage of the Equal Rights Amendment in 1972. It focuses on the discovery by contemporary feminists of a way to justify ERA and contrasts this discovery with the pivotal response by the amendment's most effective critic in the U.S. Senate, Sam Ervin. The third and fourth chapters tell the story of mobilization by women lobbying for and against ratification; with chapter 4 the chronological narrative ends. The next three chapters explore what the amendment meant to pro- and anti-ERA women and to the North Carolina General Assembly. Social conflict is never a clear story line told from one vantage alone; it is always many stories. The parties to conflict never fully understand their opponents; they rarely agree on the issues, or at least on precisely what the issue is. The stories never begin at the same time; nor do they end simultaneously. It is to be hoped, however, that they are all interesting and that, from the many stories told here, readers will fashion their own understanding of the Equal Rights Amendment in the 1970s.

Because this book is about women and politicians, its authors owe both a special debt for their help in writing it. We began research when the outcome of the ratification process was still uncertain, a situation that helped us understand the intensity of feeling involved in the issue and appreciate the authenticity of that feeling for partisans on both sides. North Carolina ratificationists—activists and legislators alike—not only consented to be interviewed but placed at our disposal legislative papers, personal correspondence, organizational material, and press clippings. That they did so three years before the deadline for ratification of the Equal Rights Amendment reveals a sobering faith in the integrity of historical research and the people who do it. Former North Carolina Senator (now Justice) Willis P. Whichard and North Carolina Representative George W. Miller, both of Durham, were especially generous on this score, as were campaign activists who literally weighed us down with cardboard boxes of material, among them: Wilma Davidson, Nancy Dawson-Sauser (formerly Drum), Florence Glasser, Tennala Gross, Elisabeth S. Petersen, and Miriam Slifkin. Former Speaker of the North Carolina House of Representatives Carl Stewart provided valuable counsel on interview strategy. Former ERAmerica representatives Joan McLean and her North Carolina co-worker Kay Sebian provided bed and board as well as information and insight. Anti-ERA activists such as Toni Suiter, Alice Wynn Gatsis, and the Reverend Kent Kelly, to name only three, and anti-ERA legislators such as Marshall Rauch and Benjamin D. Schwartz also opened their doors to investigators as acts of faith. Other state senators such as Wilma Woodard and

Russell Walker provided impromptu consultations that both enhanced our understanding of North Carolina politics and saved us from the folly of our own personal bias. These individuals are but a few of those who interrupted busy schedules to talk with us, for we actually participated in more than one hundred conversations, listed in appendix 4.

The Center for the American Woman and Politics at Rutgers University, New Brunswick, New Jersey, and its director, Ruth Mandel, provided critical financial and psychological support in the early stages of this project as well as important advice upon its completion; we are especially grateful also to Marilyn Johnson, who helped at an especially difficult time. As co-recipient (with Jane De Hart) of a center grant to study the mobilization of North Carolina women in the ratification campaigns of 1973, 1975, and 1977, Roxie Nicholson contributed substantially to the work on the organization and leadership of the pro- and anti-ERA coalitions in these campaigns. The energy and insight that she devoted to this initial phase of the project as well as the effectiveness of her interviews with a number of activists, especially opponents, entitles her to co-authorship of sections of chapter 3. John Scott Strickland, Harlan Gradin, William Auman, and Jennifer Lettieri also contributed valuable research and critical assistance, bringing their impressive intelligence to what were often less than exciting tasks. The University of North Carolina at Greensboro, where the study was initiated, and the University of North Carolina at Chapel Hill, where it was completed, each provided a semester's leave of absence; the latter also defrayed a portion of expenses incurred, as did the Duke-UNC Center for Research on Women. The Institute for Research in Social Science, the Institute of Government, the Southern Historical Collection, and the North Carolina Collection—all at the University of North Carolina at Chapel Hill—are graced with superb staffs that make them hospitable places to do research. This was equally true of the North Carolina State Archives in Raleigh and the Schlesinger Library at Radcliffe College, Cambridge, Massachusetts. Especially valuable were the Schlesinger Library's rich and as yet uncataloged papers of the National Organization for Women (NOW), to which Eleanor Smeal and Alice Cohan graciously arranged special access. The Henry A. Murray Research Center, also at Radcliffe, made available to one of the authors ideal quarters for writing over the course of two summers.

Any book, especially one so long in the making as this, bears the imprint of scholars too numerous to name but who nonetheless helped to shape the final product. We are grateful to participants in the 1978 Berkshire Conference on Women's History who commented on an early statement from this project after it was presented as the keynote address. Members of the Homewood Seminar in American Religious History at the Johns Hopkins University, Baltimore, Maryland, helped refine analysis on religious opposition to ERA, and colleagues in the Social History Workshop in North Carolina criticized the project in its early stages; so did members of a graduate history seminar in the University of North Carolina at Greensboro. Participants in the 1983 Seminar on Cultural Constructions of Gender at Brown University, Providence, Rhode Island, were especially helpful. Joan Wallach Scott, Donald Moore Scott, and Mary Frances Berry responded to preliminary papers with their customary acumen. Jacquelyn Dowd Hall, Joan Hoff, Linda K. Kerber, and Jane J. Mansbridge also made helpful comments on scholarly papers and drafts of individual chapters. William H. Chafe, Cynthia Harrison, Linda K. Kerber, Alice Kessler-Harris, Arthur S. Link, Frances Fox Piven, Dorothy Ross, and Harry Watson read the penultimate draft. The criticism of each nicely complemented that of the others, bringing a range of insights and perspectives that enhanced the book. In committing valuable time to our project they demonstrated the communal nature of scholarship and writing. If they have not saved us from embarrassing mistakes, imprecise language, murky allusions, and idiosyncratic glitches, it was not for want of trying. For remnants of this list that remain, we bear responsibility.

Nancy Lane at Oxford University Press has been supportive of this book since its inception, believing in its completion during times when we found it difficult to do so. We are grateful for her sustained counsel, support, and friendship. As copyeditor, Ann Hofstra Grogg displayed not only the extraordinary care and precision that is characteristic of her work but also a welcome tolerance and sympathetic understanding of the constraints on our time that we especially appreciate.

Elizabeth Farrior Buford and Benjamin J. Cohen lent invaluable support to this enterprise, and most especially its completion, with precisely the right blend of patience and impatience. They know best the significance of their contribution and the depth of our gratitude.

We are also grateful to each other. Contributing complementary strengths and tasks to the common enterprise, we developed the argument, which bears our equal imprint, as if there were but one author. Because there were two authors, there were difficulties as well as rewards in the creative process, but we were able to surmount them through a common commitment to the book. The result we hope conveys the extent to which sex, gender, and ERA were not only scholarly subjects, but also issues that inspired politics and enriched lives.

Raleigh, N.C. D.G.M.
Chapel Hill, N.C. J.S. De H.
February 1990

Contents

Sex, Gender, and the Politics of *ERA*

1

Parable: Woman Suffrage and "Manhood" in North Carolina

Bill Whichard was pleased. The youthful representative of the Eighteenth District from Durham had just watched the North Carolina Senate ratify the Woman Suffrage Amendment to the Constitution of the United States: 44–0. He had introduced the bill of ratification almost a month earlier in the North Carolina House of Representatives, where it had passed by a vote of 92–0 amid generous comments about the many "contributions" of women to the Tar Heel State. The language of ratification betrayed traces of condescending gender etiquette while conceding the importance of women's rights. But it was not the wording of the resolution that was so disconcerting as its timing, for ratification had been introduced into the North Carolina General Assembly on August 13, 1920, and it was now May 6, 1971.

Whichard had tried to make a symbolic point while making amends for the past. He was to learn, however, that the values preventing ratification of woman suffrage in the North Carolina of 1920 were not dead. Then, Tar Heel legislators had refused to act out of anger at women's participation in public life, motivated by complex psychosexual and racial anxieties as well as by simple political calculation. Conservatives renounced the good-humored hypocrisy of exaggerated compliment to rail at threats to traditional definitions of private self and public sphere. That the emotion thus expended revealed something more complicated than access to the ballot box is suggested in the common acknowledgment that woman suffrage was part of the long process that had birthed a "New Woman." In asking for the ballot as if it were her "right," she had challenged the very definition of "sex" because voting was a public act identified with the "virile manhood" of the state. Because her enfranchisement did not debauch the New Woman or destroy the American family, it is easy to dismiss her opponents who warned these things would happen. That their values persisted was evident in the 1970s when their preachments were repeated in attacking a new generation of feminists who also wanted to amend the U.S. Constitution. Suffragists and partisans for ERA were linked through the common goal of equality and the elemental anxiety that this goal aroused among many of both sexes. That Whichard should have been both the successful sponser of woman suffrage (1971) and the unsuccessful sponsor of the Equal Rights Amendment (1973, 1975) is not surprising. He understood the historical linkage that draws us from the 1970s and 1980s back to the early years of the twentieth century and the fight for woman suffrage.[1]

Birthing a Female Public

The first achievement of the enfranchising process in North Carolina was a public female constituency that developed after the Civil War. Before secession North Carolina women, like their sisters throughout the South, had been distant observers of Northern feminism, which was identified with the radical wing of abolitionism. Slavery had held Southern women back, North Carolina suffragists believed, because control of slaves dictated control also of white women. Slaveholding had dulled men's sensitivity to individual freedom and created the hypocritical "chivalric ideal" that had chained women to a pedestal. The ideal had not disappeared with abolition. It lingered in laws that restricted married women's autonomy, but it had been weakened. In the following generation, women slipped out from under the restraints of common law and domesticity with ever-greater frequency to create their own public life, to do what the Methodists among them called "woman's work for women." The phrase was traditional enough—"woman's work"—to reassure men who feared the growing independence of women. It was ambiguous enough—"for women"—to bond women who would never have dreamed of transforming "duty" into "right" together with those who were ready to do precisely that. "Woman's work" indicated a process of unfolding possibilities realized through organizations that provided frameworks of social support, networks of communication, and a style of public discourse.[2] When changes in work and technology made possible the independence of women and revealed new opportunities for "usefulness," women used the the traditional language of religion and domestic obligation to justify changing the character and scope of their social role. In doing "woman's work" they were simply "improving" their traditional roles as wife and mother. For conservatives the frustrating thing about this formula was that women's activism was not exactly what activist women said it was. Unmarried women—of whom there were more than a few in the new women's organizations—were obviously not perfecting traditional roles. Instead of becoming wives and mothers they were becoming activists. Other women who had been both wives and mothers were obviously not behaving as men had been told to expect wives and mothers to behave. They were going to meetings, passing resolutions, setting up agendas, and appointing committees. The future was open, reported a group of Methodist women in 1920 who quoted Christ Jesus approvingly: "Behold, I have set before thee an open door and no man can shut it."[3]

In the vanguard of opening doors were women's clubs. Sallie Southall Cotten of Pitt County in eastern North Carolina presided over the first state conference of club women in 1902. For more than a decade she and they had been preparing for that day. Like the church women, they "avoided politics"; that is, they did not lobby for woman suffrage—at least not at first. Their disclaimer was part of a litany of reassurance that Cotten intoned for years. No—she was not going to claim men's prerogatives—but she never said precisely what they were. No—she was not interested in wearing bloomers—that was a personal choice. Yes—women had a special calling to do things that men would not do. Her personal dignity in the face of questions from men who needed to be told that she would not invade "their world" carried her through many awkward moments.[4] She understood the pressures upon women not to act as she did. She was sensitive to the almost nauseating tension experienced by women if it was even obliquely suggested that they were suffragists. Acquaintances who attempted to recruit women to "beautify" classrooms for schoolchildren had frequently found themselves in the most delicate of social situations. They patiently assured women that such activity was natural. To be sure, the home was the primary wifely responsibility, but even the most devoted housekeeper needed a change in routine. The change would not compete with religion—this was an especially important observation—because caring for children was the "handmaiden to church extension." The negotiations seemed always to turn on a familiar theme: "Woman can do much in a modest and suggestive way without seeming to encroach on man's prerogatives."[5]

Nonetheless, a public constituency of women began to take shape, and in 1904 the Raleigh *News and Observer* devoted an entire edition to it. As with discourse usually designed to sanction gradual change, that of the special edition was a mixture of tradition and innovation. Sallie Cotten celebrated the home as a joint creation through reciprocal "patience and self-sacrifice" of both husband and wife. Other writers applauded the creativity of Sunday school teachers, religious writers, and missionaries whose moral vision spilled over into the "world." Their energy seemed to guarantee the civilizing, uplifting, cleansing, and reforming of a world ruled by men. If the essayists meant to share their moral code with others, it was out of a politics not of resentment but of hope. The few women invading the business and professional world were touted as representatives of a new generation building upon respected predecessors. A Raleigh physician, for example, praised the female doctors whose example had led her into medicine. Her tone, like that of her collaborators, was positive, because, wrote the enthusiastic editor, "Every woman is now reaching out and striving after that which she believes to be a higher and better condition."[6]

Reaching out was associated with public education. During the 1890s women schoolteachers began to address problems that they faced as women as well as teachers. This dual concern was natural in a state where concessions to educational needs were extracted from the legislature with great effort. The fact that education was associated with women did not enhance its value as a public resource until teachers made common cause in 1912 with the North Carolina Federation of Women's Clubs in a "revolt," as Edith Royster called it, against "irritating restraints" and "disabilities." An assistant superintendent of Wake County schools and outspoken member of the North Carolina Teachers Assembly, Royster had been charged by both the federation and assembly to report on the legal status of women in the state. Simultaneously both organizations petitioned for laws to allow women to help make policy in public education. Legislation was not all they wanted, but it was all they asked at the time in hopes of stretching the traditionalists' concession of a nurturing role to women into an argument for greater public responsibility. The organizations did not intend to reshape society any more than they intended for women to remain captive in traditional roles. They were simply trying to get things done that needed doing and in the process to emphasize the great loss to North Carolina resulting from "legal limitations to women's services in this state."[7]

If the organizations could not get the General Assembly to relinquish its view of women, they could and did get it to allow them to sit on school boards. To activists, the law to put them there was no more important than the lobbying process itself; it had mobilized them. Club women who met in May 1913 were encouraged by the words of their genteel founder, who promised that women were about to assume "power and responsibility." The fight to sit on school boards had made Cotten bolder. The report to the federation on the status of women revealed the penalties exacted by an archaic legal system. What could she have been thinking as North Carolina's chief justice, Walter Clark, urged his audience to their next victory—woman suffrage? What had happened to her reassurances to uncomprehending men? Her diary is suggestive. As she worked with Edith Royster in the study of discrimination, Cotten began to confront the frustrating fact of her political helplessness. "We can never serve there," Cotten had written about women on school boards, "until we are enfranchised." Political helplessness was now more than an annoyance; it was unfair, it was injust. As it rained, she had thought about her frustrations and hopes: "Must we all become suffragists?"[8]

The question was unsettling. So was the answer.

Pure Woman and Virile Men: Sexual Symbolism and Power

As North Carolina women stretched toward an increasingly sophisticated understanding of their position, they were confronted by a threat from within the troubled Democratic party. The issue was power. The time was 1898. The pawns were white women and black men. The weapons were the hysterics of pathological racism and the symbolic purity of white women. The latter were transformed by Democratic rhetoric into potential victims of black rapists, the symbol of helpless Carolina in the hands of Republicans and Populists after the election of 1896. To dislodge them from power, the Democratic party leadership joined white supremacist Democrats throughout the South in sounding the alarm against "Negro Domination!"[9] The few blacks who had reeceived minor state offices after 1896 scarcely lent empirical support to such charges. But Furnifold Simmons, the state party chair, insisted that where public officials were susceptible to political pressure from blacks, they were insensitive to the needs of civilized whites. Throughout Western culture this racist celebration of white "civilization" against the barbarism of "lesser breeds without the law" captured the imagination of elites. It is not surprising, therefore, that editors and preachers should have joined politicians in becoming evangelists of white supremacy.

With Simmons and the Democrats, press and pulpit attacked universal suffrage as a prelude to "sanctionized mongrelization" of the races. They called upon white men to "assert their manhood." "The white womanhood of North Carolina," they cried, expect white men to "do their duty." To prove this, White Government Unions enlisted young women as "cheerleaders" in a deadly contest. "The white good women of North Carolina," wrote one excited man, "are unusually aroused." They have been doing "all in their power to arouse the White Men." The result of this "arousing" was the "Revolution of 1898." For more than a generation afterward, the Democratic party of North Carolina would proudly fly the flag of white supremacy that Simmons and his storm troopers raised. Young party stalwarts would remember with nostalgic pride the night they bludgeoned black people into political oblivion. They could recall the "agony" of being forced to discount more than one hundred thousand black votes in order to protect white women's virtue, their own "honor, and our Christian civilization." To emphasize what this "protection" entailed, Democratic vigilantes in Wilmington murdered fourteen protesting blacks, wounded and banished others, and forced the administration of the port city to resign. It was an efficient, broadly symbolic coup d'etat. Just as efficient was the campaign to disfranchize blacks in 1900. The politicans to whom suffragists would appeal, therefore, had come to power a little more than a decade earlier by killing people, restricting suffrage, and identifying successful politics with arrogant masculinity and terror.

The political role of women in the Revolution of 1898 was as a symbol to justify such acts. Their own interests and sensibilities were suppressed, a situation that was altered slightly in the fight for prohibition. When temperance forces entered politics in the 1880s, the Democratic party had opposed them. Women had wanted to prohibit use of beverage alcohol to protect themselves and their children from the abuse and negligence of men who retreated from their homes to the male companionship of the saloon. At first Democratic leaders believed the agitation to be dangerous because it created conflict on issues they themselves had not defined. If people other than party leaders could dictate issues, dissidents might possibly gain power through the liquor question. After a generation of failing to silence antiliquor forces, they switched sides and dried up the state for the sake of the Revolution of 1898. By 1908 Democratic politicians and the Anti-Saloon League could submit a referendum to the voters making it illegal to sell or manufacture intoxicating liquors in the state. Rhetoric so useful before was spat out once again: "The Rum-crazed Negro" became a menacing symbol, free to drink himself into a licentious rage and "prey upon our property, our homes, and to wreak ruin upon the most helpless members of our households." Unlike their role in the Revolution of 1898, women served as more than evocative symbols. They spoke at rallies and paraded under banners and at last surrounded the polls to plead, cajole, pray, chant, and sing for prohibition. Their children, too, joined to help strengthen—or perhaps weaken—their fathers' resolve. By a vote of 113,600–69,400, the white men of North Carolina redeemed the state from Demon Rum as they had redeemed it from Negro Domination.[10]

The logic of women in politics was by 1908 very clear indeed. They were allowed there only so long as they served the interests of Democratic party leaders and behaved as if their sex was the most important thing about them. They had to be willing to be categorized by how different they were from men rather than by the ways in which they were similar. This physical difference was freighted with cultural meanings shared with other Southern states as well as with the North and Great Britain—the purity of the race, the moral integrity of society, the leadership of the future. Gender difference was a potent source of political possibilities, where racist fantasies could be linked with the sexual overtones of oppression to transform the sexual difference between men and women into a symbol of white manhood. The helplessness, innocence, and susceptibility associated in men's minds with the concreteness of anatomical difference between male and female demanded that men act in certain ways, to protect certain things, for certain reasons if they were to be taken seriously as "virile" men. If women should behave in ways that were perceived to challenge sexual distinctiveness, the personal and political definition of many men would be challenged.

Enter Woman Suffrage

Women thus had to act within a political system that manipulated them as puppets in a morality play men had written. Although their morality was linked with gender rather than political vision, the association was nonetheless finding form. Indeed, women were beginning to believe that their gender demanded they participate in politics. In 1913 a group of middle-class white women and men met in Charlotte to form an equal suffrage association and apply to the National American Woman Suffrage Association (NAWSA) for membership. The social process that had brought them to Charlotte wed their own experience with a broader trend outside the state as the equal suffrage movement gathered momentum with majorities in Washington (1910), California (1911), and Illinois (1911). Activist North Carolina women came to suffrage through their intensifying sense of female solidarity against men who had ignored their wishes in education and the law. They now intended to challenge their status on issues ranging from the age of consent to policy decisions. They were fed up with politicians who did not take them seriously.[11]

The idea that North Carolina women should participate in politics on their own behalf was so outrageous that even North Carolina's reforming Progressives dared not damage their appeal in the People's Platform of 1914 by including woman suffrage. And, when later in the same year one Progressive actually did attempt to introduce woman suffrage into the Democratic party platform, he was almost literally laughed out of the convention. The women, however, were not laughing. By March 1914 seventeen equal suffrage associations sent delegates to Charlotte to elect officers: Barbara Bynum Henderson of Chapel Hill was elected president; her sister, Suzanne Bynum, was in charge of publicity; and her sister-in-law, Mary Henderson, was in charge of lobbying the legislature. They submitted three bills to the General Assembly of 1915. The first was a bill making women eligible to be notary publics. The second was a bill to raise the age of consent from fourteen to twenty, and the third was an equal suffrage bill. Suffragists expected no miracles. The fall election had given the new General Assembly a conservative mandate by rejecting constitutional amendments to reform education, state government, and a regressive tax system.[12]

In February 1915 before a joint committee of the General Assembly, Barbara Henderson assumed a role better suited to a magician. She had to persuade the heirs of the Revolution of 1898 to think of women as something other than the sex symbols of white supremacy. She began her statement to the sons and nephews of Confederate veterans by quoting from the Gettysburg Address. "I am speaking to you today," she began, "on the assumption that you are democrats, that is that you believe in a government of the people, by the people, for the people . . . , the right of every human being to have some voice in the laws which govern him." If men continued to withhold suffrage from women, she observed, the motto of government should be: "of the men, by the men, and for the men." The words were not appealing, perhaps, but they were more appropriate. Since her audience could not think about women without thinking about sex, she met Southern chivalry head on. The protection associated with it was identified with a woman's physical distinctiveness, but Henderson shifted "protection" from sex to politics and argued that women should be enabled to "protect" themselves. Her implication was that men would not always do so. Without the vote women were ignored in politics and thus not really protected. Although this fact might not be harmful to women as "protected"—as middle class—as she, she wondered what the General Assembly would do for "the woman whose man does not protect her, and who had been deprived of the power to protect herself."[13] The best-protected women in the country, suffragists insisted, were those living in states where they could vote.

In this and other statements suffragists repudiated being categorized by their relationships to men: "wives, daughters, sisters, cousins, aunts, as if they were not human beings in their own rights" but "relations to a human being." If they had been male they would not have been ignored as individuals "in their own right"; but their sex made them an appendage and they did not like it. Even in a debate over political rights, women could not avoid striking out at cultural constraints because they so affected the former; but those constraints prevailed. Equal suffrage bills failed in both houses. To express their contempt a group of drunken legislators invaded suffragist headquarters in the Yarborough Hotel in Raleigh and destroyed a large map of the United States.[14]

North Carolina suffragists were not alone in defeat. Suffrage referenda lost in Pennsylvania, New Jersey, New York, and Massachusetts. But 1915 was nonetheless a good year because, in December, Carrie Chapman Catt accepted the presidency of NAWSA. Her enthusiasm and drive were contagious: North Carolina "sufs" gained converts and self-assurance. In late 1916 they were given a new sense of solidarity with the national movement by a change in strategy. After the Democratic National Convention adjourned with a platform encouraging the enfranchisement of women by state legislatures, Catt called an emergency convention. There she outlined her "winning plan" of achieving enfranchisement by federal amendment. This strategy would require pressure on senators and congressmen as well as state legislators, but greater demands seemed to promise greater rewards. The nuclei of women in North Carolina cities began to reach out into the countryside, and when the General Assembly convened in 1917, lawmakers seemed to be a little more pliable than before; the margins of defeat were not so large. All that suffragists could win in the new House of Representatives was a bill on divorce that, as one headline read, "Provides That Male Adulterers Are as Guilty as Female."[15]

Suffragists fared better in the North Carolina Senate. On February 19 Senator Kelly Bennett, a pharmacist from Swain County in the western mountains, submitted a bill to allow the women of Bryson City, the county seat, to vote; on February 26 it passed its third reading. It was the second successful bill for the freshman senator and the first suffrage victory in the North Carolina General Assembly. Almost sixty-three years later, Teela Bennett could proudly remember her husband's first success in politics. They had both traveled more than three hundred miles of railroad track to Raleigh. She had been there before as a teacher—she was most impressed with Edith Royster—and now she was there as the wife of a young and progressive, suffragist politician.

"Why was 'Dr.' Kelly a suffragist?" she was asked.

"I think," she smiled, "that his wife had something to do with it." She recalled a suffrage meeting near Raleigh in an old plantation house; she was one of the very few wives of assemblymen present. Most of the women there were with the local equal suffrage association. What so impressed her about that meeting so long ago was its "intensity," the "electric" atmosphere sparked by women who knew what they wanted and were determined to get it. The feeling had been strong; it animated her still. And because she lived so intently in the present, she mentioned the failure of the General Assembly to ratify the Equal Rights Amendment.

"I am," she confessed in genteel anger, "ashamed of North Carolina."

Senators could allow women to vote in the isolation of Bryson City where they could do little harm, but presidential elections were another matter. One man objected to a bill enfranchising women at the federal level because it would mean that his black cook would be able to vote whereas his own wife could not do so because of her more refined sensibilities. Besides, he insisted, voting women were contrary to the teaching of Saint Paul. The sponsor of the bill whose patience with such arguments had worn thin carefully replied that if his colleague's cook "was more patriotic than his wife, that was no reason why the cook should be deprived of the vote." As for Saint Paul, What did he know? He was merely a "miserable old bachelor." Someone demanded retraction on the slur on the sainted apostle. "I will," the senator replied, when the men of North Carolina admit that they "have less regard for the women of this state than the men of other states." He never apologized, and his bill failed 20–24. A few days later, the House killed Kelly Bennett's bill. Even the women of Bryson City would not be allowed to vote.[16]

Entwined with suffrage mobilization in the spring of 1917 was the war effort. Even before war was declared on April 6, Carrie Chapman Catt had called state suffrage leaders to Washington, D.C., to discuss the role of NAWSA after Congress had finally acted. "War to the Nation," reported Otelia Carrington Cuningham to her sisters, "means first and foremost a democracy for men and women alike." Our best effort, she insisted, is to spend time, money, and energy in "Suffrage service, suffrage organization, and Suffrage Education." Organizing the North Carolina division of the Women's Committee of the Council of National Defense was therefore not a diversion from the suffragists' goal. They participated in Liberty Loan drives, pushed programs to eliminate waste and increase production of food, helped teach farm women about nutrition, and tried to prevent curtailment of the state's social services. They investigated working conditions for women and saw to it that children were healthy and in school instead of sickly and in factories. It is not surprising that such action should have reinforced the view of women as reformers.[17]

As American women acted the part of good citizens, antisuffrage sentiment diminished. On January 10, 1918, the U.S. House of Representatives passed the so-called Susan B. Anthony Amendment 274–136 to enfranchise women. The only North Carolina representative to vote for woman suffrage was Zebulon Weaver of Asheville. The mountains of the Tenth Congressional District that he represented had sent suffragist men like Kelly Bennett to the General Assembly as part of the pattern of democratic individualism seemingly endemic there, but neither of the state's U.S. senators joined Weaver. One was Lee Slater Overman, who could relate to women only by telling them how pretty they were, and the other was Furnifold Simmons, whose views on women had been amply demonstrated in the Revolution of 1898. The state's suffragists organized a petition campaign, knowing that it was futile. The surge of suffrage sentiment in the nation, wrote one woman, "will not make a particle of difference" in North Carolina because "the politicians are as afraid as death of letting anything come to disturb" them.[18] In reply,

Carrie Chapman Catt wrote, "I know it is difficult to move North Carolina," but she did hope that her Tar Heel colleagues could understand the psychological satisfaction of indelibly impressing upon the minds of adversaries that woman suffrage was inevitable. "Don't be discouraged," she wrote; "The time will come when you will be proud that you supported it." And so they canvassed villages, towns, counties, and colleges. Almost all the women at the state normal college in Greensboro signed petitions; so did those at Presbyterian Peace College and Baptist Meredith. Men at the University of North Carolina joined their fellows at Trinity (now Duke University), Wake Forest, and Davidson. But the senators, while surprised at the outpouring of prosuffrage sentiment, were not with the majority when the U.S. Senate passed the Woman Suffrage Amendment on June 4, 1919.[19]

Despite rejection by their senators, North Carolina suffragists had by this time discovered that they were much more popular than they had been just four years before. The North Carolina Republican party endorsed woman suffrage in April 1918 and by the end of the year had been joined by Methodists, Baptists, nurses, farmers, labor, and club women. The motivation of the latter was clear. They wanted the vote in order to fight discrimination against women and to get a "square deal" for children.[20] And the legislative session of 1919 suggested that the legislature just might do so—at least later if not sooner. The Senate voted 35–12 to enfranchise women in municipal elections, and in the House sufs counted an eight-vote majority. But they watched their advantage disappear before a combination of dirty tricks, misleading statements, ignorance, and perfidy. Opponents were giving suffragists another object lesson in just how "dirty" politics could be: They lost 49–54.[21]

Referral of the Nineteenth Amendment to the states for ratification in the spring of 1919 transformed even lukewarm prosuffrage women into jubilant partisans. By March 22, 1920, the vote of only one state was required for ratification. The conventional wisdom had it that Delaware (or Vermont or Connecticut) would surely ratify before the fall elections. By the end of March, the assumption of inevitability affected the calculations of North Carolina politicians. To be sure, Senator Overman never relented his opposition; but Senator Simmons was weakening, as was Governor Thomas W. Bickett. One of the three men who wanted to succeed Bickett was even willing to be identified publicly with woman suffrage. In February, Lieutenant Governor O. Max Gardner was named to the honor role of NAWSA for "distinguished service" to the cause.[22] Since Gardner's two opponents for the Democratic nomination were opposed to women suffrage, and possibly because he thought its inevitability might give him a ready-made constituency, he decided to make it an issue. Gardner needed a friendly state convention to write a platform tailored to his "progressive" views. His men—and as it turned out, women—worked at the precinct and county levels to claim a majority of delegates to the state convention.

Precinct leaders in the central and western parts of the state were so sure that ratification was inevitable that they treated potential women voters as they had traditionally treated men under the age of twenty-one who would be

President of the N.C. Equal Suffrage Association, Gertrude Weil, (first left) and her fellow activists had a network of experienced and committed suffragists in place long before Congressional passage of the Nineteenth Amendment—an advantage ERA ratificationists lacked.

eligible to vote by election day: they invited them into party deliberations. Some women, however, came uninvited. On the evening of March 27, 1920, precinct meetings throughout Raleigh were invaded by squads of women. They so shocked the men—some of whom were their own husbands—that they met almost no opposition as they passed prosuffrage resolutions and elected delegates to the county convention. Cornelia Jerman had so effectively organized Raleigh's Second Ward that her women outnumbered the men three to one. She later confessed that she had not really meant to laugh, but the men were so dumbfounded that she simply could not help it. In some precincts women were named "committeemen" and delegates to their respective county conventions. Since Cameron ("Cam") Morrison of Charlotte was opposed to woman suffrage, the women were natural allies of Max Gardner, although one of their co-conspirators, coaches, and cheerleaders was Josiah William Bailey, a Democratic party activist who supported both woman suffrage and Morrison.[23]

By April 8, when the State convention met in Raleigh, the presence of women in county conventions had suggested to the sanguine a slight shift in North Carolina politics. Because ratification of the amendment was a Progressive issue, and because Morrison had been so outspoken in his condemnation of Progressivism and woman suffrage, and because the women and their male allies had announced that enfranchisement meant the end of machine politics in North Carolina, woman suffrage was identified with innovation and progress. Although neither of these was an asset in North Carolina, the party was Gardner's—for a moment. The women delegates were obvious but quiet—it was not they who forced the suffrage issue on the convention. Too much public oratory, they believed, would damage a ratification campaign should one be required, and by early April it seemed to be a real possibility. After the precinct meetings, Delaware had unexpectedly failed to ratify the Anthony Amendment.

Carrie Chapman Catt then decided that North Carolina would have to be the last state in the ratification process. She told North Carolina suffragists' president, Gertrude Weil, to make "every person in the state eat, sleep, and drink ratification." The miracle seemed almost possible after Senator Simmons's public concession that women suffrage was now "inevitable" and that the convention should adapt itself to this reality. Suffragists were not overjoyed by his tepid comments, but they were pleased that he had acknowledged the respectability of women suffrage, thus making Max Gardner's own advances seem legitimate. Against opposition from eastern black belt counties and the pleas of his antisuffrage allies, Gardner decided to force woman suffrage onto the floor of the convention and into the platform. His eloquent brother-in-law and convention keynoter hailed equal suffrage as a matter of justice, and amid cries of "NO NO" asked the convention to yield gracefully to the inevitable. Party conservatives refused. Senator Overman refused—he was being challenged in the primary by a suffragist attorney from Greensboro. Morrison also refused. He was fast transforming woman suffrage into a symbol of opposition to the Revolution of 1898.

Gardner's delegates were too much for Morrison. They ignored antisuffrage appeals to states' rights and white supremacy and by a vote of 585–428 endorsed woman suffrage and instructed the General Assembly in special session to ratify the Nineteenth Amendment. "The convention was not converted," wrote one political reporter, "it was biffed over the head with a fact." Josephus Daniels, former editor of the Raleigh *News and Observer* who was now Woodrow Wilson's secretary of the navy, had been following the convention closely. He was ecstatic. "It is all over," he telegrammed Carrie Chapman Catt, "but the shouting!"[24]

Suffragist Ideology and Classification by Sex

Daniels was wrong, as Catt suspected. Failure of the General Assembly to ratify the Nineteenth Amendment in 1920 resulted from legislators' response to the radicalism they inferred from suffragists' ideology and the threat they thought it posed to their brand of politics. But there was something else, too. Woman suffrage was far more complex than recording a vote. It was about human relationships and personal identity—whether or not women would be defined or classified essentially by traditional views of behavior thought to derive from their sex. The word "sex" is used here instead of "gender" because it is the former word by which the struggle for equality between men and women has been identified. "Sex" has a broad range of connotations and is often used as a synonym for "gender," even though the latter word really refers to the cultural meanings assigned anatomical differences. The word "sex" is ambiguous even for people who maintain the distinction in their own minds because the ambiguity is itself valued. (It is difficult indeed to think of "gender appeal.") Differences between the sexes were at issue when suffragists argued that women were a public constituency with legitimate interests and goals. Differences between men and women should not be used to deny citizens of their rights or the Republic of their talents. Antisuffragists focused on the wrong things when they insisted on gender differences, suffragists were saying, because the former saw them as restrictive and prescriptive rather than as the basis for a richer politics built on an appreciation for the varied experiences of women.[25]

Suffragists understood the politics of their situation very well. They lobbied not to achieve utopia but political concessions for girls and women. They wanted better educational and sanitary facilities, equitable pay for women teachers, and the appointment of women to policy-making positions. To gain these goals they had to appeal to men who did not think of themselves as suffragists. The appeal therefore had to be phrased according to values that a broad spectrum of voters could approve: not necessarily justice, logic, or equality, but a woman's natural interest in children and family. Avoiding confrontation with the exclusive virility of North Carolina political life, suffragists spoke of the special sensitivity of women to issues of little interest to men. They printed the appropriate symbols in capital letters: CHILDREN, PROTECTION, BETTER BABIES, WISE MOTHERS, SOCIAL PURITY. No one should doubt that mothers had to have the vote in order to be good mothers.

If the suffragists' public discourse was affected by traditional views of womanhood, it was relatively unaffected by the pathology of white supremacy. Given the ease with which rhetoric of the Revolution of 1898 had been resurrected in the prohibition campaign of 1908, it is surprising that it was not broadly employed on behalf of woman suffrage. It is true that suffragists shared the racist assumptions of men, and they were not above using what one Louisiana suffragist called a "race prejudice appeal," but they preferred to hem and haw. "It would be impossible," they insisted, "to raise the color issue in the equal suffrage amendment because there is no mention of color, and color cannot be [included] by the utmost strain of construction." The Anthony Amendment, they insisted, "does nothing but eliminate the word MALE from the constitution." In private correspondence, too, they could agree that there ought to be "higher qualifications for voting than at present." If such comments were equivocal enough to reassure racists, they reflected the situation of reformers who may not always understand the implications of their own innovations. Such ambiguity was meant to deflect the opposition of men wary of equal suffrage without indicating that suffragists approved the rhetoric of white supremacy. It was not easy to fight for the principle of equality in a racist society; but they had to do just that. If their words suggest that the suffrage movement was an extension of white middle-class hegemony, that was precisely what powerful white middle-class men were supposed to think.[26]

The most carefully phrased statements of political pragmatics did not disguise the suffragists' belief, however, that they were in the vanguard of change. The precise nature of that change was probably evident to only a few, but in retrospect the general tendency was clear: women were changing the way in which they thought of themselves. Inherent in suffrage ideology was a commitment to independence and autonomy for women. Rhetoric easily went beyond the political; both sexes, they emphasized, should have equal opportunity "for the development of the personality and the attainment of true liberty."[27] No matter how much they spoke of women's special contributions to civilization and emphasized the continuity of citizenship with motherhood, suffragists could not deny that they wanted a new way of looking at women. The fight for suffrage had already made them different: they could never be the same after challenging men who thought of them as "only females"—in the genteel sarcasm of Edith Royster. With her, they must have seethed when traditionalist men used the sublimated sexual innuendo of traditional banter to dismiss their intellects and deny their rights. They were now telling men that to treat them merely as the other sex was to demean them.

To treat women merely as the other sex dictated that women achieve their goals by attempting to flatter or cajole—sexually defined acts—rather than by voting, which was a sex-neutral act demanding attention to rational interests. "Sex" as men defined it meant being used as political bait in the Revolution of 1898. It meant listening to politicians tell women how pretty they were to emphasize that the female sex meant beauty and the male sex meant power. "Sex" as men defined it was the game that Marjorie Shuler, for example, refused to play when confronting an antisuffrage politician after he had broadly waved the manly banner of states' rights. A political reporter enjoyed the confrontation. " 'But how about the federal prohibition amendment?' Miss Shuler asked . . . as tenderly as a Boston philanthropist talks to a poor relation! 'Oh! Well, I ain't a-goin' ter argue with you,' said Mr. Hammer in a whisper that could be heard eight blocks, and Miss Shuler agreed. She wasn't arguing; she was telling him." Hammer's game kept him in power and was reflected in slogans that had little to do with citizenship insofar as Shuler was concerned. His anger erupted from being unable to dictate his relationship with a woman; her sex should have made her deferential and silent. "Sex" as men defined it meant not listening to Marjorie Shuler.[28]

The extent to which North Carolina women resented being dismissed as "only females" was suggested by a minority's persistent and enduring relationship with the Congressional Union. This group withdrew from NAWSA, which thought the union's tactics were needlessly confrontational. The union—which later became the National Woman's party—continued to send organizers into North Carolina, where they were praised for their effectiveness. The union's crisp arguments were sometimes used verbatim to express Tar Heel suffragists' position. Julia Alexander, an attorney from Charlotte, angrily observed that the union's methods should be used against the General Assembly. We would then, she wrote, probably accomplish more in one month than we have in the past year.[29] The union broke nicely with the past; certainly it insisted that women had to confront the issue of classifying them by sex. But Alexander's enthusiasm for confrontation was not shared by most North Carolina suffragists, and frustration with being classified as "only females" did not necessarily drive suffragists to the full feminist implications of their ideology.

North Carolina suffragists were reluctant to push too far because of their own personal conservatism and the way in which their opponents sketched out "full implications." These "implications" were so apocalyptic that they invited sarcasm. To the warning that woman suffrage would weaken the family by transforming it into a political arena, Mrs. Josephus Daniels replied, "There has been no lack of love and marriage in the states where women have been voting for twenty years." Mr. Josephus Daniels agreed, but he and she both knew that "the missus" was also Addie Worth Bagley and that her value as a person—so important to her husband as "My Dear Blessing"—was quite beyond her rôle of wife and mother. His enormous pride in her could not be expressed in traditional categories. When asked to fill out a questionnaire about himself, he was at a loss as to how to list his wife's occupation. One of their sons suggested "Mother," but that was not quite fitting because he had not, as he pointed out, listed his own occupation as "Father." Moreover, she had not been in the "mothering business" for about sixteen years. He finally concluded that she was a "politician."[30] Although the Danielses knew that public activism had made their lives different from those of other couples, they never really pushed the full implications of suffragist ideology to subvert family-based gender roles. If traditional categories were strained, suffragists believed that men and women could simply shape more serviceable categories as they had in the wake of college education for women. Since opponents of the latter had promised the same chaos now prophesied by opponents of woman suffrage, and since it had not come, opposition to both seemed alarmist.

There was—as some opponents sensed—a paradox at the core of the suffragists' ideology. The dual achievements leading to sexual equality were to be the empowering of a public constituency of women on the one hand and the relinquishing, on the other, of the traditional classification of women by their sex. The first was explicit, the second was implicit. Suffragists were part of a long process that politicized women's identity as persons defined by sexually related roles and transformed them into a political group working for specific public policies and holding government responsible for them. Tradition had defined women by sex and exploited and rewarded them by sex, so they were thrown together and excluded from power by sex. Their politics had been developed within the sphere men conceded to them—religious and other voluntary organizations, even lobbying groups. But now women were pushing against spherical limits by politicizing sex. In doing so, suffragists had changed sex from the definition of family responsibility into a new public agenda. Largely developed by women, the definition expressed personal independence, self-interest, and social responsibility. The new, public role was at odds with the exclusively domestic role, although women seemed to have embraced both with less anxiety than logic suggested. Classifying women by sex was unacceptable when men did so to deny women their rights; it was acceptable when women used it to explain their common experience and special contributions—when it was used to bond them together for their own goals.

The Threat of Woman Suffrage

Many women, of course, did not want "equality" any more than some men wanted to grant it. Having lost the battle, having been proved to be false prophets, and having expressed their fears in cliché, stereotype, bromide, and prayer, antisuffragists have not fared well among historians. They fared no better among suffragists. North Carolina's chief justice, Walter Clark, thought women who opposed suffrage perhaps a little distracted. They reminded him of the "little lady" from Marblehead whose family was in the whaling business and who asked, at the introduction of kerosene lamps, "What will become of the poor whales?"[31] Anna Howard Shaw, president of NAWSA, referred to her opponents as the "Home, Heaven, and Mother Party," and the feminist writer and lecturer Charlotte Perkins Gilman thought them a little too romantic about motherhood. Even a kitten, she had observed in a combination of sacrilege, lese majesty, and scientific detachment, can be a mother.

The implied attack on "motherhood" has suggested to some historians that opposition to woman suffrage was really deep-seated anxiety about the future of the family. Woman suffrage, it was charged, would change the home into a mere restaurant and hotel by enticing women out of the domestic and into the public sphere, depriving children of their nurture and husbands of their haven. Ignoring the economic and personal reasons that had already driven or drawn women into work outside the home, antisuffragists attempted to reaffirm and thus presumably to reinforce the traditional locus and rationale of woman's work—in the home and for the family. An "aggressive minority" of suffragists, wrote Mary Hilliard Hinton about women with whom she had worked in women's clubs, was trying to force all women into politics when they would rather have remained in the privacy of their own homes—something Hinton had, of course, refused to do.[32] That woman suffrage should have evoked laments about the destruction of the family is not surprising; they came naturally after a century of discussing women's roles. That many women who deviated from those roles should have felt themselves placed at odds with family responsibilities was a natural enough result of the way in which domestic responsibilities had been divided. But those roles and not the family were the subject of discussion; "family" was a mystifying symbol. Opponents of woman suffrage who bewailed its impact upon family life did not talk about how familial functions had been transferred to the state or community. They did not focus on the quality of family life and how husbands could improve it. They did not suggest how women's life within the family could be made better. Opponents did, however, warn that woman suffrage would bring feminine political and professional competition with men.

The complaint that enfranchisement would turn the family into a political arena at regular intervals focused on men and women in conflict, not on familial solidarity. If the latter had been the essential priority, maintaining the traditional relationship between men and women would have been irrelevant because the relationship could have been adjusted according to an agreed-upon higher value, that is, the family. But that was not the priority. The priority was the subordination of women. Expressing concern about the family was a way of maintaining traditional sex roles, not a way of defending the family. There is a real temptation, of course, to say that both "concerns" were held equally. But that solution would ignore the persistent emphasis on what sex meant rather than on family life and what antisuffragists meant by that. Focusing on the family rather than definition by sex is a little like insisting that the Civil War was over states' rights rather than slavery.

The basic reason for insisting on what may appear to be a minor point is to place in proper perspective the suffragists' own dissatisfaction with being defined essentially by sex. Their opponents understood them very well indeed, so well in fact that male antisuffragists worried about the "superior positions women will hold" after enfranchisement and "man's subordinating himself to women." The substitution of female for male domination was an almost universal theme. If Southerners accused suffragists—in a particularly revealing and suggestive metaphor—of intending to "unhorse the manhood of the South," the same fear had also been expressed in the North as well as in Great Britain. The theme of female domination was as familiar as the folk myth of the henpecked husband and as defensive as virility would permit. "If you think," read one pamphlet "that the legislation which for several years has been steadily hostile, in every respect and at every point, to husbands and fathers, should be carried still further against them . . . you will vote for this nineteenth amendment."[33] True men—men who could exercise authority without apology—would not yield to such an assault.

Politics was, of course, the ritual and liturgy as well as the substance of power, and the sexual terms of political discourse revealed a fusion of political power and sexual prowess. The virility of politics was so persistent a leitmotif in North Carolina that it was expressed easily and naturally. The Revolution of 1898 had been a victory of white manhood on behalf of women; the public conversion of Cameron Morrison to prohibition had been expressed in a "virile argument." And the justification of Max Gardner's candidacy in 1920 specifically bypassed an appeal to race or class for one addressed to the "virile manhood" of the state. Even the man who courted women's support and invited them into party politics could not slough off "virile" posturing. His opponents, after all, had said that men who supported woman suffrage were sissies.[34] The implications for women were clear: women who did not want to be virile would not want to vote. Moving from the domestic sphere in which sexuality was controlled into a sphere where it was celebrated would unleash the restraint that made civilization possible. In a culture where social problems of ignorance, want, and crime were perceived as a breakdown of self-discipline, the loss of Mother, whose moral teaching and example made discipline possible, was traumatic. That loss was symbolized by the act of voting because, in the ethos of virility, voting was perceived as a public display of self. Good women were not supposed to display themselves; only prostitutes and actresses did that. Good women would plead, as did the president of the North Carolina branch of the Southern League for the Rejection of the Susan B. Anthony Amendement, for "protection from political responsibility." Otherwise the future was bleak. "Western Races," announced a headline in The *Woman Patriot,* "on Verge of Collapse through Feminism." The idea was part of the cultural system of England as well as the South, of New York as well as North Carolina, of Boston as well as Raleigh.[35]

Beyond celebration of motherly sacrifice was a political calculation. Whereas English antisuffragists fretted about the votes of working-class women and their Yankee counterparts frowned at the anticipated increase in the immigrant vote, Southern antisuffragists warned of the return of blacks to the polls. Thus, when hearing in 1919 that the U.S. Senate was about to submit the suffrage amendment to the states, a North Carolina assemblyman blurted out, "Have we forgotten the political Revolution of 1898?" He and his allies hoped that no one would forget it. The Susan B. Anthony Amendment, they screamed, "is a Force Bill!" It would force Negro Domination. Look at Carrie Chapman Catt's words, shouted the bold print of evocative broadsides: "Suffrage Democracy Knows No Bias of Race, Color, Creed, or Sex." Catt had said that she favored equal suffrage for all—foreign born, women, Afro-Americans. Antisufs handed out photographs of white and black suffragists working together for common candidates; they also included a portrait of "Mrs R. Jerome Jeffray (Negro)," who was a friend of Susan B. Anthony herself! Remember, legislators were reminded, Frederick Douglass, who forced the Fifteenth Amendment upon the South, was Afro-American; he favored woman suffrage, and he had actually married a white woman. A novelist warned that the Afro-American press discussed the Anthony Amendment in terms not only of "equality of the sexes, but the equality of the races!" It is not surprising that one opponent should have scolded Gertrude Weil that she, as a Jew and "knowing that the Jews were God's chosen nation," should know her Bible well enough to know that Afro-Americans were a "cursed race." Her helping them get back into politics revealed that she did not think as highly of women as he did. "If you did," he told her, you would never "fight as hard as you have for this abolitionist movement."[36]

Authority and tradition were for many antisuffragists best expressed in the moods of religion. Suffragists were damned for resisting biblical authority and especially that of Saint Paul, whose writing supported submissiveness of wives to husbands (as well as slaves to masters). Suffragists, however, knew the Bible, too, and they knew that Paul had not always been authoritative or clear. Chief Justice Clark was insistent on this point as he quoted 2 Peter 3:15, 16 with approving good humor: "Our beloved brother Paul . . . hath written some things hard to be understood, which they who are unlearned and unstable wrest as they do also the other scriptures, unto their own destruction."[37] Widespread as Saint Paul's authoritative mysogyny may have been, it was never fanned into the flames of a religious crusade such as the profamily movement of a later generation.

Woman Suffrage in the General Assembly

Antisuffrage rhetoric exposed the anxieties of many Americans in 1920, but its logic was not by itself sufficient to explain why a majority of North Carolina legislators refused to ratify the "inevitable"—as Senator Simmons had called woman suffrage. Resistance to the "inevitable" became identified with political orthodoxy in the second gubernatorial primary election of 1920; it was a disappointing lesson for the man who had steamrollered the state convention into a proratification position. Max Gardner found that voters were furious with him for the way in which he had brought women into the party even if defended as a political accommodation made for its own good. Siding with the inevitable was not a virture in North Carolina, as Cameron Morrison made clear in his defense of the Revolution of 1898. The lieutenant governor, whispered Morrison's partisans, is a "mixed marriage advocate" who—they then shouted—would bring back Negro Domination. They warned that "Max Gardner and miscegenation, suffrage and socialism, ratification and eroticism are the same thing." There was no way to have a civil discussion of issues facing the state. Finally on July 3, after what one woman called "an intense appeal to the worst prejudices of our people,"[38] the Democratic primary came mercifully to an end. Cameron Morrison won 70,332–61,073. The Revolution of 1898 had been vindicated; its logic had nominated Morrison and would elect him over modest Republican opposition in the fall.

On July 7, Governor Thomas W. Bickett called for an extra session of the General Assembly in August to discuss tax reform and woman suffrage. Women activists knew that even with a friendly Senate they faced a formidable task. "All the world," wrote Catt to Weil, "loves a good fighter and a good fight." But her brave front could not counterbalance her earlier confession that it was "unfortunate that North Carolina" was so crucial to ratification. Annoyed by Josephus Daniels's stubborn optimism, she was also dismayed at Governor Bickett's penchant for being "childish" and "petulant" as he had been when told by President Woodrow Wilson to carry the state for women. Maybe they could do it by themselves. Weil, Cornelia Jerman, and their colleagues were by now seasoned campaigners with access to politicians and newspapers.[39] And they were encouraged by the changing mood beyond the state. When Jerman returned from the Democratic National Convention, she had been refreshed by contact with men who were actually comfortable with women as colleagues. Among them, she recalled, "there was no flippancy, no evasion in their attitudes towards women." But now reality confronted her in North Carolina, and she knew, as she told her troops, that they all must be busy "EVERY MINUTE!" They obliged her. They wrote personal letters to each legislator and organized referenda from women attending summer school in Chapel Hill and Raleigh—who favored suffrage 5:1 and 3:1. They recruited male allies as well. Josiah Bailey was making one of the best investments of his political life by throwing himself into the campaign with his wife's friends, castigating "old minds, old bachelors, and [old?] politicians." "We put up a big fight," wrote a Raleigh activist, and "we try to make everyone believe the amendment will be ratified—BUT WE CAN NEVER TELL AND WILL NEVER KNOW until the roll is called."[40]

But Jerman did, in fact, know that the fight was over before it began. On August 11, 1920, the day after the General Assembly had begun its work, 63 of the 120 members of the House sent a telegram to Tennessee legislators assuring them that Tar Heels would not ratify. They urged the Tennesseans not to "force" woman suffrage on North Carolina by ratifying in their stead. Suffragists were shocked. Among the signatures were names of men who had given their word to support ratification. Antisuffragist men had often told suffragists that women were more moral than men, and the assemblymen seemed bent on proving it. "The men have got to lie to the women," wrote a reporter, "to convince the women that the men are not on their level."[41] And yet the public rituals of political life continued to be celebrated as if words and gestures, meetings and debates were needed to impress on celebrants and constituencies alike the solemnity of the issues. Suffragists even hoped to change minds. They had come so very far in five years. They had won a major victory at the North Carolina Democratic Convention. Their own ideological awakening had been nurtured by awareness of the new and exciting things happening to women all over the nation, indeed, the world, and they could not believe, despite abundant evidence, that most North Carolina Democrats were as angry with politically active women as they actually were.

Suffragists appealed to Cameron Morrison to deny that his victory had resulted from his opposition to suffrage; he refused. They appealed to men among the sixty-three to hold true; and they also refused. They appealed to Senator Simmons who believed woman suffrage "inevitable," and he, too, refused them. To be sure, on August 13 the Raleigh *News and Observer* ran stories under impressive headlines: "Simmons Appeals to Legislators"; "Senator Simmons Stands Squarely for Ratification of the Amendment." But the headlines reflected the newspaper's editorial policy instead of what the senator had actually said. "I have no desire," he had written, "to press my views upon the members of the legislature now in session." He did, however, repeat his personal opposition to woman suffrage even as he acknowledged its "inevitability." He saw no reason to help the women even though their enfranchisement could not now have hurt Cam Morrison.[42]

The Southern Rejection League, from its headquarters in the Raleigh Hotel, sent encouragement to its champions to defend home, family, and white supremacy. Decked out in pale red roses to contrast with the yellow rose of suffrage, members happily passed out leaflets in the rotunda of the State Capitol Building. They opposed equal suffrage, they explained, because the "forces" behind it denied "the divine inspiration of the Bible" and set "human reason as the criterion of worship in the overthrow of Church authority."[43] With such issues at stake, it was fortunate that they knew how to influence assemblymen. William O. Saunders, a newsman from Elizabeth City, was impressed. The closest approximation of a "village atheist" in the General Assembly, he observed, "They are having the time of their lives, lobbying and playing politics in their endeavors to defeat the suffrage amendment." Indeed, Saunders thought the average antisuffrage woman "more skilled in lobbying" than suffragists, "holding on to the hand of a legislator, and looking soulfully into his eyes, wooing him with the pulsing pressure of her very body as well as eyes, lips, and hearts." Saunders warmed to his subject: "Some of them are peaches, and don't need the vote as long as there's a drop of the blood of Adam (or should I say David?) left in man." The comment about David was the kind of textual exegesis Saunders delighted in. But his scandalous insight into the dynamics of traditional interaction between the sexes and the dramatic politics of sublimated sexuality outraged his opponents because, as a legislative resolution of censure said, it implied "general improper conduct."[44]

At 11:15 A.M. on Friday the Thirteenth of August, Governor Bickett led a procession of women into the House chamber to address the General Assembly on the proposed Nineteenth Amendment.[45] He had wanted the Tennessee legislature to ratify in order to relieve him of the responsibility of facing the issue, for he had a higher priority. He had worked hard for tax reform against what one observer called "political assault and battery, mayhem, arson, burglary and highway robbery." Because of this opposition, Bickett thought woman suffrage more "inevitable" than tax reform. It was not that Bickett had no sympathy with suffragists; he did. He was married to one: Fanny Yarborough, who had studied at the universities of North Carolina and Chicago. He walked into the chamber with his wife and her good friends, Cornelia Jerman and Addie Daniels; two rejectionist women trailed behind. Out of place, they were nonetheless first to applaud Bickett's speech. He knew, he explained, that sentiment in the General Assembly was "decidedly against" ratification. "With this sentiment," he announced, "I am in deepest sympathy. . . . I have been profoundly disturbed about what politics would do to women." Suffragists were miffed; their opponents responded with tumultuous applause.

Then Bickett shifted gears. Woman suffrage was inevitable; voters throughout the nation were already inviting women into their political "homes." He put it to Southern gentlemen: "Shall we receive them with a smile or with a frown?" He had used variations on this theme in several speeches earlier in the year; Southern gentlemen should be gracious, he thought. And then he betrayed his true feelings: "Judgment and justice, mercy and humanity all cry out that women have the first right to speak when the issue is whether or not the world shall henceforth be ruled by righteousness or by blood and iron." He was referring to the League of Nations and the hope that women who birthed the casualties of battle would have a voice in the issues of war and peace. Nodding in agreement, Senator Alfred Scales, a partisan suffragist, fanned himself with an antisuffrage fan. Twenty minutes after the governor had finished his speech, Scales reported favorably out of his Constitutional

Amendments Committee his own proratification resolution by a vote of 7–1. The General Assembly then went home for a long weekend.

On Tuesday at 11:30 A.M. debate on ratification began in the Senate.[46] Scales was floor leader. His opponent was Lindsay Carter Warren—who would one day be Franklin Delano Roosevelt's comptroller general and was, a reporter wrote, the smartest young politician "this side of purgatory." The first speaker had just received a rousing response from the suffragist west gallery when Warren rose to speak. Pressing the lapels of his jacket and trying to appear older than he was, he looked up at what a reporter called "the Battalion of Death." He would not, could not go against the wishes of his constituents for the sake of "expediency." He spat the word out in disgust at the governor. "I denounce such an insidious appeal from Governor Bickett to our baser natures." After all, he explained with the grave seriousness of a sophomore who had just read his first book of ethics, "Right is right and wrong is wrong." Antisuffragists cheered. Addie Daniels, down on the floor, was momentarily nonplussed, then she laughed and began to clap, urging the west gallery to follow her example. For two minutes she led them on while sweat rolled down Warren's flushed face.

Then—as if to mock Cam Morrison's appeal to the Lost Cause—the last Confederate veteran to serve in the General Assembly followed Warren. The women, he remembered—it was long ago—had made sacrifices for us all, "and can we say to them now that they have no right to vote?" But it was Powell Glidewell from the central portion of the state who lashed in contempt at Warren. The latter had chaired the Platform Committee at the state convention, and it was his platform that the convention had rejected. "When you lost the fight in the state convention," Glidewell addressed his youthful colleague, "are you not sport enough to take your licking and stick by the party you claim as yours." It was not a question. As for Warren's obsequiousness before public pressure, "If your people are not willing to let you vote your judgment, it is no affair of mine." It was a shame "to see a young man absolutely refuse to allow an idea to enter his head." Glidewell could not leave Warren alone. "A wise man may change his mind," he said, referring to President Wilson, "but—" he turned to face Warren directly, "a fool never does." The Senate dissolved in an uproar.

Warren had the last laugh. He submitted a substitute to postpone consideration of woman suffrage until 1921, and it passed 25–23. Senators Titus G. Currin of Granville and Wilkins P. Horton of Chatham ignored their written commitments to ratification and joined opponents of woman suffrage. The substitute was a way of being against woman suffrage without being against woman suffrage. Senator Horace E. Stacy of Robeson was confused; he thought he was in fact voting for suffrage as he had promised. Senator Marmaduke J. Hawkins of Warren County was absent, and Taylor O. Teague may have been forcefully restrained by antisuffragists so that he could not vote.[47] There was no time to find Hawkins, rescue Teague, explain the facts of life to Stacy, or shame the two defectors. Late that night Max Gardner was still trying to find a way to get a vote to reconsider.

The next day, North Carolina women were enfranchised by the Tennessee legislature. Catt wired Gertrude Weil: "RATIFICATION ACCOMPLISHED PROCEED WITH STATEWIDE CELEBRATIONS." Two days later the North Carolina House of Representatives rejected the Anthony Amendment 41–47. As one assemblyman insisted, reality was irrelevant: North Carolina should contend for principle. Harry Grier, a Republican representing Statesville who had once scolded suffragists for their impiety and their suggestive (at least to him) clothing, and who took his religion and tobacco in strong doses, urged his fellows to pass a special resolution rejecting the Anthony Amendment as a "violation of the principles for which our fathers in gray fought and died." Antisuffragists raged out of control at having their victory over North Carolina women nullified by Tennessee. They tabled a Senate resolution introduced by Alfred Scales to provide the machinery for registering and voting women in the fall elections. "Good God!!" cursed a Gardner man to the lieutenant Governor, "Is the House trying to commit suicide?" Gardner rushed to the other chamber, where his allies refused to vote enabling legislation if Morrison or Simmons received credit for passage. Finally they arrived at a settlement and, conditioning their legalization of woman suffrage on the U.S. Supreme Court's approval of Tennessee's ratification, in effect arranged for the registration of North Carolina women by a vote of 35–27 out of 120 members. The action of the House in tabling Scales's resolution, said suffragists, was the "stupidest, most indefensible act of the special session."[48]

The crusade for woman suffrage in North Carolina ended without the sweet resolution of victory. The women had hoped so very much that theirs would be the first southeastern state to ratify and thus challenge the myth of Southern parochialism. They had come so far and won so much; the victory in the North Carolina Democratic Convention had been so encouraging. But the party refused to play by its own rules. The women resented this refusal and the arrogant way in which they had been treated, and they would remember the slights for a very long time. But they could laugh about it a week later at a banquet for the "immortal sixty-five," as Cornelia Jerman called the sixty-four assemblymen and Max Gardner who had stood by them. They conducted a brief, mock session of the General Assembly under "Warren's Rules of Disorder." One woman impersonated a laughing Thomas W. Bickett as the "assembly" debated whether to enfranchise men: "Resolved by the House of Misrepresentatives and senility concurring," in the spirit of the great philosopher who said, "Right is right and wrong is wrong." The female "Governor Bickett" intoned that she "did not fear the effect of the ballot on men, but she did fear the effect of men upon the ballot." It was good to laugh, perhaps to remember how Glidewell had baited Warren, what Addie Daniels had done, what Josiah Bailey had said outside the legislature. It was good to release tension; but it was better to leave this all behind, as Otelia Carrington Cuningham, former president of the Equal Suffrage Association of North Carolina, said. The ballot, she wrote, was a "means" to untold possibilities, to improve citizenship, to pass sound and intelligent laws, and to move beyond the parochial and "prejudiced politics" of the special legislative session.[49] American women had gained the right to vote; it was the end of the beginning.

From the perspective of a later conflict at the junction between gender and politics, the fight for woman suffrage in North Carolina is illuminating. It began within an intensifying and accelerating process of women's activism in both paid and volunteer work; from there, a few middle-class, comparatively well-educated, relatively young white women in the urban industrial center of the state created networks to lobby for woman suffrage. This mobilization was a natural "next step" in the process of forming a female constituency that, suffragists insisted, ought to be able to hold government directly responsible for its actions on issues that mattered to women. Emphasis on the specific act of voting begged the question as to what women's interests actually were, but even this limited grant—implicit in the idea that women were citizens—was nonetheless elusive. This was true partially because in attempting to broaden the franchise after 1913, suffragists were going against the grain of a section of the country that had restricted suffrage a little more than a decade earlier. If suffragists as a result were guarded about the impact of their enfranchisement on white supremacy, and if they were ambivalent about race relations, they nonetheless made few if any appeals to the Revolution of 1898. That event—although undoubtedly justified in the minds of most women—seemed to be part of a past from which suffragists were trying to escape.

Suffragists' opponents were primarily men. To be sure, conservative members of the women's clubs such as Mary Hilliard Hinton opposed woman suffrage as a transformation of the political world. And some of her "sisters" employed traditional feminine "wiles" to reassure antisuffrage legislators that all women had not been "unsexed" by the mad rages of the twentieth century. Among them were wives of politicians and members of the United Daughters of the Confederacy, but antisuffrage women in North Carolina were neither broadly organized nor active. As for legislators, men from the rural eastern part of the state that had the highest density of blacks—now disfranchised—and the lowest rate of industrial development were most opposed to woman suffrage, that is, to any relaxation of their control. Generally in areas least disturbed by intrusion from the outside world, woman suffrage was most despised. The same was true of areas in which white supremacy dominated political rhetoric. If suffragists were characterized by ambivalence and embarrassment with regard to race, their opponents shared none of this civility.

Race and isolation were only two aspects of antisuffrage feeling. There were others. One was the way in which opponents avoided the merits of woman suffrage and subordinated them to short-term political gain. A more basic matter was opponents' anxiety at the implications of woman suffrage for gender relationships. The two matters appear to be contradictory because they put opponents in the position of either ignoring woman suffrage as a serious issue or conceding to it a greater cultural impact than it possibly could have initiated. The Revolutionaries of 1898 knew that woman suffrage was "inevitable" but dramatized their principled resistance above the "mere expediency" of supporting it. They did so for the "expediency" of electing one of their own as governor, that is, for short-term advantage. Their actions were at the level of electoral politics, but they also revealed something at stake in another level of consciousness. For men sensitive to political effect anti-suffrage legislators acted in a strangely immature way after enfranchisement. Their petulant adolescent furor was consistent with the "virile" posturing of Senators Overman and Simmons. Their accusations of "sissy," aimed at Max Gardner in the second Democratic primary in 1920, suggested what was personally at stake: manhood. They sensed a gender revolution. The bottom line for male opponents of woman suffrage was—as it was for opponents of universal manhood suffrage—fear of losing power.

As for women suffragists, they did not make a rhetorical frontal assault on what a later generation would call social roles based on gender. They laughed at what they thought were silly caricatures of such anomalies as manly women and sissified men; they turned their sarcastic wit on exaggerated predictions of domestic carnage. And yet they had indeed taken deadly aim at the myth of a masculine citizenry. If it was precisely their point that women did not have to become men in order to be citizens, they nonetheless were depriving men of one of their masculine privileges. In arguing that women offered participation essential to the improvement of republican government, suffragists were separating the meaning of citizenship and of the Republic from masculine gender. Both women and the community would benefit from this change. There was nothing that women did or were supposed to do that justified their being deprived of their rights. There were no aspects of community life, they believed, which would not be improved by bringing women's experience to bear on policy. In this position they were indeed doing what their opponents said they were doing: challenging traditional ways of thinking about what it meant to be a man because they were challenging conventional ways of thinking about what it meant to be a woman.

2

"Physiological and Functional Differences": Sam Ervin on Classification by Sex

After winning the vote, women activists were confronted by an uneasy choice not entirely of their own making. Should they seek equality or special treatment? Lobbying for suffrage, they had emphasized equality; justifying themselves, they had argued that the polity needed a woman's touch. The touch was ambiguous: it could imply that as women entered the work force and appeared before the law they could rely on traditional gender values to protect them. This reliance meant asking for special treatment. At the same time, women understood the evocative power of egalitarian rhetoric and the moral status of the equal protection clause of the Fourteenth Amendment. Whether to affirm equality, which seemed to imply the same treatment, or seek special consideration, which acknowledged sexual difference, posed a dilemma that was emotionally and morally awkward because women sensed that one did not necessarily exclude the other. They also knew that the popular logic of one might be used to confound that of the other. The result was a tension unresolved until the women's movement of the 1960s repudiated the rhetoric and politics of "protection."

This resolution was resisted by many Americans who believed that special treatment was a corollary of anatomically defined characteristics—what U.S. Senator Samuel James Ervin, Jr., called "physiological and functional differences." Although it might appear to be relatively unimportant what a conservative North Carolina senator would say about such things, his views were broadly representative of the nationwide conservative resistance to feminism and the Equal Rights Amendment. He was its primary male opponent, using his office, speeches, and prestige as the defender of traditional values—in league with Phyllis Schlafly—to frustrate ratification. In his resistance to the use of government to achieve equality for blacks and women and in his steadfast and profound belief that women by their physical nature were to be reserved special status in the law, he represented male opposition to ERA. He believed that since women by nature deserved special treatment, equality threatened them. Ervin had not benefited from the process through which feminists moved from a position similar to his to one against which he fought. To place him and the fight for ERA in perspective, therefore, we need to review that process.

Classification by Sex: Rights and Liberation

Soon after enfranchisement, the tension between equality and protection created a rift between the two organizations most active in the struggle, the National American Woman Suffrage Association and the Congressional Union. The latter became the National Woman's party, and NAWSA became the League of Women Voters. Differences between the two groups birthed in the suffrage campaign widened as they debated how to achieve women's rights. The league's strategy was to teach women to advance their interests in a nonpartisan way. In North Carolina, for example, former suffragists worked for the secret ballot and juvenile justice. They fought laws that discriminated against women and lobbied for legislation to protect them in the workplace and support them in the courtroom during trials for rape. State by state the league pressed for equal access to jury duty, guardianship, and fair wages, arguing that women should be treated equally. It also lobbied for laws protecting the health of female industrial workers by treating them specially. In pursuing "specific bills for specific ills," the league embraced the legal classification of women by their sex. This the National Woman's party refused to do. It also regarded restricting feminist action to specifics as a dilution of women's activism. Worse—insisted the party's president, Alice Paul—attention to detail prevented women's achieving the equality that could only be guaranteed by amending the U.S. Constitution. Since the equal protection clause of the Fourteenth Amendment was applied specifically to blacks, the party in 1923 proposed an Equal Rights Amendment to remove all support for legal discrimination on the basis of sex.[1]

In retrospect, both the league and the National Woman's party appear to have been right. The issue of sex-based protective legislation was far more complex than either side acknowledged, as the historian Nancy F. Cott has pointed out. In a sexually segregated labor market in which women were usually relegated to substandard jobs, they benefited from protective laws that shortened their work week and improved working conditions. In the long run, however, such laws barred women from positions where skill levels and pay scales were higher. By restricting economic opportunity, protective legislation consigned women to work that had necessitated protective legislation in the first place. The position of the league, which was concerned about women's mothering roles as well as the quality of their workplaces, seemed to help perpetuate existing inequalities. The National Women's party, which emphasized women as paid workers and focused on a utopian rather than an actual workplace, refused to concede the vulnerability of women workers and ignored the need for any protective legislation for either sex. Emphasizing different aspects of the dualism between equal treatment and special treatment, the two sides remained deadlocked.[2] The Equal Rights Amendment was embraced by only a minority within organized feminism.

By the 1940s and 1950s, the conventional wisdom of opposition to the Equal Rights Amendment had set themes that would recur in the 1970s. This wisdom was nurtured by experience with a Supreme Court that had not yet established a reputation for broadly interpreting the law to benefit petitioners whose rights had been violated. The Court—indeed, the entire system of judicial review—was conservative, a fact that affected those who were concerned about the fragility of what women had wrung from a reluctant, male-dominated political system. The Equal Rights Amendment could conceivably be used to deprive rather than extend the rights of women—at least that is what the league said in 1943. The distrust was obvious in a leaflet warning members of the alarming results of ratification. It was a litany of suspicion. With regard to: the laws of consent, Would women benefit from ERA? "Who knows? Nobody! Who must decide? The Courts!" Divorce laws, Would women benefit? "Who knows? Nobody! Who must decide? The Courts!" Property rights, Would women benefit? "Who knows? Nobody! Who must decide? The Courts!" The amendment was too vague, the courts overburdened, the possibility of governmental centralization too great. "The real discriminations against women," insisted the league, "result from custom and prejudice—not from laws. A constitutional amendment won't change these attitudes. It will aggravate them."[3]

Constitutional scholars were equally suspicious. In 1953 Paul Freund, a distinguished constitutional lawyer at Harvard Law School, wrote a statement endorsed by forty-three lawyers and law school deans whose arguments would become standard fare for opponents a generation later. The amendment, warned Freund, could make every legal issue relating to women into a constitutional question because its "range" was of "indefinite extent." What seemed to be especially troubling was the proposed "constitutional yardstick of absolute equality between men and women in all legal relationships." The meaning of this imposition, he claimed, was unclear, and court decisions would be required to apply an absolute standard where flexibility was desirable, as in the case of protective legislation for women. Freund believed that woman's place in society could not be defined by an "abstract rule of thumb" but by statutes and court cases relating specifically to the roles women played as wage earners, citizens, individuals, and family members. In each of these roles, however, sex defined women as a special kind of worker, citizen, and family member. The implications of the stereotype were suggested in Freund's discussion of citizenship. There he reserved an exemption for women from certain duties such as service on juries. The assumption that women should require an exemption from laws applying to men was widespread at the time, having taken sustenance from the Supreme Court's decision in *Muller* v. *Oregon* (1908) that special protection for women workers was justified by their special role as mothers. Indeed, after listing physical and social attributes presumably applicable to all women, the Court concluded that woman "is properly placed in a class by herself." Since that classification as an important part of Anglo-American jurisprudence was "reasonable," Freund concluded that it should be maintained. The abstraction of " 'equality' is a worthy ideal" when mobilizing legislation to "remedy particular deficiencies of the law," Freund observed, "but as a constitutional standard, it is hopelessly inept."[4]

Necessity for exemptions was and continued to be firmly fixed in the assumption that classification by sex was required to protect women. When the U.S. Senate first debated the Equal Rights Amendment in 1946, colleagues agreed with Claude Pepper (D-Fla.) that they could not pass it unless assured that it would not deprive them of the power "to make reasonable classifications in the protection of women." The same reservations were expressed in 1950. Then, Carl Hayden (D-Ariz.) added a rider to the proposed amendment, stating that nothing in its language "be construed to impair any rights, benefits, or exemptions now or hereafter conferred by law upon persons of the female sex." For twenty years the Hayden rider prevented favorable action in the Senate.[5]

By the end of the 1960s, however, increasing numbers of women would challenge classification by sex. Taking their lead from anthropologists, critics would attempt to separate "sex" (anatomical differences) from "gender" (the meaning a particular culture attaches to such differences). They would argue that assumptions unsubstantiated by anatomy had penalized both sexes but especially women and called into question the very meaning of gender. In the meantime, practice was operating more effectively than theory as changes in the workplace undermined traditional gender roles in the postwar years even as women continued to pay them lip service. By 1960 some 40 percent of American women were employed in full- or part-time jobs. Unlike earlier counterparts, these women were not young, single, or poor. Nearly one-half were mothers with school age children, and many were middle class. They had entered the labor force because of changes in their personal life as well as in the economy. They restricted the number of children they bore and benefited from a longer, healthier life than previous generations. They also found new white-collar and clerical jobs increasingly available at the same time rising economic expectations required an additional paycheck to achieve the status to which they aspired. To be sure, when asked why they worked, women frequently responded that, as wives and mothers, they had taken on jobs outside the home as part of their family responsibilities. They often chafed at the double burden of housework and wage work, the unfairness of unequal pay, and the low value placed on their jobs, but frustration was expressed in personal terms as a reflection of the "way things are." Most were not yet ready to define themselves in terms other than those dictated by family roles. The contradictions between inherited definitions—old classifications—and the reality of their own experience were there nonetheless, perceived if not yet understood.[6]

To understand fully the implications of these developments, women needed a feminist consciousness. The few suffragists who had sustained feminist commitments beyond the 1920s had not been able to convert large numbers of women; even those activists working to improve the status of women rejected the feminist label. By the early 1960s, however, America had changed. Women were beginning to participate in a process that would help generate that consciousness. The first steps began inauspiciously enough as work on official commissions that systematically compared the situation of the sexes in American society. Initiated at the presidential level in 1961 by John F. Kennedy and replicated by governors in all fifty states, these commissions consisted in part of women who had already established their credentials as professionals and activists. Their reports indicated widespread sexual discrimination. Disturbed by mounting evidence of what the North Carolina report termed "the enormity of our problem," members of state commissions began to pool their knowledge and suggest reform. Much of "the enormity" had already been publicized by Betty Friedan's attack on domesticity in *The Feminine Mystique* (1963); data from the commissions detailed it. When commission members gathered in Washington in 1966 for the Third Annual Conference on the Status of Women, they encountered the catalyst that transformed many of them into activists. That catalyst was the Equal Employment Opportunity Commission (EEOC) the federal agency charged with implementing the Civil Rights Act of 1964.[7]

Despite the fact that the act had proscribed discrimination on the basis of sex as well as race, EEOC refused to take sex seriously. The word "sex" had, after all, been inserted into the bill by conservatives to prevent its passage; consequently the first head of EEOC regarded the "sex provision" a "fluke." Representative Martha Griffiths (D-Mich.) thought otherwise. Having rallied women legislators in an unlikely alliance with conservative Southerners to keep the "sex provision" in the bill, she was determined to see that legislation enforced. Allies in EEOC agreed. They, along with a small group of Washington feminists, argued that the agency would take gender-related discrimination seriously if women had a civil rights organization to lobby as effectively for themselves as the National Association for the Advancement of Colored People (NAACP) did for Afro-Americans. Before the conference was over, twenty-five women had paid $5 each to join the National Organization for Women (NOW), which after 1969 was complemented by Women's Equity Action League (WEAL). The two organizations launched a major attack on sex discrimination in employment and education while a third group formed to fight discrimination in politics. The National Women's Political Caucus (NWPC), established in 1971, hoped to increase the number of women holding elective and appointive office.[8] Soon women in the professions were fighting for equality in such overwhelmingly male groups as the American Medical Association and the American Historical Association. Equality was contagious.

That so many of these new feminists should have been white, middle- and upper-middle-class women with professional training was not surprising. The very fact that they had sought advanced training meant that they were less likely to find conventional roles satisfying, as the political scientist Jo Freeman has explained. Moreover, when measuring themselves against men whose professional and personal lives paralleled their own, they found the greater rewards enjoyed by the latter seemed especially galling. Privileged though they were, such women felt more "deprived" in many cases than did women who were in reality far less fortunate. The particular combination of circumstances that forced them to confront their condition was as diverse as the individuals themselves, but they could all understand the awakening of a Southern legislator. She had, she recalled, "succeeded in getting Mississippi women the right to sit on juries (1968)" against "appalling" arguments. She confronted the widespread assumption about the incompetence of women once again when she tried to help women obtain credit from reluctant businesses. But she became enraged when, after her divorce, she was denied a home loan. "The effect," she remembered, "was one of the worst traumas I've suffered."[9] For many women like her, the indignity of gender-based discrimination could no longer be ignored.

The process of awakening consciousness that generated renewed demands for women's rights simultaneously provided the impetus for "liberation" among a younger generation of women. Many of them veterans of either the civil rights movement or the New Left, they infused feminism with a radical new perspective and instigated a cultural revolution. Social change had been the goal of the Southern-based Student Nonviolent Coordinating Committee (SNCC) in the struggle for racial equality and of the Students for a Democratic Society (SDS) in efforts to mobilize the Northern urban poor. Drawn to such groups by a commitment to equality, college-age women left their middle-class families to do things they would never otherwise have thought possible. In the South they joined picket lines, created freedom schools, and registered blacks to vote. In the North they trudged door to door in grim slums to mobilize women whose struggle for survival had made them understandably suspicious of intruders. Summing up the feelings of hundreds, one young woman wrote, "I learned a lot of respect for myself for having gone through all that."[10]

If being in "the movement" brought a new understanding of equality and self-worth, it also brought new problems. Men professing racial equality were not necessarily committed to sexual equality. Women frequently found themselves assigned the domestic and clerical chores of movement work and refused a part in deciding policy. They also found that the sexual freedom of the 1960s could feel suspiciously like exploitation. A draft resisters' slogan was a male fantasy: "Girls Say Yes to Guys Who Say No." Summing up the situation, the historian Sara Evans concludes, "The same movement that permitted women to grow and develop self-esteem, energy and skills generally kept them out of public leadership roles and reinforced expectations that women would conform to tradition as houseworkers, nurturers, sex objects."[11] By 1967 sexual stereotyping by male radicals had created tensions so acute that women left the movement to organize on their own behalf. As radicals, they were accustomed to challenging prevailing ideas and practices. As activists, they had acquired a deep-seated conviction that personal problems could be "social" enough to require political solutions. They had also learned techniques to help them transform an awareness of discrimination into a determination to fight it—"consciousness raising." They sought liberation from ways of thinking and behaving that had stunted their growth, limited their options, and fettered them to men. As feminists, they proposed to change the world as they had changed themselves.[12]

An integral part of this resurgent feminism was the assault on legal classification by sex. As one of the first attacks on such classification, an article in the December 1965 *George Washington Law Review* by Mary O. Eastwood and Pauli Murray stands out. There Eastwood, a veteran feminist civil servant, and Murray, a multitalented black lawyer and feminist, argued against Paul Freund's position that sex was a valid classification of persons in the law. They repudiated the *Muller* decision for having added to the problems of women by focusing on their attributes (gender) as a class of persons and fixing upon all women the social function of wife and mother even though many were neither. The injustice of this decision could be understood through the analogy of treating all men before the law as if they were husbands and fathers and assuming that legal doctrine with respect to men should be elaborated from this classification. That women were wives and mothers was, of course, true, and the social functions they thus performed were essential, but "social policies that are genuinely protective of the familial and maternal function," Eastwood and Murray argued, could be based on sex-neutral law that focused on those functions and not on the sex of parties to a case. The rights of women as individuals were threatened by the assumption that they could appear before the law only as wives and mothers. In treating all women differently from all men the law perpetuated the "separate but equal" doctrine that had afflicted blacks in American society. As it had been unfair to blacks, so it was to women.[13]

Eastwood and Murray conceded that classification by sex had been an important means of protecting women earlier in the century. By the end of the 1960s, however, it was clear that such classification had discriminated against women. Proof came in unanticipated results of the Civil Rights Act of 1964 and the enforcement of Title VII, which prohibited discrimination on the basis of sex in employment. In administering Title VII EEOC had found that state laws protecting women from hazards in the workplace had actually "protected" them from high-paying jobs. In 1970 the Labor Department's Women's Bureau reported that "such laws and regulations have ceased to be relevant to our technology or to the expanding role of the female worker in our economy." These work rules did "not take into account the capacities, preferences, and abilities of individual females" and therefore tended to "discriminate rather than protect." Thus one of the reasons for opposing the Equal Rights Amendment had disappeared with the passage of time. Accordingly, the bureau reversed its previous position and endorsed ratification.[14]

Amending the U.S. Constitution seemed logical to feminists not only because classification by sex penalized women but also because the Supreme Court seemed unlikely to uphold the principle of sexual equality without a constitutional change. Women's rights litigants had appealed to the Fourteenth Amendment, hoping the Court would extend to sex the clause forbidding a state to "deny any person within its jurisdiction the equal protection of the laws." In extending that clause to cover laws penalizing persons on the basis of their national origins, the Court had made race and nationality "suspect" classifications. Laws so classifying persons would no longer be subject to "ordinary" scrutiny—which meant only that they bear a reasonable relationship to some legislative purpose. Rather, they would automatically be regarded by the Court as "invidious" (based on unreasonable group antagonism) until proved to be justified by a compelling legislative purpose.

As late as the 1960s, however, the Court had declined to make sex a suspect classification. Even when applying ordinary scrutiny, it did so in such a manner that the "flimsiest" of reasons were regarded as adequate to justify classification by sex. The case of *Hoyt* v. *Florida* (1961) is an example. Upholding a Florida law exempting women from jury service on the antiquated basis of separate spheres, justices held that "woman is still regarded as the center of home and family life," hence the reasonableness of Florida's law. In 1969, still using the minimal standard of scrutiny in a case involving prisoners' rights, the Court held that statutory classifications would be set aside "only if no grounds can be conceived to justify them" as having some rational relationship to a legitimate goal of the state. As the case against sexual stereotyping mounted, the Court would soon begin a major review of the standard of reasonableness, transforming the equal protection clause into a potent weapon with which to challenge state laws discriminating against women. At the time, however, feminists had only the record of the Court to go on, and the record was not good for women.[15] Amendment advocates concluded that waiting upon the Court and the Fourteenth Amendement was like waiting for Godot.

Classification by Sex: Senator Sam to the Rescue

That women would wait no longer was indicated by a public demonstration. On February 17, 1970, a group of women interrupted a hearing of the U.S. Senate Subcommittee on Constitutional Amendments to demand that it consider an Equal Rights Amendment. A few weeks later, in March, the Citizens' Advisory Council on the Status of Women sent a report to President Richard Milhous Nixon to enlist his support for ratification. The council explained with appropriate citations to specific cases that neither the Fifth nor the Fourteenth Amendments could redress grievances against women. Filling an egregious void in constitutional law, an Equal Rights Amendment, argued the council, would guarantee equality of rights with respect to such things as inheritance, property rights, employment, divorce, and governmental action. Passage would not force women to do things for which they were physically unsuited, or weaken laws against rape, or reshape personal attitudes. It would not abrogate the right of privacy or deny dependent spouses special consideration in divorce settlements. It would benefit both sexes.

Actions of demonstrators and councillors had their desired effect. On May 5, 1970, Senator Birch Bayh (D-Ind.), chair of the Subcommittee on Constitutional Amendments, began hearings on ERA. With 72 senators and 223 representatives sponsoring the amendment, it seemed to enjoy the kind of support usually reserved for the three-day weekend. The hearings repeated in more detail much of what the Citizens' Advisory Council had written to President Nixon, ranging over myth, stereotype, general conditions, and the vestigial discrimination of common law. Women expressed outrage: first, from feminists who believed it ludicrous to have a hearing on conceding something so obviously just as "equality of rights"; and then from a woman who attacked this "mad whirl toward equality and sameness." Like the League of Women Voters so many years before, she wanted specific laws for specific wrongs. Most of the seven hundred pages of testimony was favorable to ratification, however, and the amendment was referred positively to the Judiciary Commitee without a dissenting vote on June 28.[16]

In the House the amendment was blocked by a longtime opponent, the chair of the Judiciary Committee, Emanuel Celler (D-N.Y.). Representative Martha Griffiths, however, used a discharge petition with 218 signatures to move the amendment to the floor of the House. It passed 352–15 on August 10 and was sent to the Senate for action. There, Senator Samuel James Ervin, Jr. (D-N.C.) took up where Celler left off and became the amendment's chief antagonist. Since 1954 he had been an eloquent spokesman for a restricted constitution and minimal government in defense of the white South's constitutional rights and in opposition to the civil rights acts of the 1960s. Consistent with this position, he had warned in 1957 that the Equal Rights Amendment would do more to centralize federal power than any action since passage of the Fourteenth Amendment.[17] He was afraid that an ERA would grant to federal courts the jurisdiction to review state laws covering marriage, divorce, property rights, and work. Suspicion of federal power and the nefarious changes wrought by the Fourteenth Amendment were not, however, the basis upon which Ervin attacked ERA when it came to the Senate in August 1970.

The issue was gender. The amendment, he believed, would abrogate all "rational" distinctions between men and women in the law; there were no "irrational" distinctions between the two. These distinctions were based on "physiological and functional differences" between the sexes so natural and sacred as to have had a moral economy based upon them. Men possessed muscular strength; women, unless they were feminists perhaps, possessed intuitive power with which to distinguish between "wisdom and folly, good and evil." Men thus responsible for protecting women, who in turn, provided moral order and the motivation for defending it, both of which together made civilization possible. Since in the fusion of physical necessity and moral obligation women bore and reared children, they were to be excluded from "arduous and hazardous activities." Ervin's argument was the same as that of the Supreme Court in *Muller* v. *Oregon*. "Any country," he warned, "that ignores these physiological and functional differences, is woefully lacking in rationality."[18]

Considering Ervin's position helps highlight both sides of a complex issue in their first collision. Much anti-ERA material was based on Ervin's public statements. Understanding him helps understand the way many male legislators thought of the amendment when they read his anti-ERA speeches, which he sent to them as part of a nationwide lobbying campaign. Ervin believed he was defending timeless values against ephemeral fad, superficial posturing, and political pressure, all of which he identified with feminists. Their refusal to concede the rationality of legal distinctions based on physiological and functional differences imperiled society. He believed, therefore, that further hearings were called for, this time before the Senate Judiciary Committee with himself in the chair.

The renewed hearings began September 9, 1970.[19] In them opponents expressed the same reluctance to surrender conventional gender roles that motivated Ervin. To be sure, the testimony was often presented within the framework of legal objection, but the structure itself was sustained by appeal to traditional views of gender. The "absolute" grant that "admits of no exception" continued to worry Paul Freund. ERA allowed no legislative discretion, he found, conceding to an insistent Ervin that the amendment would prevent passage of laws for the "necessary protection of women," even those justified by physiological and functional differences. This draconian inflexibility meant that protective legislation would be repealed. And, given his understanding of capitalist *real politique,* Freund did not believe that ERA would extend protection to men. The business lobby would prevent it. Like working women who observed men on the loading platform and assembly line, he was wary of tinkering with protective legislation even though it was then in the process of being dismantled. The concrete and troubling problems women faced could not be solved by application of the principle of equality, Freund believed: the concept was too vague. His colleague from the University of Chicago, Philip Kurland, agreed, scolding feminists by implication when he deprecated "the temper of our times that demands instant and simplistic solutions to complex problems."

Uncertain though they professed to be before a future shaped by the Equal Rights Amendment, Kurland, Freund, and other lawyers opposed ratification because it called into question roles that had been attributed to women by tradition and law. Thus "equality" meant that women would be subject to compulsory military training. It meant that divorce, child custody, and family law in general would have to be scrutinized; it might mean that single-sex colleges would be illegal. Opponents did not consider whether such traditional patterns did in fact deprive women of equal protection; rather, the very fact of change—regardless of need—was thought to be sufficient reason for opposing ratification. Change was linked with sex- and gender-based anomaly to such an extent that the possibility of unfair treatment of women under current law was rarely addressed. Throughout the hearing wisped the insistent voice of the senior senator from North Carolina saying that ERA would forbid Congress from ever making a legal distinction between women and men. He paid no attention to the law professor who told him that that interpretation was not accurate. Ervin was somewhat surprised when the man who told him this, Leo Kanowitz of the University of New Mexico and one of his own witnesses, presented a brief on behalf of ERA. Kanowitz admitted to a chagrined Ervin that he had indeed once been suspicious of ERA but that he had changed his mind. That was not what Ervin wanted to hear, and he did not ask why Kanowitz had changed his mind. Consistent with this attitude, Ervin was often absent while proponents of ERA testified. Wilma Scott Heide, a sociologist and chair of NOW's Board of Directors, accused him of playing charades.

Statements from witnesses were not the only material upon which Ervin had to rely in staking out his position. He asked states' attorneys general throughout the country to evaluate the impact of passage. Responses to his inquiries are interesting for what they reveal not only about the opinions available to Ervin but about attitudes toward the amendment before it was submitted for ratification.[20] Some attorneys general, such as Slade Gorton of Washington, objected to the "drastic and wholesale inhibition" upon the state legislature but pointed out that his state had a law against sexual discrimination in employment anyway. Others replied that ERA would affect protective legislation for women but added that their states had already repealed these laws. A few replied laconically that ERA would have little significant effect. Louis J. Lefkowitz of New York thought there would be little direct effect because the state legislature had already changed protective legislation. He feared that there might be some short-term disadvantages for housewives and homemakers but did not explain. He found it regrettable that women had faced such "delays and recalcitrance" in their "search for equality" as now to require "the difficult task of working out a constitutional amendment which would be fair in its application to all women." He did not say it ought not to be done.

Most letters anticipating changes in protective legislation suggested that the issue was insignificant. Most predicted change in family and property law, in divorce and child custody cases; some thought laws relating to crimes associated with sex offenses would have to be reworded. Attitudes toward this change varied from the nonchalant to the skeptical to the alarmist, although there were few of the latter. The reply from North Carolina's attorney general, Robert B. Morgan, was among the most complete and indicated a thoughtful survey of technical changes that would make laws more precise and evenhanded. True, there would probably be changes in the conveyance of property and the wording of family and divorce law, and in the provision forbidding "lewd women" from living within three miles of colleges. (One can imagine Morgan chuckling at that.) It was also possible that North Carolina income tax laws would have to be rewritten—something that subsequently happened without passage of ERA. But Morgan's report was like that of colleagues from other states. Few expected great change from ratification. Changes in family and divorce law then already under way were expected to be sustained by ERA and not thought to be very significant. Many of Ervin's respondents shared the views of the president of the Texas Bar Association, who thought that the Equal Rights Amendment was "meaningless" and merely "a political gesture."[21]

Ervin concluded, however, that changes in the law could be *vast* if ERA were ratified. He interpreted the probability of change as indicating not a need but a danger. He interpreted the double standard of the law as not shameful but sacred, the archaic assumptions and language of laws relating to "sex" crimes not as damaging to women but as protecting them. He paid no attention to arguments in favor of ratification; he never acknowledged the impact of the Civil Rights Act of 1964 on protective legislation even though he was unalterably opposed to the act. Since he implied support of protective legislation when he attacked ERA, he should also have attacked the act for placing women at risk; but he failed to do so. He never considered rulings of EEOC that such legislation was discriminatory. He need not have agreed with the conclusions of the commission if an honest doubt was raised by ambiguities or conclusions based on ambiguous evidence—and experience has revealed such doubts justified. Perhaps he so detested the commission's activities that he could not bring himself to rely upon it for information. More likely, however, his hostility to ERA was based on the noticeable increase in anomalous gender relationships that he identified with feminism. This conclusion is suggested by the way in which he warned that ratification would lead to changes already taking place: former husbands' claiming alimony, wives not taking their husband's name, married couples living apart.

Following Ervin's hearings, the Judiciary Committee reported ERA positively (15–1) to the full Senate for debate in October.[22] There, Ervin attacked "militant women who back this amendment. They want to take rights away from their sisters" and pass laws "to make men and women exactly alike." He emphasized the differences between himself and proponents by discussing *Hoyt* v. *Florida,* in which the Supreme Court had failed to extend to women the equal protection clause of the Fourteenth Amendment. In the House, Representative Griffiths had cited the case to show why women could not rely on the Court's interpretation of this clause to extend them protection. Ervin observed that he had as much right to comment on *Hoyt* as did Griffiths. Hoyt had been found guilty of second-degree murder by an all-male jury and appealed the verdict on the ground that women had been excluded from the jury, thus denying her a trial before her peers. The Supreme Court denied her appeal. In thus refusing to extend the equal protection clause of the Fourteenth Amendment to women when it should have done so, feminists argued, the Court had demonstrated the need for a constitutional amendment specifically designed to guarantee women's rights. Ervin defended the Supreme Court. It had actually protected Hoyt, he said, because placing women on the jury would have guaranteed her conviction of murder in the first instead of the second degree. Why this was so, he did not explain. Shifting the focus from Hoyt to the privileges of women, he observed that Florida law already allowed women to serve. They could simply call the clerk of the court to insist they be placed on the jury list. Ervin thought this a minor effort constituting sufficient protection for women if not for the accused.

Senator Bayh asked Ervin a question: "Why should she not be treated as men, especially under this particular statute?"

Ervin: "Because there is an old Spanish proverb which says that an ounce of mother is worth a pound of priest. She should not be required to leave her children, who need her far more than the court does. This statute says: 'Treat me like a mother, whose first obligation is to her children.' That is what the statute says. It treats her like a mother."

Bayh responded: "Mothers also happen to be citizens."

Ervin: "I believe in doing something special for mothers. The senator from Indiana may believe that mothers should be treated like everybody else, but I do not believe that." Any mother, he insisted, could now "serve on a jury even if she has to neglect her children."

Ervin was determined to resist the encroachments of feminism. The torrent of angry, scolding, and pleading letters and telegrams pouring into his office from business and professional women indelibly stamped on his mind not their arguments—which some developed at great length—but the impression that women who wanted passage of the Equal Rights Amendment were a minority who worked outside the home. Exceptions to the general rule of what women ought to be, they did not know how to use available laws to fight sexual discrimination. Worse, they were willing to sacrifice the rights of women for their own selfish interests. They were undoubtedly sincere, he acknowledged with the kind of condescension one reserves for thoughtless children and impulsive adults, but wrong. Given his solicitude for mothers, it is not surprising that toward the end of debate in the Senate he attempted to amend ERA with his own version of the Hayden rider. Condemning feminists' attack upon "the wives, mothers, and the widows of this country," he submitted an amendment exempting laws designed to enable women "to perform their duties as homemakers and mothers."

Supporters were unprepared for the positive response to Ervin's rider, which, like that of Senator Hayden before him, doomed the amendment once again. Both friends and enemies of the amendment finally agreed to defeat it.

Ervin versus Rights: Civil and Equal

Birch Bayh remembered his respected adversary with affectionate chagrin. "Sam did not in his heart of hearts understand the kind of problems women were facing. . . . He was very bright and yet could not see what was going on under his nose; he just couldn't see it." In historical perspective, however, it is not surprising that Sam Ervin should have opposed the Equal Rights Amendment. He had, after all, fought every piece of civil rights legislation presented while he was in the Senate.[23] If he did not believe that black Americans had grievances that should be redressed by federal legislation, he could not be expected to concede the same to women. He had been in the Senate for almost two years when he and John Stennis (D-Miss.) helped Richard Russell (D-Ga.) write the Southern Manifesto signed by nineteen senators and eighty-two representatives in response to the historic Supreme Court decision *Brown* v. *Board of Education* (1954). The Court had acted, they insisted, in an unconstitutional manner. The next year Ervin fought the Civil Rights Act of 1957 and then in 1960 another civil rights act and again in 1963 and 1964. He insisted that the issue was resistance to governmental tyranny and denied—although heir to the Revolution of 1898 for which his father had stumped the state—that blacks had ever been deprived of the right to vote in North Carolina. Much as he attempted to cast the issue in the dispassionate language of the law, he could not hide his defensiveness as a white Southerner. The attempt to guarantee the voting rights of black Southerners was, he charged, "an insulting and insupportable indictment of a whole people." It is therefore not surprising that he was an important figure in the longest Senate filibuster in history against the Civil Rights Act of 1964 or that he should have tried to destroy the Equal Employment Opportunity Commission. His targeting it in 1971 and again in 1972 lends perspective to his insistence that ERA was not needed because women could already obtain redress through the EEOC—precisely that agency he was most determined to abolish.[24]

The senator, whose father had told him vivid stories of white suffering during the Civil War, was profoundly hostile to the ideal of equality—especially as unfurled by Yankees and blacks. The ideal was too ambiguous to be a sure guide in law and was especially mischievous when the power of government was placed behind it. As Ervin told the Tennessee Bar Association in 1964, Americans must choose between "equality coerced by law and the freedom of the individual"—they could not have both. Or, put another way in the same speech, "No men of any race can law or legislate their way either to economic or social equality in a free society." It can be safely assumed that Ervin felt the same way about women. He thought the issue facing federal lawmakers was freedom of choice, although he had never been, as a law school dean pointed out to him, "similarly exercised over numerous state statutes" supporting segregation and restricting freedom of choice in North Carolina. But Ervin was not so concerned with conditions that prompted civil rights legislation as he was with the people against whom it was aimed. He quoted Associate Justice John Marshall Harlan to the effect that every American has the freedom to "use and dispose of his property as he sees fit, to be emotional, arbitrary, capricious, even unjust in his personal relations."[25] By intervening to secure voting rights, integrated public accommodations, and integrated schools, the government was preventing individuals from choosing their associates.

The individual's range of action was far more important to Ervin than the social context which limited that action. When civil rights activists challenged racist politics, Ervin denounced their blaming blacks' shortcomings on whites. When they tried to change human behavior, he accused them of trying to remake a world impervious to human manipulation. Individual responsibility was important to Ervin; he had little sympathy for those who shirked it or whined about it. "I certainly agree with you," he wrote a college professor, "that 'Snivel Rights' is a pretty good name for the so-called Civil Rights Bill."[26] Reformers were snivelers, and the politicians who aided them, irresponsible sycophants. They should "quit telling Negroes that all of their problems are due to racial discrimination and that they can be lifted to a more abundant life by an act of Congress." Worse than sycophants, they aspired to a tyrannical "thought control" which even his Presbyterian God had not ordained. "God depends upon the persuasive power of the Gospel," he lectured a minister with whom he was particularly angry, "and not upon the coercive power of man-made laws for the regeneration of the heart of human beings. . . . If God had intended to coerce all men into doing right under all circumstances, he would have made them puppets." He thought theologians should stay out of politics; the Protestant Reformation, of course, had allowed politicians in theology.[27]

The "tyranny" of trying to change culture through governmental "edict" was dramatically spelled out for Ervin after 1970. And if anti-ERA letters to him are any guide, people felt the same as he throughout the United States; their response to the Equal Rights Amendment was affected by federal policies desegregating public education. Ervin's most intense work on the Equal Rights Amendment from the spring of 1970 to March 1972 was carried on at the same time he was confronting busing crises in North Carolina. The most publicized incident was that instigated by a federal court judge who in 1971 ordered busing between Charlotte and its surburbs to achieve racial integration of public schools. On behalf of the Charlotte-Mecklenburg Classroom Teachers' Association, Ervin filed a brief with the U.S. Supreme Court arguing that the court order denied "thousands of children of their equal protection of the laws." The Supreme Court did not agree with Ervin; it rarely did. Neither did the Department of Health, Education, and Welfare (HEW). Its directive to the Wake County Board of Education in July 1971—three hours' drive east of Charlotte—seemed to surpass other busing policies in tyranny. The schools of Cary in that county had been integrated in 1967, but HEW thought the county if not the town remiss in its plan for integration and ordered it to move students throughout its jurisdiction to achieve a 75–25 racial mix in each school. Cary's parents were perturbed. In the fall children who had walked to school in the spring would have to board buses for a twenty-mile ride. Letters poured into Ervin's office by the hundreds. Displaced Yankees and native Southerners, integrationists and segregationists, black and white, professionals and working-class people objected. "HEW," accused one man for his fellow parents, "is the voice of POWER and not the voice of REASON."[28]

Such power was dangerous. Sam Ervin seemed always to have believed that aphorism with rigorous consistency.[29] To be sure, he had not been concerned with the power of individuals or corporations, but he was—as he said in the civil rights debates—concerned about defending individuals from arbitrary acts of government. In defining the dialectics of power, he understood the polarity of individual and government if not that between races and sexes. Thus he authored an act in 1966 to allow poor defendants to be released from jail pending trial. He wrote legislation to create a system of federal defenders, enlarge legal protection of the mentally ill, reform the system of military justice, and guarantee to native Americans living on reservations the same rights as other Americans. He attempted to safeguard citizens from governmental surveillance and blacklisting. He tried to protect citizens from the government's restriction of civil liberties in the District of Columbia through preventive detention, draconian sentencing, and the invasion of private homes by police without warning. He vigorously defended First Amendment guarantees of a free press and the absolute separation of church and state. There could "be no real freedom of religion," he wrote, "unless government keeps its hands off religion and religion keeps its hands off the government."[30]

In all these things, Ervin opposed government's power to restrict the rights of individuals. He would consistently, therefore, oppose granting government power to restrict the rights of the individual who chose, in Harlan's word, "to be emotional, arbitrary, capricious, even unjust in his personal relations." Any assumption that Ervin could have supported ERA because it restricted government from imposing on individual women arbitrary and stereotypical standards has to be placed against his absolute opposition to the civil rights movement and its political allies. If he saw in civil rights acts the attempts by tyrannical government to shape the personal relations of its citizens, he would have no difficulty interpreting the feminist agenda in the same way. Since he thought that physiological and functional differences were basic and unchangeable, he could hardly be expected to favor amending the Constitution to impose on American women an understanding of gender that he seems to have believed was, borrowing Harlan's words again, "emotional," "arbitrary," and "capricious." If he would not use government to prevent such a combination of adjectives in personal relations, he would certainly not use government to impose it.

Senator Sam was also a traditional gentleman with a sense of honor surpassing that of most men. While fighting in the trenches during World War I, he had led a valiant charge against a German machine gun nest in a heroic act that almost cost him his life. He later championed academic freedom in the state legislature and, still later, as a judge, faced the personal meaning of sitting in judgment on others, finally resigning his seat because every time he sentenced a felon, something good within him died. Living by a strict code of civility, he had little sympathy for a politician like Richard Nixon who gained office by lying about such a decent man as Jerry Voorhis of California. As he expected politicians to act like gentlemen, he expected women to act like ladies. The perceived lack of civility of "women's libbers" seemed to offend him, possibly because it indicated disrespect for themselves as well as other women. When he tried to inscribe their "proper" role—"homemakers and mothers"—into the Constitution, he did, to be sure, betray an insensitivity to women's position as great as that relating to blacks, but he meant it as a tribute. It was probably not tradition Ervin defended so much as his understanding of what was moral and reasonable, and feminists seemed to be especially *un*reasonable. Their failure to understand the horrors of war in dismissing the significance of drafting women must have been a case in point. His own experience in the trenches of the First World War made him unable to comprehend a demand for sexual equality that embraced terror, mutilation, and death.

ERA as Symbolic Feminism: The *Yale Law Journal*

Senator Ervin was once again chief antagonist of ERA when it was submitted to the Ninety-second Congress in 1971. He had been able to prevent passage in the fall of 1970 by appending amendments unacceptable to ERA's supporters. Until the spring of 1972, when ERA finally passed Congress, it brought torrents of mail warning of the danger to working women, wives, widows, and mothers. Once again there was support from the Nixon administration, a host of women's organizations and a broad range of voluntary and governmental associations. There was also something new in an authoritative interpretation of what ERA would mean. It was in the form of a long, scholarly article from the *Yale Law Journal* of April 1971 and reprinted in the *Congressional Record* for October 5, 1971. Written by the Yale law professor Thomas I. Emerson, with Barbara A. Brown, Gail Falk, and Ann E. Freedman, it would soon be cited by opponents and proponents alike.[31]

The Equal Rights Amendment, the Yale article said, offered a "broad re-examination and redefinition of 'woman's place.' " In their analysis of classification by sex, the authors focused on sexual stereotypes and traditional gender definitions that, when reflected in law, denied women the right to appear as individuals before the bar of justice. The assumption that there was a special place for women in society restricted their freedom. Thus women needed a "bold new direction" in redefining their "place" through a process that would first abolish the separate treatment of women from men in the law. Beyond the law, social change was the implied positive goal. "The political and psychological impact of adopting a constitutional amendment will," the authors said, "be of vital importance in actually realizing the goal of equality." The amendment will "have effects that go *far beyond the legal system.*" The meaning of that phrase, "far beyond the legal system," was not altogether clear, as it was more evocative than analytical. The multiplier effect would theoretically result from the fact that "all aspects of separate treatment for women" were inevitably discriminatory and "inevitably related." Racial segregation was the negative paradigm in the authors' minds: separate treatment created "a dual system of rights and responsibilities" which meant inevitably that one group would be dominant.[32] To solve the problem of male dominance as Americans had addressed the problem of white dominance meant that sex, like race, could no longer be "a permissible factor" in determining "legal rights." Classification by sex was to be understood as "overclassification" because it placed a person before the law as representative of an abstraction rather than as herself. This classification, the authors argued, denied the "rights of each individual to develop his or her own potentiality. It negates all our [American] values of individual self-fulfillment."[32]

The language employed by the authors was that of social ideals rooted in the liberal tradition, but the expectations were profoundly subversive. Posing the language of individual rights against stereotypes entrenched in law and custom reflected the feminist assumption that socially constructed differences between the sexes were the chief source of women's oppression. Weakening the legal supports for these constructions would help weaken the *systemic* oppression of women. In this authoritative statement and in the positive response to it among their constituencies, ratificationists were affirming ERA as a means of subverting male dominance. Separating female individuals from their anatomical distinctiveness before the law was believed to be beneficial to them. There were to be exceptions, to be sure, because not all distinctions based on sex were stereotypical, but the goal was clear: to challenge the allocation of resources, property, and privilege according to gender. The tone of the article's introduction indicated a far more wide-ranging and radical effect than that which scholars believed would have resulted. There was to be some disagreement over whether the amendment would have made sex a prohibited rather than a suspect classification, although the latter seems to have been more likely. Even that less radical change, however, would have indicated—as the authors concluded, that society was "prepared to accept and support new creative forces."[34]

Feelings evoked by these "forces" flowed freely from letters to Senator Ervin from ratificationist women of all ages.[35] In their animated expectation of change, they fused justice and equality and blurred the distinction between equality and equity. They had been penalized by traditional views of gender, they believed, and were enraptured by the discovery that there were reasonable and convincing ways of making it illegal to define persons by sex. The Equal Rights Amendment was so well received by millions of American women because it seemed to be addressed so precisely to their own experience of restrictive gender definition. It is not surprising that the Yale article should have become holy writ.

Ervin on Emerson: The Argument against Ratification

Like the Bible, where much depends on the interpretive viewpoint of the preacher, the Yale article became as authoritative for opponents of ERA as it was for proponents. This incongruous situation resulted from Sam Ervin's reading into the *Congressional Record* an attack on ERA purportedly based on the arguments of Emerson and his co-authors. While accepting the article's authority, the senator distorted its argument and conclusions.[36] After the amendment was sent to the states for ratification, Ervin's speeches against passage in the Senate were circulated to legislators and lobbyists throughout the United States. In them, Ervin pleaded for the traditional view of women in which gender (culture) and sex (anatomy) are fused. He ignored the Yale article's analysis of how sexual stereotyping deprived individuals of their rights. In a simpler context of arbitrary governmental action he would probably have sided with the individual, but, as we have seen, he did not acknowledge the power of social convention to restrict human rights. By neglecting to address the way in which classification by sex could impede an individual's access to the bar of justice, he acted inconsistently with his own ideal of individual freedom—at least Birch Bayh thought as much. Undoubtedly Ervin believed that when a female was treated as a woman she was respected as an individual, in stark contrast to the argument that sex-based differentiation ignored the individual who did not fit the stereotype.

Ervin paid no attention to the clear explanation that ERA would not affect legislation in which the biological sex of the parties was relevant. For example, laws allowing leave for childbearing would be acceptable when applied to one sex, but leave for childrearing would have to be allowed either sex. The purpose of the amendment, to abolish the "dual system of sex-based rights and responsibilities," was presented by Ervin without explanation of the difference between overclassification and reasonable classification, even though he often posed "reasonable" distinctions between the sexes against the "unreasonable" equality to be achieved under ERA. He ignored the context from which he ripped individual statements and failed to address the essential core of argument about what, in fact, was reasonable and unreasonable.

Ervin's arguments were frequently a mixture of alarmist expletives and unexplained implications; these were dramatically obvious in his discussion of the impact of ERA on the military. Of twenty separate items quoted from the Yale article in a section on the military, nine amount to one essential point: that the armed forces—because of past unequal treatment of women—would have to change. Indeed ERA claimed Ervin, would "require a radical restructuring of the military's views of women." Ervin interpreted this statement of undeniable contemporaneous bias against women as a reason for not passing the amendment; he was opposed to "radical restructuring" regardless of need. Equality of treatment for women military personnel, despite the many instances cited in which it was needed, was thought by Ervin to be absurd. He approached the issue not from the viewpoint of service women who had suffered sexual discrimination but from that of people who thought of men and women in terms of stereotypes. The idea of women warriors was so foreign to Ervin's way of thinking that he could not address the problems of the many women already in the military. More of a nuisance than they were worth, military women would have to have "athletic facilities" made available to them if ERA passed. Why that should have been an argument against ratification is not clear.

The essential point in Ervin's discussion of the military relied on the stark fact that, under the Equal Rights Amendment, women would have been subject to registration for a possible draft. Once drafted, he feared that qualified women would serve in combat zones. This issue—especially in the traumatic aftermath of the Vietnam War—was complex and provocative whether faced starkly or placed in a nest of careful explanation. Even if national interest and public safety were placed above equality as a priority in constitutional rights, and even if women in frontline combat might be relatively rare, the impact of ERA would undoubtedly have been to force scrutiny of traditional assumptions about gender and military service. That this scrutiny would have changed battlefield policy radically is uncertain. Much legal opinion held that ratification would have a limited impact on military commanders' discretion about who would serve in combat. ERA would probably not have changed rules on participation in combat so much as the system of rewards associated with it. But what "probably" would not have happened was beside the point to Ervin. He was concerned about the possibility that steps to ensure equality of the sexes would lead to the death of combatant women. With the most radical feminists, he thought it a matter of consistency. To a less radical one he wrote: "If you admit that women generally would not be eligible to serve in combat, you are in effect admitting that there are valid distinctions between men and women which should be reflected in legislative proposals—and ERA would destroy all laws that make distinctions between men and women."[37]

Ervin was, of course, wrong; "valid distinctions between men and women" were not to have been abrogated by ERA—merely those proved to be invidious in their effect on individuals. Pro-ERA publicists never argued that women were the same as men; they never denied the legitimacy of laws making reasonable classifications such as those relating to pregnancy. Conceded differences between the sexes should not be used to deprive either sex of its rights. If there were differences between the two sexes relative to their effectiveness as warriors, these differences could be stated specifically so as to weed out undesirable men as well as women. The argument was clear in the Yale article, but it made no impact upon Ervin. The idea of separating social role from sex, although familiar to feminists, was largely unacceptable to nonfeminists. Ervin seems to have considered the idea nonsense. His total lack of concern about the way women were treated in the military suggests that as anomalies they should have received no "special consideration." A man who denies that blacks have any legitimate complaint in this society would certainly not be receptive to the complaints of women unless they wanted to be treated "like women."

Women who behaved like women were protected by the law, Ervin believed; but he charged that the Equal Rights Amendment would change that. He distorted the impact of ratification upon the crime of rape; indeed, he misrepresented the Yale article in such a way as to reverse its meaning, claiming that ERA would abolish punishment of sex-related crimes. Hysterical charges that ERA would declare the Mann Act unconstitutional were based on statements attributed by Ervin to Emerson and his co-authors but ripped out of context to create an aura of sensationalism. Scrutinizing the Mann Act under cover of the Equal Rights Amendment did not mean that women could now be transported across state lines for immoral purposes with impunity—but that is what Ervin implied. Scrutiny of laws originally intended to protect women seemed dangerous; appropriate rewording of laws to have them apply to both sexes seemed frivolous. Moreover, for some reason Ervin was appalled that ERA would forbid "*finding legislative justification in the sexual double standard.*" He italicized the statement either because he was confused or because he literally did not want an end to the sexual double standard, a strange position for one holding such high standards of personal morality as he.

As for family law, Ervin's selections from the Yale article reflected the belief that ERA was a declaration of war on homemakers. After ratification,

Ervin argued, citizens could not dictate "different roles for men and women within the family on the basis of their sex"; the argument was very much like the warning of domestic warfare associated by opponents with women suffrage. What Emerson and his co-authors had actually said was that authorities would no longer be able to dictate such roles through stereotyping in court cases, laws, or judicial decisions. No one seriously believed that the amendment would allow women to haul their husbands before the court in the allocation of domestic duties—although few have probably not daydreamed about the possibility. The Yale article pointed out that the suggested amendment left couples free to allocate "privileges and responsibilities" according to "their own individual preferences and capacities." New laws could even have been passed basing "marital rights and duties on functions actually performed within the family, instead of sex." The dispassionate tone would probably not have reassured traditionalists, however.

Such personal freedom as the Yale authors envisioned, encouraged by the resurgent women's rights movement and embraced by men as well as women, was indicted by Ervin as a subversion of family life. A woman might no longer be required to take her husband's family name or to accompany her husband to a new home. Ervin's argument ignored actual practice and changes already embraced by courts, legislatures, lawyers, and individuals. Despite trends in case and statutory law that had already encouraged the settlement of divorce cases on their merits, Ervin darkly warned that ERA would have required what judges in North Carolina and the rest of the country were then doing—rejecting bias in favor of women in suits concerning child support, child custody, and alimony. The senator not only ignored the state of family law in the United States; he also conveniently ignored the Yale author's argument that domestic law should be rewritten more precisely in terms of parental function, marital contribution, and ability to pay.

As for those women—a majority in North Carolina—who worked outside the home, Ervin charged that legislation protecting them would have been invalidated by ERA. He claimed, for example, "a school board regulation imposing maternity leave at least four months prior to the expected birth of her child would be invalidated" under ERA. The Yale article had indeed said so, but what Ervin neglected to mention was that the invalidation was not of the maternity leave but the imposition of an arbitrary four-month period regardless of physical condition that could unfairly deprive women of needed income. The point had been to show how protective legislation could be discriminatory when based on stereotype—in this case, a stereotype of pregnancy—rather than the facts of each individual case.

In charging that ERA would leave women unprotected in the workplace, Ervin was appealing to what proponents believed were anachronistic assumptions about protective legislation. Prevailing wisdom since the 1920s had been that the stereotype of women's frailty helped protect them in heavy industry. The Civil Rights Act of 1964, however, had in effect voided sex-based protective legislation and with it the objections of organized labor and women's groups to the ERA. The Yale authors had argued with charming optimism that under ERA protection would be extended to all workers without the stereotyped classifications that discriminated against those persons originally thought to benefit from them. The argument became a mainstay of the pro-ERA argument, but it never really captured the imagination of blue-collar women who knew the ways of bosses at least as well as did the Equal Employment Opportunity Commission and Yale lawyers. Persistent and pervasive resistance to governmental regulation of health and safety in the workplace suggests such skepticism was warranted. It was probably reasonable for Emerson and his co-authors to say that they thought "labor legislation which confers clear benefits upon women would be extended to men," but it was not reasonable to believe that employers would actually comply. Working women who had endured the perfidy of employers firsthand therefore could be and were frightened by Ervin's warnings, which were probably based more on traditional views of womanhood than on concern for women workers.

Ervin's indictment of ERA, when stripped of its alarmist tone, was accurate with regard to the draft although the issue of women in combat was much more complex than he acknowledged. Statements on protective legislation were based on concerns that had motivated feminist opponents of the Equal Rights Amendment in the 1920s and were still relevant in an industrial system in which employers could cynically "support" equality by denying protective legislation to both men and women rather than extending it from one to the other.[38] The drafting of women and the invalidation of protective legislation were arguments that convinced by emotion—especially by revulsion at female corpses and the exploitation of "helpless women and girls." This emotion—as well as an appreciation for his homey, commonsense approach—suggests what Senator Ervin was really concerned about. In his statements upon which anti-ERA literature is based are two themes: a defense of traditionalist views of womanhood and a suspicion of the concept of equality itself. When placed within the context of other public statements about ERA, these are reinforced by a third theme derived from the second: antipathy to governmental action to assure equality.

The response to Ervin's arguments in the U.S. Senate was to pass the Equal Rights Amendment 84–8. The amendment had already passed the House of Representatives 354–23 on October 12, 1971, and it finally passed the Senate on March 22, 1972, after nine amendments submitted by the North Carolinian were rejected. Ervin had wanted to exempt women from compulsory military service (defeated 18–73), and combat units (defeated 18–71); to exempt protective legislation (defeated 11–75); to exempt wives, mothers, widows (defeated 14–77); to exempt laws imposing upon fathers' responsibility for supporting their children (defeated 17–72); to exempt privacy from being affected (defeated 11–79); to exempt sexual offenses (defeated 17–71).[39] In view of the overwhelming defeat of his efforts in the Senate, it might have seemed as if Sam Ervin were engaging in a form of Calvinist masochism, but Senator Sam was not feeling masochistic.

By submitting his exemptions and provisions to the Senate, Ervin had created a record with which to mobilize opposition in the ratification process.

In speeches to his colleagues and especially in his exposition of the Yale article, he had written some of the major pamphlets of the antiratification campaign. With rejection of his amendments by the Senate, he could now claim that ERA would require military service for women if the draft were reinstated and that women would serve in combat units. He could also charge that the amendment would withdraw protection from women workers, wives, mothers, and widows; force the sexual integration of public rest rooms; abolish the crime of rape; and relieve men of responsibility for supporting their families. The record was clear; the Senate had had a last chance to prevent such results but had failed to do its duty. If Ervin's arguments had not convinced the Senate, he would turn to the people, and especially the women of America, for vindication and victory.

The Ervin-Schlafly Concordat

When Ervin made his last-ditch effort in the Senate on March 22, 1972, he was already a hero to conservatives. His celebration by the media during the 1973 Senate hearings on the Nixon administration's connection with the infamous Watergate break-in followed by one year his celebrity as an opponent of feminism and the Equal Rights Amendment. When he lost the votes on ERA in the Senate, he had already been made an honorary member of HOW—Happiness of Womanhood—an organization of antifeminist women headed by Jacqui Davidson of Kingman, Arizona. Davidson had met Ervin on a trip to Washington, D.C., in the fall of 1971 and had worked with him to coordinate telegram campaigns against ERA. She was not alone in her admiration of the grandfatherly figure. Telegrams and letters poured into Ervin's office from all sections of the country: from New Jersey—praising his fight for the white men of America; from North Carolina—could he now do something about trash along the highways; from New York—it is "incredible how anyone in their right mind could vote such legislation in." From Louisiana he learned that FOE—Females Opposed to Equality—had named him "Father of the Year." Women from all over the United States were writing to thank him, as a North Carolina woman put it, for his "efforts in opposition to the Women's Liberation Crew."[40]

By February 1972, Ervin had heard from Phyllis Schlafly. "I wish you success!" she wrote, and included a copy of her own attack on ERA in the *Phyllis Schlafly Report.* She had written five books by this date, one of which was the widely circulated ode to Barry Goldwater and the Republican Right, *A Choice Not an Echo.* When Ervin lost in the Senate, Schlafly was ready to fight in the states. Using the readers of her ultraconservative newsletter as a base, she launched a nationwide campaign against the ratification of ERA. In July, fresh from successful fights against ratification in Illinois and Ohio, Schlafly was boundlessly optimistic: "I am sure we can beat ERA nationally, if we organize and use our wits." She was planning a confidential workshop for two key leaders from each state at St. Louis in September and was already pushing opponents of ERA in each state to demand equal time on the airways to gain publicity. Her goal was a national anti-ERA organization, one which could attract well-known women to lend prestige to the cause. By April 1973 she had in place a network of women from Alaska to Massachusetts, and she would very much "appreciate it," she wrote to Ervin, if he would send to the "girls" on lists she had provided copies of his speeches opposing ratification. A year later she had achieved all her goals but one—she was unable to find well-known women opposed to the amendment. "We have all the arguments on our side," she wrote Ervin, "as well as common sense, logic, and the majority of women. But we need visibility at the top." She needed a little glamour and publicity, she confessed. "Can you please do something to help?"[41]

It was natural to turn to Ervin. He had already done a great deal to help his newfound friend from Illinois—even placing his franking privileges at her disposal. In the early days of their association Schlafly had suggested sending at public expense Ervin's speeches as anti-ERA pamphlets to state legislators and precinct committees throughout the country. Almost ten years earlier, she explained, she had developed a similar working arrangement with Senators Everett Dirksen (R-Ill.) and Styles Bridges (R-Vt.). Dirksen had given her 100,000 reprints of his speeches and the privilege of sending them wherever she wished. To recipients and postal authorities, the franked envelopes seemed to have come directly from Senator Dirksen instead of Schlafly's home. Bridges had cooperated in the same fashion, she claimed, and both Senators had given her enough franked mailing labels to reship large packages of materials to like-minded people in other parts of the country who could distribute them where they would do the most good. To help Ervin make the right decision, she had "taken the liberty" of enclosing a copy of one of his speeches—edited to make it more readable—and a "suggested draft" of a letter to accompany it.[42]

Ervin liked Schlafly's idea. He had already been using the *Congressional Record* of his speeches and legislative maneuvers as the basis of antiratification lobbying. Within a few weeks of the amendment's approval by the Senate, he was sending materials to interested constituents, putting North Carolinians who wanted to work against ratification in touch with each other, explaining to others how to influence the North Carolina General Assembly, and referring non-Carolinians to a National Coalition for Accountability, a right-wing group based in Mount Vernon, Virginia. Schlafly's plan was, therefore, welcome, although he altered it slightly to maintain actual control of the mailings from his own office. Schlafly sent lists of addresses of state legislators as the amendment was scheduled for consideration, and Ervin's office dispatched antiratification packets as directed. By the fall of 1972, Ervin's office had become a national clearinghouse and communications center for the opposition.[43] By late winter of 1973, his anti-ERA speeches had gone out to STOP ERA activists in twenty-four states and to legislators in twenty-five states. Opposition women had found their champion. "I feel so important," wrote a woman from Pensacola, Florida, "when I speak and say this is the latest from Sen. Ervin's office."[44]

Unable to prevent the passage of ERA in the U.S. Senate, Sam Ervin used his speeches, office, and prestige as the defender of traditional "womanhood"—in league with Phyllis Schlafly—to resist ratification.

This Florida activist celebrated her importance against people who, she believed, denied it. Ervin's office symbolized faith in herself. It seemed to lend authority to her actions and feelings about feminism; it gave her a sense of participating in a significant public event; it may have provided a harbinger of victory. There was, after all, one man who stood against the fanatic, mindless, angry temper of the times. Ervin was the personification of male resistance to the Equal Rights Amendment. A feminist activist, upon learning that historians were studying the defeat of ratification, insisted that there must have been a connection between Ervin and "the insurance companies"; he must have represented them. And he did, of course, but not as she would like to have believed. He was not their attorney; he was not bought with a fee. He represented them in that both based their anti-ERA positions upon a traditional understanding of what classification by sex meant in law and custom. Although his exceptions were rejected in the Senate by overwhelming votes, Ervin's views represented the etiquette and commonsense wisdom about what sexual distinctiveness meant. Based on formulations of gender roles authenticated by personal experience, these views were resistant to feminist analysis of how stereotype had restricted women. Indeed, what some would call stereotype was stipulated by Ervin to be natural, obvious physiological and functional differences.

Ervin's resistance to governmental intrusion into family life through busing was something with which people throughout the nation could identify. The irrationality of bureaucratic and court-based government would be more than mischievous if loosed under the aegis of a feminist agenda upon the women of America. People who smarted under the lash of affirmative action and the Equal Employment Opportunity Commission could find in Ervin's attack on ERA a satisfying compensation for previous hurts. They could find in his appeal to nature-based responsibilities a commonsense repudiation of the principle of equality. His proposed amendments to ERA became the classic statement of that common sense; Phyllis Schlafly made him and his office useful in her bid to become one of the most well-known women of America. They both understood that a minority of women could dramatize the amendment as feminism itself and in so doing use the stipulation that three-fourths the states concur in amending the Constitution to discredit the ideal of equality.

By the fall of 1972, Ervin began to believe that he could defeat ratification. "Time," he wrote his fans throughout the United States, "is our best friend and weapon in defeating ERA." "Educating the public . . . is going to take time but we are starting to make some inroads." If only "3% of the women in America were able to bring about passage of the ERA by Congress," he observed, "then another 3% working just as hard, should be able to defeat it." "We are holding in more states than we need," he encouraged his dedicated allies, "and I am hopeful that we will enlarge that number."[45] Among that number he thought would be North Carolina.

3

"We Just Didn't Realize It Was Going To Be That Difficult": Political Socialization the Hard Way

When the Equal Rights Amendment was introduced in North Carolina in 1972, ratificationists hoped for a swift victory. Neither they nor their opponents realized that they were enlisting in a decade of struggle; neither anticipated the way in which their lives would change through an unfolding process of organizing, lobbying, and electioneering. Proponents expected to build on the political promises of the 1960s but were thwarted by an intensifying conservative trend. To suffer defeat as they did—narrowly and repeatedly—while at the same time developing a collective consciousness of empowerment was a profoundly paradoxical experience. Compounding the frustration was the fact that each of the six ratification campaigns seemed to generate ever-greater demands as heady visions of victory gave way to sober assessments of all the things they had failed to do. Under such self-scrutiny, assumptions that ERA would be ratified on its intrinsic merits as a legal issue yielded to a new appreciation of political demands made upon partisans of a controversial issue. As they struggled to meet those demands, they learned to be expert players in what was for most of them a new game; they also learned that knowing the rules and executing them well did not guarantee victory.

The experience of women opposing ratification was more rewarding. Outside the formal networks of voluntary and professional organizations, they worked tirelessly at the grass-roots level to gather signatures, build telephone networks, and encourage letter writing. Offended at the movements for women's rights and liberation and frequently allied with ultraconservatives in both religion and politics, antiratificationists were to become identified with the STOP ERA forces of Phyllis Schlafly and the mass-mailing computers of the New Right. Although lacking neither expert organizers nor persuasive speakers, they were less structured as a movement than were proponents. They were not politically "socialized" in the same way as proponents: they did not enter politics in the same way, and they did not lose. Consequently they saw little reason for agonizing reappraisals. As defenders of tradition, they were spared the burdens of innovation in a political culture guided by the adage, "If it ain't broke, don't fix it."

The six campaigns for ratification involved a variety of "stories" about a range of ratificationists, opponents, and legislators. Since ratificationists were innovators, their stories dominate the narrative that follows. Ratification, they believed, was something that they could accomplish through their own efforts, and it is upon these that the first focus rests. Opponents responded by telling legislators that they spoke for women who challenged the feminist revolution. Understanding the Equal Rights Amendment is not based on that narrative alone, however, or upon analysis of the expectations and self-consciousness of ratificationists. Equally important were the experience of women who opposed them and the character of the General Assembly. These need to be understood in their turn; but the place to begin is with the high hopes of activist women in 1972.

Moving Toward Equality

In 1972, the confidence of the ratificationists seemed to be legitimate. Equality seemed to be on the move in North Carolina, where for more than a decade black people had been pushing back the barriers of racism. Despite frustrating resistance, old relationships seemed to be surrendering to new ones under pressure from the civil rights movements and the federal government. The cadence of dissent intensified as the war in Vietnam joined the war against racism in civic consciousness. In the university town of Chapel Hill, conservatively dressed citizens kept vigil in silent protest before the downtown post office. On campus, not-so-conservatively dressed students listened to teach-ins and in 1970 brought the University of North Carolina to a grinding halt in the wake of the killing of fellow students at Kent State. As the struggle against racism and war became a struggle against sexism, women added their own demands to the insurgencey. They began to meet in small groups, exchanging insights, reassessing options, and gaining confidence. It was time to redefine womanhood and self and society. It was a time of possibility and change, with relationships renegotiated, work refocused, and career commitments reinforced. It was a time for forging female solidarity and activism: founding consciousness-raising groups, publishing newsletters, establishing nonsexist kindergartens, and writing nonsexist children's books.[1]

Women also worked to change institutions. At the former Women's College of the University of North Carolina in Greensboro, feminists created a Women's Studies Program to introduce campus and community to ways in which gender shaped personality, social roles, and thought. On the Chapel

Hill campus, predominantly male until a recent equalization of admission standards, local activists in the National Organization for Women pushed for equal access to scholarships regardless of sex. With others, they sought equity in the hiring, pay, and promotion of faculty members. From the town they demanded caring treatment by female personnel for rape victims. For domestic workers they urged inclusion under minimum wage legislation. In Charlotte, NOW attacked discriminatory practices in schools and financial institutions. In the capital city of Raleigh and in Winston-Salem, NOW members also struggled to implement their national agenda. Their efforts were complemented by the North Carolina Women's Political Caucus (NCWPC), which, under the guidance of Martha McKay, proposed to recruit and mobilize women for political action. McKay was a natural politician: in the 1960 gubernatorial campaign she had mobilized support for a victorious Terry Sanford and remained active in the party after he left office. Later the former Democratic national committeewoman moved to the New York headquarters of American Telephone and Telegraph to devise an affirmative action program. While there, she became acquainted with Congresswoman Bella Abzug (D-N.Y.), Betty Friedan, and other founders of the National Women's Political Caucus.

Eager to mobilize North Carolinians, McKay called more than a thousand women from across the state to Duke University on a cold Saturday in January 1972. Striking in their diversity, they included blacks, whites, and native Americans, housewives and students, as well as lawyers and labor organizers. Bonded in newfound solidarity, they scrutinized responses of gubernatorial candidates to questions asked them by McKay. Some made a mental note that one of the leading Democratic contenders had been "unable" to attend: Hargrove ("Skipper") Bowles of Greensboro tried to avoid being linked with the national party by running his campaign independent of the state Democratic party. Others noted that the answers of the Republican candidate—James E. Holshouser of Banner Elk, subsequently elected in the Nixon landslide—suggested an attractive sensitivity to the women's movement. Those in attendance applauded women running for lieutenant governor in both parties and hammered out a political agenda. Juanita Kreps, who would subsequently become secretary of commerce in the cabinet of James Earl Carter, provided a model of intellectual clarity and poise. McKay provided a model of controlled anger and determination. Attacking Senator Sam Ervin's stand on the Equal Rights Amendment, she pledged women of the state a new senator, even it if meant running herself. The feeling of the moment was an exhilarating "empowerment."[2]

The feeling was reinforced by the effective manner in which ratificationists pushed passage of ERA in the U.S. Senate and House of Representatives. Congressional staffers, mobilized by Michigan Congresswoman Martha Griffiths, joined forces with feminists in organizations such as NOW and the more-established women's groups. The result was a tightly structured network subsequently enlarged to include a wide array of women's groups, liberal organizations, and party organs. An ad hoc coalition that exemplified interest group politics at its best, ERA supporters proved to be resourceful and effective. While diligently maintaining contact with key members of Congress and their aides, ratificationists mounted an intensive lobbying campaign suggestive of broad grass-roots support. So effective were they that activists could produce up to ten thousand letters with twelve phone calls. The public seemed to have risen up to demand something as thoroughly American as the decade of civil rights legislation that preceded it. Coming from a Congress that generated more women's rights legislation than all its predecessors, passage of the Equal Rights Amendment indicated that women's issues had finally acquired popular and political salience.[3]

When the amendment was submitted for ratification, twenty-two states fell immediately into place. The rush to ratify was consistent with the experience of women at the 1972 Democratic National Convention in Miami. There the new rules for selecting delegates had made women 47 percent of the delegates and helped them insert a woman's plank in the platform. With the base of delegates, too, women could win support for a female vice-presidential candidate and elect a woman party chair. Martha McKay was pleased: "[We] were admitted in greater number than ever before to what Carrie Chapman Catt called the penumbra of the political process—the shadow which surrounds the dark center. But the umbra of the process was not yet penetrated."[4] The swift positive response to ERA seemed to be similar to the exciting drama at Miami; it confirmed Washington activists' assumption that a national lobbying effort on state legislatures similar to their victorious assault on Capitol Hill was unnecessary. But in McKay's words, the umbra was not yet penetrated. Ratificationists did not understand that each constitutional amendment ratified in the twentieth century had been based on a carefully constructed consensus created at both national and state levels. Nor did they know that their suffragist predecessors were far better prepared to campaign for ratification than they. Years of suffrage agitation had produced a relatively strong central organization ready to mount a national drive through state organizations. North Carolina suffragists had been active for only six years before the Anthony Amendment was referred to the states, but that was five and one-half years more than pro-ERA activists had had to prepare. The national ERA Ratification Council had no staff or resources; it functioned through its constituent groups, which included NOW, the caucus, and traditional women's groups such as the League of Women Voters and the American Association of University Women (AAUW). These groups could emphasize the importance of ERA to state and local affiliates, which were nonetheless essentially on their own.[5]

"A Liberal, White, Middle-Class Coalition"

In North Carolina, caucus leaders took the initiative. Sharing the assumptions of national activists, they expected ratification in the 1973 legislative session and saw no need for an all-out effort that detracted from other projects. Confident that the issue was primarily legal, they entrusted the ERA portfolio to Elisabeth S. Petersen, a young graduate of Vassar College and Duke University Law School. In keeping with the bipartisan character of the caucus, Patricia Locke, a Republican activist from Charlotte, was appointed co-chair. It was assumed that McKay would be available for needed consultation, but her commuting between New York and Chapel Hill meant that the burden of launching the campaign fell to Petersen. It was a lot to ask of a novice lawyer in her first job and before she had taken the law boards.

As in other states, ratificationists undertook a low-key campaign to explain the meaning of legal equality. Women lawyers, whose expertise, demeanor, and dress dispelled any linkage with radicalism, recruited knowledgeable women like themselves from organizations that had already endorsed the amendment. Members of the National Federation of Business and Professional Women (BPW), which had lobbied hard for congressional passage of ERA, were presumably committed, but no one, not even NOW and the caucus, were taking chances with their own rank-and-file. Newsletters argued the need for an explicit constitutional guarantee of sexual equality, relying on excerpts from the 1970 report of the Citizens' Advisory Council on the Status of Women. Petersen and her colleagues also tried to show how the principle of equality would apply to matters such as sexual offenses, overtime work, military service, alimony payments, and child custody and support. In doing so, they provided information and arguments to women who would have to "sell" the amendment against objections crafted by Ervin and disseminated by Schlafly.[6]

Although it was assumed that McKay and other caucus veterans would lobby the General Assembly, their work was to be made easier by addressing members through the mails. Legislators had already received Ervin's comments on the *Yale Law Journal* article. Because the full report of the Citizens' Advisory Council contained a statement-by-statement rebuttal of Ervin's exposition, it was essential to have it in members' hands before January 1973. The council's secretary—correctly scrupulous about such matters—explained that copies of the report were available but that it was "illegal to use the postal frank for lobbying purposes." North Carolina ratificationists would have to pay postage themselves, thus forbidden the sweetheart arrangement that saved opponents so much money.[7] Petersen also sent letters to candidates for the General Assembly asking if they would vote for ratification.[8] The responses were ready for release before election day; the task had been relatively easy.

So was that of finding sponsors for the bill of ratification. State Senator Charles Deane, Jr., son of one of the few Southern congressmen to have refused to sign the Southern Manifesto, wrote offering his service. Petersen accepted. Three women representatives also volunteered, but she persuaded them to co-sponsor, leaving the bill in the hands of Willis ("Bill") P. Whichard. He was, she explained, also sponsor of an abortion bill that he would keep in committee until ERA had passed, thus preventing linkage of the two issues. A man with a sense of history, Whichard was an appropriate choice. The fact that his prosuffrage great-grandfather had once served in the legislature that failed to enfranchise women prompted him to introduce the woman suffrage resolution of 1971. This legacy, and concern for his daughters' future, had inspired his sponsorship of ERA.[9]

Petersen's other tasks proved to be more difficult, especially that of recruiting people to coordinate pro-ERA activities. McKay herself had organized North Carolina women by contacting key individuals in each congressional district who, in turn, mobilized others. Petersen followed her example but found that veteran activists often had other priorities. Opponents of the Vietnam War believed the election of George McGovern more compelling a cause than ratifying ERA. Women with full-time jobs lacked the time to speak, recruit, and raise money. Women who had just lost primary races for electoral office offered to do what they could, but desire and time were not commensurate.[10] Such problems could not be solved by the caucus's creation of an ERA Ratification Council to coordinate lobbying. At first, it included the League of Women Voters, NOW, BPW, and AAUW, but it later attracted other women's groups and the AFL-CIO, which kept a low profile because of the state's antiunion bias.[11] A state-based structure thus existed that theoretically could be replicated at the local level but in fact was not. Problems of time and priorities remained. The mountainous area in the western part of the state and the flat, equally rural, politically conservative eastern part of the state proved especially resistant. Women who were willing, even eager, to work felt very much alone. "I am so isolated here from the ideas that make our movement possible," lamented one ERA supporter, "I will need help to continue even temporarily."[12]

Predictably, local coalitions developed in the urban areas of the piedmont, where there were a variety of women's organizations. But even there the record was mixed. Charlotte, Greensboro, and Winston-Salem organized quickly, pulling in not only constituent organizations of the Ratification Council but also such local groups as the Young Women's Christian Association, (YWCA), Church Women United, the North Carolina Women's Clubs, and the Republican and Democratic Women's Clubs. Under the leadership of Jane Patterson—"in politics" since campaigning for her father at the age of ten—the Greensboro group mobilized its own lobbying program. Durham also should have had a strong local coalition based on the presence of two universities—Duke and North Carolina Central—and a strong black middle class; but it did not.[13] The goal of recruiting the broadest possible support among North Carolina's black women, males, and traditionalists simply eluded pro-ERA activists. They could not break out of their own world. In retrospect, Petersen recalled, we were "a *liberal,* white, middle-class coalition of a few groups with little N[orth] C[arolina] strength, talking only to ourselves."[14]

This problem would not have been serious if the coalition had been the only group in action. It was not. Dorothy ("Dot") Malone Slade had seen to that. Phyllis Schlafly recruited her to help mobilize North Carolina women to STOP ERA. Drawing on members of the John Birch Society and other right-wing groups, Schlafly had organized an anti-ERA workshop held in September 1972 in St. Louis. Attending that workshop had been the gray-haired Slade, who worked at the American Tobacco Company in Reidsville, small town near Greensboro and the Virginia border. She had first become acquainted with Schlafly at a meeting of the National Federation of Republican Women. Having written to Ervin on a variety of issues, Slade had emphasized that she did not wish to be "liberated" from her privileges as a woman.[15] Other letters suggested a deep concern about a broad range of issues that had led her to join the John Birch Society. Thus she was receptive to the purpose and style of the St. Louis workshop. There, participants were given packets explaining the background of ERA and warning of its danger to family and country. From these materials they prepared and delivered three-minute talks like the ones they would deliver back home when they returned to mobilize friends and neighbors.

How well Slade learned her lessons was soon apparent. Unable to refuse Schlafly's request to head the state unit of STOP ERA, she invited fellow Birchers to help. The invitation was readily accepted in part because ERA tied in with so many issues about which these women felt strongly, especially the increasing power of the federal government and the changing tone of public life. Convinced that the idea of discrimination against American women was ridiculous, they looked for the hidden agenda behind the proposed scrutinizing of sex as a classification in law. They found it: ERA was to be the legal means of imposing feminists' ideas about gender on American women. Absent the word "imposing," it could be argued that they understood feminist goals fairly well. Deeply offended at the prospect, these women began talking about the ramifications of ERA to every group with which they were associated. At the same time, Slade put together a list of women whom she thought could be recruited. Although less than a dozen women showed up at the first meeting, they brought names and addresses of others and left armed with packets containing reprints of Ervin's speeches, Schlafly's reports, and instructions on how to defeat ratification. As new names kept coming in, Slade sent out more packets, scribbling across the top a note referring to the acquaintance who had united them: "Jane Doe said you'd help with this."

The General Assembly of 1973

On the first day of the legislative session in January, Bill Whichard introduced a bill of ratification into the House of Representatives. A similar resolution in the Senate was endorsed by twenty-five of the fifty members, a number that usually assured passage. Ratification also had the good wishes of James E. Holshouser, the first Republican governor since 1900, and the state's leading newspapers. It was, as one editor said, a matter of "simple justice."[16] Before the first gavel had convened the House, material opposing ratification had begun to pile up in legislative offices. In addition to Ervin's speeches were pleas from Homemakers' United of Scottsdale, Arizona, and members of the Tennessee legislature. From Slade's hometown came a petition from more than one thousand antiratificationists. Letters from constituents urged representatives to hold firm against "RADICALS bent on destruction of the family." In addition to all the paper was more personal lobbying. Wearing STOP ERA badges affixed with pink diaper pins, busloads of opponent women scurried in and out of legislative offices, clutching Bibles and offering to pray for legislators. Fascinated by women who claimed that "women weren't meant by the Lord to be equal," newspeople were quick to exploit the anomaly. By the end of January one thing was clear to everyone: not all women supported the Equal Rights Amendment.[17]

Those women who did, however, stepped up their efforts. The General Federation of Women's Clubs armed its constituent groups with carefully researched responses to accusations against ERA. The North Carolina Women's Political Caucus used its second annual meeting in Raleigh to emphasize the seriousness of opposition. On hand was Elizabeth Koontz, a black North Carolinian who, as head of the Women's Bureau of the U.S. Department of Labor, had led it to endorse ratification. Also there was Betty Friedan, founding member of both NOW and NWPC, who arrived from New York with word that "outside" money and political forces were behind opposition in the Tar Heel State—a view shared by McKay. Women legislators asked members to talk with their representatives—but "don't get emotional" warned one of them. She was wary of the political ambience within which ratificationists had to function: the pervasive masculinity that had impressed suffragists more than half a century before.[18]

By February 1, when the Senate Constitutional Amendments Committee began public hearings, ratification had become a major battle. Dispassionate analysis by the head of the Institute of Government at the University of North Carolina at Chapel Hill gave way to a blanket indictment of ERA by a professor of law at Wake Forest University. Robert E. Lee had been invited as a disinterested expert, but he assumed the role of antagonist when he claimed that there was probably "no place in the world" where women had "greater rights under the law than in North Carolina." Ratification of ERA would destroy those rights that, he added indignantly, women would never have had "if men had not wanted to release them." The amendment would invite further interference in the lives of North Carolinians, he warned, from a Supreme Court that had already done too much in busing children and banning prayer. ERA partisans, packed by the hundreds into the hearing room, laughed and hissed at Lee. One supporter recalled that "we acted out too much."[19] Their fervor if not their demeanor was matched by Phyllis Schlafly. She and her North Carolina counterparts carried warnings of danger. One witness read aloud a letter from Justice Susie Marshall Sharp that expressed her complete accord with Sam Ervin's views. Becoming in 1975 the first woman to serve as chief justice of a state supreme court, Sharp was already well known and respected by members of the General Assembly. Her letter provided a weapon that, as Slade later remarked, opponents used "to good advantage." Supporters lacked a similar advantage, but they did insist on the importance of legal equality. Both groups emerged claiming victory.[20]

In the weeks before the vote, debate and lobbying predictably intensified. For ratificationists, the emergence of anti-ERA women activists was scarcely acknowledged and little understood. Friedan reiterated her earlier charge that "big money" was behind the right-wing groups that fueled the opposition—a charge she admitted privately that she could not substantiate. She also took the media to task for depicting supporters of ERA as bra-burning man-haters. Schlafly and Slade responded quickly. The North Carolina coordinator of STOP ERA acknowledged her affiliation with the John Birch Society but vehemently denied that large sums of "outside money" financed her organization. She did not insist, as she could have, that Birchers had as much right to be heard as anyone else. As for "outside money," she wished she in fact had some: STOP ERA had only $25 in its bank account donated in small sums by personal friends. Looking like a "gray-haired school teacher," Slade, according to one reporter, spoke with the "fiery zeal of a tent evangelist" as she recounted how she had initially financed anti-ERA mailings from her own meager resources.[21] The publicity helped. Small contributions began coming in from individuals and churches, swelling Slade's list of contacts.

These were ordinary people, Slade insisted, bound together not by elaborate organizational structure but by deep moral conviction. In this posture antiratificationists projected themselves to state legislators as a grass-roots movement that expected legislators to do their duty. They were also an effective lobby, capable of generating petitions, letters, and delegations of women needed by anti-ERA legislators to legitimate their opposition. The relationship with "their legislators" was a formal one of deferential respect. Leading ratificationists assumed a different role. Relying on members of constituent groups of the Ratification Council for local lobbying, they attempted to develop a "partnership" with legislative leaders. They were not reluctant to telephone Bill Whichard with information and suggestions about what to do with "their" bill although they had asked him to make it his own.

These relationships were tested when a bill to refer the Equal Rights Amendment to the people of North Carolina was introduced in the House. Designed to delay a legislative vote and protect representatives wary of taking a stand on the issue, the bill was resoundingly defeated. Ratificationists were gratified. The margin of victory (32–83) seemed to evidence their political effectiveness and the incipient power of an organized female constituency. Other observers thought that it reflected Whichard's careful organization and planning. To pro-ERA lobbyists and lawmakers alike, the vote suggested smooth passage of ratification in the House.[22] Sizing up the situation, opponents responded immediately. The House Constitutional Amendments Committee was to have met two days after defeat of the referendum to vote on sending the ratification bill to the floor. Suddenly the chair, Claude Kitchin ("Kitch") Josey, who opposed ERA, delayed action by reporting he was ill. In the meantime the Senate Constitutional Amendments Committee agreed to send its own ratification bill to the floor on February 28. ERA supporters were aghast. They had been preoccupied with fighting the referendum in the House and had assumed that initial action on ratification was to occur there. Now they faced a Senate vote less than a week away. A preliminary count indicated an even split favorable to ratification because a tie vote would be

The appearance of STOP ERA lobbyists in 1973 fascinated reporters and photographers and stunned ratificationists who had not anticipated organized female opposition.

broken by the lieutenant governor as presiding officer of the Senate. And James ("Jim") Baxter Hunt was staunchly in favor of ERA.

On February 28, the visitors' gallery was filled with hundreds of onlookers, most of them women wearing buttons emblazoned with STOP ERA or ERA YES. As debate began, discussion of the legal implications of ratification bogged down in emotion-laden oratory. An eloquent opponent cited an eight-year-old boy who begged, "Please don't send my Mama to war." A senator from the mountains charged that ERA was being used by "militant and abnormal people for selfish motives." A colleague was more precise about "women's libbers" who "hated men, marriage, and children." Debate shifted as the primary sponsor, Charles Deane, urged the General Assembly not to wait fifty years as it had on the suffrage amendment to act on behalf of women. When opponents responded that the Fourteenth Amendment already guaranteed equality, someone in the press gallery muttered that "the Fourteenth Amendment couldn't pass this body!" Seasoned political observers responded not to rhetoric but the whispered exchange between Deane and a man who had promised to vote for ratification. Fears were confirmed when Gordon Allen voted no. It was one of two defections. When the votes were tallied, ratification lost 23–27, with 19 Democrats and 4 Republicans for it and 16 Democrats and 11 Republicans voting against. With no possibility for reconsideration, ERA was dead until 1975.[23]

Opponents were jubilant. Writing to Ervin, then busily corresponding with officials in ratified states about the possibility of rescission, Slade thanked him for his help. Insisting that time and a "majority of the people" were on their side, she nonetheless urged Ervin to help ensure the reelection of their legislative allies. Schlafly, who had just sent the senator a batch of labels with West Virginia addresses, was even more optimistic. The North Carolina vote meant that eighteen states had now refused to ratify. "If you had not shown us the way," she wrote, "we would have been beaten before we began."[24]

The only consolation of supporters was that a change in only two votes would have made the difference. Yet the very narrowness of defeat made it all the more frustrating. In the postmortem that followed, proponents agreed that Sharp's public opposition and private telephone calls to legislators had been damaging. They also observed that Governor Holshouser had not made ratification a top priority. Caucus leaders argued that one of the powerful Senate "insiders" should have served as sponsor, something they had not said earlier when the decision was made to accept Deane's offer. They complained that Whichard should have refused to permit postponement of the meeting of the House Constitutional Amendments Committee. They did not suggest how anyone could have prevented the Senate from acting first. No one explained how to prevent defections resulting from the political process. Candidates had pledged support early in the fall when ERA was thought to have been backed by women in general. They learned otherwise. Allen, for example, had in all likelihood switched because of conversations with his pastor and women in his hometown. Looking back over their own role in the first ratification struggle, a few astute women confessed their own naïveté. They had actually assumed that ERA was a legal issue requiring only logical explanation by experts. They also believed that once legislators gave their word to do something they would not change their minds—although that is precisely what they wanted opponent legislators to do. They had also assumed that one or two women could handle all the necessary lobbying as well as their personal and professional commitments. They had hoped that organizing from the top down could generate grass-roots pressure throughout the state. One caucus leader remembering the caucus convention, wryly observed: "It had been marvelous in 1972 and we were still riding high. . . . We just didn't realize it was going to be that difficult."[25]

Learning about Politics: Preparing for the Next Round

The realization had a sobering impact in Washington as well as Raleigh. In 1973 only eight new states ratified the amendment. Worse, Nebraska legislators rescinded ratification, confronting proponents with a new challenge: they had to hold as well as gain. To say that the opposition had appeared as suddenly as "a villain in a melodrama" was journalistic overstatement; to say that "bewildered backers" were caught "off balance" was not. Ratificationists tried to make their opposition illegitimate by associating it with the Radical Right. Fascinated but confused by alarmist rhetoric, feminists conceded the opposition's appeal to "dependent wives who have no identity outside the family and fear that the ERA will threaten their status." They also thought it appealed to "men for whom equal rights was a threat to their masculinity" and to legislators "opposed to anything that smacks of women's lib."[26]

To meet the opposition, national ratification leaders resolved to educate the public and strengthen their coalition. Creating tools through a massive research effort was congenial to women in such organizations as the League of Women Voters and the American Association of University Women. So thoroughly did they document the plight of women and the expected benefits of ratification that no one could have researched and explained the data without having her consciousness raised even further. But the feeling had to be shared, which meant that coalitions had to be built. Doing research was easier than forging constituents of the Ratification Council into a single-issue lobby against the countervailing weights of differing agendas and resources. The Connecticut and Ohio state coalitions were successful,[27] but for awhile at least, North Carolina ratificationists seemed paralyzed. After the January 1974 meeting of the Ratification Council, Elisabeth Petersen wrote, "Everyone thinks we should move but nobody wants to help."[28]

Desire and knowledge did not coincide: "everyone" could not agree on what to do. Constituent groups tried to impose the style of each upon the whole. Caucus members who prided themselves on nuts-and-bolts pragmatism differed from league women wedded to detached nonpartisanship; both favored a less assertive posture than that of NOW. During the spring and summer of 1974 one activist recalled, virtually everybody's personality seemed to be a "problem," and the meetings were endless. Feminism—as Oscar Wilde had said of socialism—took all one's evenings. Gradually, however, structure and policy emerged. The coalition divided its work into fund raising, grass-roots mobilizing, and legislative lobbying and disseminated information packets throughout the state. But basic problems persisted: activists still faced a dissonance between local situation and national "vision"; members still resisted working for national goals they did not quite understand. As recruitment inched forward it was still easier to get official endorsements than volunteers: by April 1974 twenty-seven groups had joined North Carolina's ERA United.[29]

Hiring a lobbyist and appointing a coalition representative to work the legislature were top priorities for veterans of the 1973 campaign. But action required debate—those endless nights again. To rely on a man to control lobbying struck not a few feminists as an affront to women. To pay a fee of $10,000 seemed extravagant, but by mid-October the coalition had hired attorney Howard Twiggs of Raleigh. To keep things going it also appointed a salaried full-time coordinator: Nancy Drum, a NOW member from Winston-Salem. Anti-ERA women who received no salaries would later scoff at pretended grass-roots movements with "paid lobbyists," coordinators, and secretaries.[30] They were partially wrong: there had been little money for clerical help. Since inadequate funds had crippled the first campaign, Petersen believed that the expanded budget implied hiring a professional fund raiser, but the coalition opted instead for contributions from constituent organizations. National headquarters at NOW, BPW, and AAUW promised to release money to state campaigns presenting reasonable budgets; later, the beleaguered coalition treasurer in North Carolina concluded that the reasonableness of her budget had been suspect from the very beginning.[31]

As in 1973, ratificationists planned to mount a "low-key," "well-organized," tight campaign.[32] Repeating the low-key style in newsletters, press conferences, and workshops, they hoped to dispel the rowdy image acquired from their hissing Professor Lee. The image was to have contrasted favorably with opponents' emotionalism as they launched the new campaign in the summer of 1974. This time the leadership was based not on the caucus but on the league and NOW. Titular head of the campaign was to be Gladys Tillett of Charlotte, a former suffragist and Democratic national committeewoman; she had chaired the Women's Division of the Democratic party under Presidents Franklin D. Roosevelt and Harry S. Truman. Tillett was to be the proponents' icon of legitimacy; but suffragists had been perceived as radicals in their own time, just as Nancy Drum was to be perceived in hers. It was Drum and not Tillett who came to be viewed as the ratificationists' leader. Drum tried to dress the part by abandoning her "shades" a la Gloria Steinem and her trendy wardrobe for clothes more suitable to working with men and reassuring women. If only she could have shed the movement's problems along with its clothes, but she faced many of the same difficulties Petersen had faced: unorganized areas of the state, incomplete files on legislators, and the activities of opponents. But there was good news, too. Ratificationists were becoming better organized, and their ranks were growing among nurses, secretaries, and Federally Employed Women. They sent publicity kits to newspapers across the state and alerted supporters to be prepared at a moment's notice to write "polite, low-key, and pleasant" letters.[33]

Opponents were galvanized as well. By October, Dot Slade had placed

STOP ERA material in the hands of every member of the General Assembly. Two months later opponents gathered in Raleigh to form North Carolinians against ERA (NCAERA). Supporting Slade's work, they were nonetheless convinced that more organization was needed, especially outside the urban areas of the central part of the state. A mixed group of fifty people, some of the founders were associated with the John Birch Society, the Church of Jesus Christ of Latter Day Saints (Mormon), or Protestant fundamentalism. Sonia Johnson, a feminist who was excommunicated from the LDS Church, was sure that there was a Morman network at the core of the organization. Marse Grant, editor of the Baptist *Biblical Recorder,* agreed. Although it is true that a few wards in the state did actively work against ERA, it is difficult to say how extensive or significant such action was. Officials of NCAERA were not forthcoming on the matter.

It is clear that NCAERA established networks to facilitate grass-roots mobilization. The eastern and western sections of the state were assigned to coordinators, who worked in turn with county and local organizers. It was a leaner and more efficient operation than that of the proponent coalition. Communicating primarily by telephone, coordinators relayed instructions to implement lobbying strategy. Like their ratificationist counterparts, opponents of ERA at the grass-roots level focused on their legislators. But unlike pro-ERA activists, they supplemented letter writing and one-on-one contact with a massive petition drive. In order to augment finances, dues were set at $5 per person and a search begun for effective media spokeswomen. Leadership of NCAERA rested with four officers: two co-chairs and two regional coordinators. Headquarters was the comfortable, well-furnished Rocky Mount home of Alice Wynn Gatsis, who served as co-chair with Bobbie Matthews. Gatsis's sister-in-law, Toni Suiter, served as secretary and—in private conversation at least—as an eloquent interpreter of what NCAERA meant.

With press reports that the network was ready for business in thirty areas of the state, volunteers began to enlist by telephone. It was all immensely heartening, recalled Sherry Hicks, an NCAERA regional coordinator. Her organizing efforts in the western part of the state had brought responsibilities far more rewarding than those associated with any of the jobs she had held prior to her marriage. Such unsolicited expressions of support, she believed, were welcome indications that antiratificationists had barely "scratched the surface." But the scratches were nonetheless impressive. Activists who collected signatures, sent thousands of them, twenty to a page, to the General Assembly. Many had been harvested from shopping centers during the holiday season; but many more came from fundamentalist churches. Antiratificationists sent Sam Ervin's speeches, Schlafly's reports, and appeals from the profamily movement to congregations; they enlisted ministers who happily worked with "family-centered people." Converts talked to people in mills, at check-out counters, at church and civic meetings, and even in the sauna at the local YWCA. From conversion came action. One woman walked into the General Assembly carrying five thousand letters to each legislator in her district. She came back later with fifteen hundred more.[34]

The General Assembly of 1975: With All Deliberate Speed

ERA supporters around the nation followed the 1975 North Carolina General Assembly with special interest, disheartened by rejection in Oklahoma and Virginia and attempts to rescind ratification elsewhere. Although momentum had flagged, with only three ratifications in 1974, the Tar Heel State was one of six that strategists thought promising. Supporters had made gains in the 1974 elections and were ready to go in the House by mid-January. Herbert Hyde of Asheville introduced the ratification bill on January 17. His co-sponsor was H. M. ("Mickey") Michaux, Jr., a fellow Democrat from Durham and the most prominent black politician in the state. House Speaker James ("Jimmy") Collins Green, a veteran legislator and skilled strategist, assigned the bill to the Committee on Constitutional Amendments and fellow Democrat A. Hartwell Campbell, who was committed to using his powers as chair to defeat ratification.[35] Antiratificationists introducted a public referendum on ERA to divert attention, while Campbell stretched out the deliberative process. He knew that the supporters' narrow margin in the House could be eroded with time—which was at his disposal. A businessman representing the tobacco counties around the eastern town of Wilson, he observed in his gracious, soft-spoken manner that ERA was "not quite so simple a proposition as Hyde suggested." A subcommittee would have to study North Carolina laws that would be affected by ratification. He then announced a public hearing on March 4 when supporters could testify; opponents would come a week later. Senator Ervin would have an extra session all to himself.[36]

When hearings finally began before an audience of six hundred "mostly white middle-class women," testimony was taken from a very few proponents. Carefully selected to convey support for the amendment regardless of race, sex, or party, they addressed the committee through reportedly "logical arguments" in an atmosphere of "order and decorum" consistent with the low-key campaign ratificationists had run.[37] Low key meant low visibility. Ratificationists were not what newspapers called the "best show in town." That award went to Phyllis Schlafly, who had, as the Raleigh *News and Observer* reported, "left her fashionable house in Alton, Illinois, her six children, lawyer husband, and two secretaries to tell North Carolina women that their place is in the home." Articulate and poised, Schlafly electrified the faithful with an attack on feminists who—if they felt penalized by their sex—ought "to take up their argument with God." Her sentiments were repeated in testimony by opponents on March 11. More than twenty speakers, including housewives, working women, lawyers, and preachers, listed the dangers of ERA in what the *Greensboro Daily News* thought were "strident emotional pleas." But the legislative summary for that day reported that the crowd sporting bright red STOP ERA signs was "somewhat subdued compared to the hordes of campaigners and shrill rhetoric of two years ago."[38]

A week later Senator Ervin repeated his by-now familiar arguments. Not wanting "his majesty to have the whole show," proponents produced their own expert, William Van Alstyne, a distinguished Duke University professor of constitutional law. They placed his response to Ervin's arguments in the

Perfectly poised, groomed, and organized, Schlafly energized North Carolina allies who shared her conviction that a woman should be treated "like a woman, not a man, and certainly not a sex-neutral person."

hands of each legislator; no one reported how many read it. In testimony of a different kind, a group of anti-ERA Raleigh housewives presented members of the General Assembly with loaves of home-baked bread and an appropriately symbolic note: "To the breadwinners from the breadbakers." Proponents also provided appropriately symbolic refreshments for legislators—a cocktail party.[39]

Finally the subcommittee studying the effect of ERA on state law and employment practices reported that passage would have no serious impact on state government. It would affect glaring discrepancies in salary and technical modifications in about two hundred laws. For example, the phrase, "a jury of twelve men" would have to be rewritten to include women; the law awarding compensation to widows would have to include widowers; rape laws would have to apply to men as well as women; the statute requiring that married women who wished to sell their property be privately interviewed to ensure that they were not being pressured to do so by husbands would have to be revised or dropped. The report was similar to, but longer than, Attorney General Morgan's reply to Ervin in the late summer of 1970. After the report was accepted, the committee voted ratification to the floor.[40]

With the date finally set, both sides sized up the prospects of victory. According to a recent Gallup poll, a majority of North Carolinians supported ratification: 58 percent for, 24 percent against. More men than women fa-

Loaves of bread tagged "To the Breadwinners from the Breadbakers" were distributed to N.C. legislators as a reminder that a vote against ERA was a vote against changing gender roles.

vored the amendment; younger women expressed greater support than their elders. Legislators were more evenly divided. Prospects for Senate approval were favorable provided the bill passed the House. Hyde predicted that it would squeak through, possibly by a margin of five votes, although most observers conceded that the margin was closer. A few anticipated a tie that would require Speaker Green, a shrewd and powerful opponent, to vote.[41]

Lobbying on both sides intensified. Then, on April 15, debate in the full House finally began before an audience of women bearing their respective insignia. Campbell tried unsuccessfully to pass a new rule requiring a three-fourths majority for passage. Hyde, assuring his colleagues that the amendment would not damage the family or make roosters cluck like hens, argued the necessity for ratification. Antiratificationists repeated the observations of Senator Ervin. The House then reported a 59–59 division on a roll call vote—one person was missing. To spare the speaker from having to break the tie, Ronald Mason, an anti-ERA ally of his, voted on its behalf. It was, after all, only the first reading, with two to follow.[42]

When the House reconvened to consider ratification, the speechmaking resumed. An address by Patricia ("Trish") Stanford Hunt was notable for the way in which it stated the position of most women legislators. Hunt, a 1950 graduate of the University of North Carolina at Chapel Hill, had become a wife, a mother, a widow, and an educator. In 1968 she won election as a Democrat to represent Chatham County poultry farmers and Orange County professors. She now addressed her colleagues in language that was direct and to the point.[43] Women had been "completely and totally upstaged" in debate over ERA, she pointed out. Opponents had made "incredible" statements about women who supported ratification. "I am not," she declared, "in favor of unisex; I am not afraid of men; I am not a man-hater; nor am I a member of any women's lib organization." She was a lawmaker concerned about her constituents. The difference between the two sides lay, she thought, in the proponents' assumption that, with ratification, courts and "state legislatures will act where changes are needed with responsibility and in accordance with the intent of Congress." "Opponents," she continued, "implicitly and sometimes explicitly assume that the courts and legislatures will act vindictively and without regard for public welfare." She accused them of "paranoia" and turned to address their arguments.

First, she pointed out that state labor laws applying to women had already been nullified by Title VII of the Civil Rights Act of 1964. Second, the claim that ERA would weaken the obligation of fathers was "not borne out by the facts." The courts did not now interfere with marriages and would not do so after ratification; a woman's right to support could be enforced only when the

Such badges highlighted the most troublesome issue in the ratification campaign—whether passage of the amendment would require not only the drafting of women but their use in combat situations.

couple had separated. Lawmakers seemed not to understand the law; North Carolina statutes concerning divorce, alimony, child custody, and child support were already in accord with ERA. What ratification would change, she insisted, was that divorced women who were not now being supported by alimony would be better able to support themselves through equitable pay. Like her suffragist predecessors, she was asking for the right of self-protection, or in this case self-support, on equal terms with men. If equal terms—carried further—meant drafting women, that would improve their economic position. Military service had always been an avenue of upward mobility, and less-advantaged women stood to gain from it, just as had their brothers.

Hunt went on: equal responsibility no less than economic advantage was at issue. "Why shouldn't young women have equal responsibility for the defense of this country?" she asked. If the U.S. Supreme Court could rule in the case of *Taylor* v. *Louisiana* (1975) that jury duty was the public duty of women, it could "also rule that laying down one's life for one's country is no less a public duty!" Women who had moral objections to bearing arms could seek noncombat or community service. Draft boards would be at least as liberal in granting such alternatives to women as to men. These arguments were not theoretical for Hunt. She recalled that she and a friend had tried to enlist in the WAVES at the end of World War II because "we wanted to do our part." Yes, she conceded to a male colleague, "War is Hell! . . . If our women share in [it], perhaps it will stop this incredible stupidity!" The issue, however, was not that ERA would put women in combat, she thought, but that it would put "us females in the Constitution," freeing us from the "highest form of bigotry [which] is paternalism." "Wherever there is *discrimination,*" she concluded, "we have lost part of our freedom and democracy. When all persons are not treated equally before the law, some one is less than a full citizen."

Passionate words, they did not change votes. When the tally was taken on April 16 for the third reading, ratification lost 57–62. Antiratificationist women in the audience burst into applause. Defections in the House made possible by Campbell's delaying tactics were disheartening. Legislators who changed votes agreed as to why they had done so. To be sure, Mason had switched his vote on the first reading to protect Speaker Jimmy Green, but there was another reason: his constituents were opposed. The same was true for others, including Myrtle Wiseman, the only woman to vote against ratification. Support in the House had always been "soft," and constituent pressure was particularly intense during the four months that ratification languished in committee. Recalling the barrage of letters and telephone calls she and her friends produced, an opponent from the port city of Wilmington concluded, "I don't think we could have worked any harder."[44]

"Nobody . . . Had Any Idea That It Was That Involved"

Drum and her colleagues could also take pride in their work: they had hired a professional lobbyist and created a larger, more effective, better-led coalition.

Since their task was greater than that of anti-ERA women, however, and because they had failed, they were more preoccupied with what they had not done. Lacking necessary resources, they had failed to educate the public. Lacking a network of skilled organizers, they had failed adequately to mobilize their constituents. Lacking the emotional intensity of their opposition, they had failed to convey the necessity of ratification. Their low-key style was, observed one legislator, "too cerebral."[45]

Indeed, style worried ratificationists very much—their own as well as that of the opposition. The contrast was clear in the discrepancy between reasoned editorials supporting them and letters to the editor shrieking at danger. It was evident during televised debates when a mixture of alarmist and seemingly untrue charges evoked a frustration that, in the words of opponents, was expressed as "anger" and "militancy." They worried about the hisses of 1973 and the authoritative manner of their best advisers. They worried about their youth and clothing and compensated by appointing Tillett and sending memorandums on decorum. They worried about the top button of Nancy Drum's blouse that, during a televised debate, had been unbuttoned. It was something, recalled a friend, "which we all do." And Drum had crossed her legs. Seated opposite in a turtleneck sweater with decidedly *un*crossed legs was the representative of STOP ERA—definitely no unbuttoned top buttons. The viewer had been dismayed; that "loose" button worried her.

Drum and her friends learned about more than image, however. They had learned so much "just in terms of organization and people and lobbying and government and laws . . . and what it took to pass an issue of this nature," as one said. And they had learned that they had been too naïve and inexperienced. They had spent too much time with people like themselves and not enough on rallying blue-collar and black women. They had spent too much time on paper organization and not enough on grass-roots mobilization; they had not been able, for example, to deliver constituent letters when they were needed in April. But they had learned, too, that "passage is not something that you do by just writing letters." One woman confessed to being "just knocked over by men who get down there with parliamentary procedures, and some of the things they were doing on the floor and the kinds of strategy that they would have to do to get things out of committee." In a remark that would have brought a snort from political activists, she added, "I'm sure that nobody in 1973 had any idea that it was that involved."

The lesson of how "involved" ratification was was learned throughout the United States even by those more sophisticated. To be sure, by 1975, supporters had won thirty-four of the thirty-eight states required for ratification; they had beaten back rescission efforts in West Virginia, Idaho, Kentucky, Michigan, and South Dakota. Majority opinion seemed to favor the amendment and feminism both. Pro-ERA coalitions existed in all unratifying states, representing almost one hundred groups. In 1976, however, not one state ratified; ratificationists' only real gains were educational—self-education, not education of legislators. NOW's veteran national legislative strategist Ann Scott observed that since ERA was a political issue it was gratifying that ratificationists were beginning to move, as a Florida proponent said of her own colleagues, "like a political machine." The battle had changed dramatically since the heady days when feminists drove ERA through Congress and into the quagmire of a few state legislatures.[46]

Campaign 1977: "ERA is for Everybody"

With the sobering experience of two defeats, North Carolina ratificationists began preparing more than a year in advance of the critical 1977 struggle. This time, they decided, they would be more "political," which meant that in 1976 they would pay more attention to electing legislators committed to ratification. This strategy required polling candidates, targeting districts, and mobilizing constituents. "Political" also meant no more low-key tactics. It meant renovating and expanding the statewide networks among those elusive grassroots. Ever ready to tinker with structure, ratificationists invested responsibility for overall policy in a new board that, when contrasted with the spare structure of opponents, was really a crowd. They elected four officers, conceded each constituent organization a delegate, and appointed chairs of eight standing committees. They also named a steering committee to implement policy and an executive committee to make emergency decisions. To avoid confusion, they insisted that the president be the "only official spokesperson." There would be a tighter ship in 1977.[47]

To command it, ERA United named a new captain—the Reverend Maria Bliss. A highly motivated activist, Bliss had settled in the small town of Asheboro on the eve of the 1975 lobbying campaign and immediately begun to organize the surrounding area where "no one knew what ERA was all about." As important as her talents was her image. She was a North Carolina native, wife, and mother of three, who had combined homemaking with an active ministry in the Methodist church. Chuckling about her image—her matronly figure, twenty-one-year marriage, years of child care, and streak of gray hair—she wryly explained why she had been chosen: she was the "everyday housewife." Her religious credentials, she thought, especially commended her to a state in which opponents had appropriated God for political purpose. That her ordination made her anomalous was not acknowledged; she was simply "mainstream North Carolina people." To underscore that belief, ERA United changed its name to North Carolinians United for ERA (NCUERA).[48]

Since goals are rarely achieved by name changes, Bliss set up her calendar for action. She sought field organizers who could recruit minority, church, and working-class women. She and her colleagues appointed committee chairs who would have their staffs ready by the end of May; the legislative committee was told to have data on all candidates ready for the August primary. If achieving these tasks demanded eighteen-hour days, Bliss and her co-workers could take comfort from the fact that this time around the North Carolina coalition would have significant help from Washington, D.C. The thirty-

As head of NCUERA, the Reverend Maria Bliss exemplified the commitment and energy which pro-ERA activists brought to a revitalized and expanded coalition effort in 1977.

member Ratification Council there had been rebaptized ERAmerica and in January 1976 reestablished in an office from which it set about raising funds with renewed purpose.[49] This renewal, like that in North Carolina, came in part from the fact that ERA had changed. Once it had been one among many means to equality, a reformist symbol of consensual American idealism calling attention to the power of discrimination but lacking evocative power. The idealism remained, of course, but as ERA faltered it became more important as a goal than as a means. Thus focused, ratificationists transformed ERAmerica from a clearinghouse for information into a source of much-needed funds and expertise for state coalitions. Accordingly, Bliss and her strategy committee received advice on everything from recruitment and mobilization to cash flow and contingency budgets.

Sobered by *Ms* magazine's account of losing a state ERA in New York, ratificationists mapped out details of the coming campaign. NCUERA's Public Relations Committee sketched out ways to convince North Carolinians that ratificationists were not "radical feminists." "ERA," they insisted, "is for everybody." The words appeared on bumpers, lapels, and coalition literature throughout the state. For its slogan, the women drew on Susan B. Anthony. They transformed her statement "Men—their rights and nothing more. Women—their rights and nothing less" into "ERA—Nothing more, nothing less." The slogan linked ratificationists to their historical antecedents and emphasized that "ERA is concerned with NOTHING MORE than the legal rights of all citizens" and "NOTHING LESS than a constitutional guarantee of equal rights under the law." Proponents carefully crafted their symbolic language—how evocative it would be depended on the emotional attraction of the words "equality" and "rights." Consistent with their slogan was a logo with the letters "ERA" and the symbol of justice. The words "nothing more" were written over one scale, "nothing less" over the other. Campaign colors were to be red, white, and blue.[50]

The committee called for action as well as symbolism. Task forces were set up on polling and information with instructions to distribute news releases, establish contact with wire services, and feed stories to targeted newspapers. They planned small dried flower arrangements and paperweights containing the coalition slogan in needlework as mementos for legislators. The committee's promotional task force issued precise instructions on how to work with other committees, detailing lines of authority and pointing out that volunteer labor not only saved money but also enhanced commitment. The committee's *pièce de résistance* was to be a media campaign designed to produce pro-ERA letters to legislators. The network upon which this campaign would rely was to be maintained by an NCUERA newsletter and frequent rallies. Assessing her committee, Peggy Carter was impressed with her co-workers but stunned at the time demanded of her. Instead of taking only a "couple of hours a week" as promised, ERA took over "my whole life."[51]

Karolyn Kay Hervey made the same discovery. Involved in ratification through NOW, Hervey believed that, as a Jewish émigré from Connecticut, she could best contribute to the campaign behind the scenes. She anticipated fitting her work as chair of the Legislative Committee into her routine as housewife and mother with minimal reliance on McDonald's or Burger King. She was wrong. She received an initiation into North Carolina politics with what Baptists would observe was total immersion. She had no difficulty establishing district voting patterns and the records of incumbents from data assembled in Raleigh. Collecting information on new candidates was more challenging, however, because it required locating people to report accurately on their financing, base of support, and associates most likely to affect the candidate's vote. Their spouse's position on ratification was a matter of more than idle curiosity. Armed with names of activists from former campaigns and building upon what they had done, Hervey's committee members began calling around the state. By June they had contacted persons in all but four legislative districts, analyzed data, organized files, and written letters asking candidates their position on ERA. Compensating for the hours of hard work was the knowledge that many of the new "contacts" had exchanged a "closet commitment" to ERA for open involvement. The experience was invigorating.[52]

Few needed a renewed spirit more than field workers. Grass-roots organizing was the one area in which proponents had always had the best intentions and worst results. As long as they worked with women like themselves, coalition activists were effective. Once they stepped beyond the familiar into the most important work of the movement, they seemed to falter. Determined to learn from past mistakes, however, coalition leaders studied the 1975 networks, collected information on the demographics of each House district and the state's one hundred counties, and pinpointed local chapters of the coalition's constituent bodies. They prepared flow charts with job specifications for district organizers, area captains, and county coordinators. That part was easy. The next step was to convene meetings, elect local leaders, and plan district campaigns. Computer printouts listed contact groups, with the names, addresses, and telephone numbers of current officers. To activate contacts, NCUERA sent kits explaining the need for ratification and what everyone could do at the local level.

The plan was laudable. The field service manual was excellent. The search for effective conveners, however, was predictably difficult while "time was," as Snuffy Smith would have said, "a-wastin.' " NCUERA leaders called the right people but got more names than volunteers. Veterans of past campaigns were frequently "interested" but "burned out" or otherwise committed. The plan itself may have been at fault: it was too elaborate, too conscious of constituent groups, too emphatic about action at every level save where it counted. There was too much "organization." The coalition admitted as much when it scrapped "democratic process"—trying to work from the bottom up—and pressed the conveners themselves into service as district coordinators. Demands on the conscripts were enormous, and some women could meet them—if they were passionately committed, knowledgeable and effective, and willing to put aside personal needs to "do whatever had to be done." But the plan seemed unable to produce enough of these ideal activists; thus NCUERA decided that it needed a state coordinator to direct local coordinators.[53]

This meant more money. Including the $6,000 in a budget of $50,000 was one thing; raising the money early enough to have an impact was quite another. The only hope was ERAmerica. If the national organization thought North Carolina likely to ratify, it might loosen its purse strings. A consultant for AAUW thought that chances were good, and after a weekend of frantic "grantwriting," an NCUERA team flew to Washington, D.C., to present its case to the heads of eight national organizations. In assessing the political climate, they pointed out that the signs were good. Jim Hunt, clearly the front runner in the governor's race, was solidly behind ERA, as was his fellow Democrat Speaker of the House Carl Stewart. It was true that Jimmy Green, the conservative Democrat who would probably be lieutenant governor and president of the Senate, was not a supporter. But his own gubernatorial aspirations presumably made him chary of alienating women in a state where poll data suggested that a majority supported ERA. Green's possible obstructionism seemed to be more than counterbalanced by the support of Stewart and Hunt. The three women left Washington with promises of funding for North Carolina (Illinois, Missouri, Indiana, Florida, and Nevada were the other states targeted). The money itself was slow to follow. NCUERA remained afloat financially into the fall through the largess of NOW, which fed the coalition monthly installments of $2,000–3,000.[54]

Assessing their efforts in the months before the fall election, pro-ERA activists thought that they had made significant progress. It seemed evident in the overwhelming numbers of requests for materials and speakers and in the many small donations accompanied by encouraging letters. It seemed evident, too, from evaluating contributions in kind. The North Carolina Council of Churches provided a model liaison with women's networks in the eastern part of the state, Tibbie Roberts; her commitment to ratification energized those around her. Roberts had been reticent to "volunteer" until memories of her suffragist grandmother's feminism thrust her into the fray. In the west she was complemented by activists from Methodist and Episcopal churches. From the University of North Carolina at Chapel Hill a Kenan Professor of Law, former Chancellor William B. Aycock, volunteered to celebrate the Bicentennial of Independence by stumping the state for ERA. His lecture to clubs and community groups so impressed newspaper editors that they frequently printed it in full. In the process of enlisting new activists, while also educating, promoting, reorganizing, and expanding, there was a sense of empowerment. There was little time, however, for the coalition's most crucial task—electoral politics.[55]

Mobilizing legislative districts was to have been NCUERA's primary responsibility before the elections. But many organizations upon which the coalition was built had not been founded on or educated in political action; even when accurate information was gathered, they were unable to use it. Even in organizations committed by state and national leaders to work for pro-ERA candidates, members were reluctant. Their hesitancy is understandable. They had not joined AAUW, for example, to work in electoral politics; if they had wanted to do that they would have joined a political party. To be asked to campaign was to be asked to change personal priorities, to do things they had not done before. They may have hoped somehow that political activity "would be done," but assuming responsibility for changing the passive into the active voice was more than they could do. The transposition would almost have demanded a change in personality, or at least in personal style. And personal inclination was reinforced by a long history of women's participating in politics through influencing those already in office, not putting them there. By tradition as well as temperament, therefore, many ratificationists were uneasy with aggressive "politicking."[56]

The different kinds of politics are seen in what the coalition was both unable and able to do in preparing for the General Assembly of 1977. It had been unable to effect a strategy for the lieutenant governor's race, in which two of three candidates were ERA loyalists: Herbert Hyde, House sponsor of ratification in 1975, and Howard Lee, an aspiring black politician. The third man, Jimmy Green, opposed the amendment. Predictably, Lee and Hyde split the liberal vote, forcing a runoff between Lee, who won a plurality in the first primary, and Green. Lee (and ERA) lost. Ratificationists were, however, able to approach and work with the next governor, Jim Hunt, and Speaker of the House, Carl Stewart, who were not necessarily beholden to them. Warmly supportive, the two men thought the prospects of ratification "tight" but agreed that it would have a good chance if it were to originate in the House with George W. Miller of Durham as sponsor. Miller, who characterized himself as a Democratic in the John F. Kennedy–Terry Sanford mode, was a veteran legislator who had a reputation for integrity and the willingness to vote his own convictions against counterpressure from constituents. (His reputation won him the unqualified admiration of voters from other districts who sent him a box of Valentine candy with the note: "Roses are red / Violets are blue / We wish our district / Had someone like you.") If proponent women had not actually put their allies in office, they nonetheless approached the legislative session with great confidence.[57]

Antiratificationist women were also confident. Their task, to be sure, had been less difficult than that of ERA partisans. If, in a political culture characterized by incrementalism, resisting innovation was easier than initiating it, opponents had nonetheless accomplished a great deal. They had defined the terms of debate and in the process transformed a legal issue into a symbol of gender revolution. They had out-lobbied ratificationists twice by the latter's own admission, and—spared the problems of coalition building and structural revamping—expanded their networks especially in fundamentalist churches. Invigorated by two victories, they were confident that with their help the General Assembly would once again do the right thing. They certainly knew that Republicans would. The party had been seized by the rightist forces of Senator Jesse Helms, who had invited Phyllis Schlafly to "keynote" the spring convention of 1976. For staunch supporters of ERA in the GOP, Schlafly's address was humiliating evidence of the totality of Helms's control of what had once been their party, too. It also provided Schlafly an opportunity to confer with her antiratificationist allies about legislative strategy.

In her New Year's message to the faithful, Schlafly claimed that the tide was shifting in their favor. State ERAs had gone down to defeat in New Jersey and New York, and public opinion polls had recorded a decline in support for feminism. But she was not heedlessly optimistic; Indiana had just ratified—the first state to do so in two years—and she thought that if North Carolina followed, proponents would develop "a momentum we cannot match." She told Ervin that "our women" reported being several votes behind but insisted, "If you will save North Carolina for us, I promise we can hold the other fifteen unratified states." There, antiratificationists had already "gone up and down their local communities repeating the arguments you so brilliantly developed in the U.S. Senate." Now, if only he could "pull out all the stops" for North Carolina women.[58]

Antiratificationist women had already sent printed material to local activists across the state and refined the tactics of past campaigns: letters, phone calls, petition campaigns, "open mike" radio shows, personal appeals. Expanding her network far beyond her Christmas card list of 1972, Slade had asked each new recruit to recruit ten more. They should come from "another section of your county," with at least one from another county. She had explained how to lobby, follow up, and keep files. "BE SURE of your facts," she had warned. "The opposition would be delighted to catch you in error." She wanted activists to be 'courteous, calm and kind," and in benediction had suggested that they "*PRAY WITHOUT CEASING!*" "Our victories in the past have been the Lord's victories; our victories in the future will be His!"[59]

The General Assembly of 1977

When the General Assembly convened in January 1977 both sides once again predicted victory. Behind the predictions, however, were careful assessments. Ratificationists could count on the active support of U.S. President-elect Carter, who had promised to do for ERA what Lyndon B. Johnson had done for civil rights. They would rely more on the governor who, it was hoped, would use his appointive power, grass-roots support, and experience on behalf of ERA. Hunt was, as NCUERA leaders reported, "vocal, visible, and steadfast" in his commitment to ratification.[60] The Speaker of the House was also an ally. Son of a mill worker and twice a graduate of Duke University, Carl Stewart had had his eye on the governorship since his days as a student; and he fully appreciated the power of women in any race for state office. The willingness of both Hunt and Stewart to put promising female colleagues into important positions simply underscored their commitment to equality. Betty R. McCain, an astute politician who gave an expansive meaning to the term, "Hunt loyalist," became the first woman to head the state Democratic party. Jane Patterson, having been apprenticed in the politics of Guilford County and ratification, moved through the 1976 campaign into the executive branch of state government after a stint in Washington, D.C., on President Carter's transition team. Trish Hunt assumed the chair of the Judiciary Committee after the Speaker would not take no for an answer. These and other appointments indicated that the governor's office and the House were "friendly territory" for ERA and women politicians.

Whether Stewart, Hunt, and George Miller could bring the General Assembly along with them, however, was problematic. Antiratificationists, too, had friends in high places. Despite the fact that they could count on strong

Lt. Governor James Collins ("Jimmy") Green (left) and Governor James Baxter (Jim) Hunt, both N.C. Democrats, were pivotal figures in ratification politics, disagreeing on ERA as on many other issues.

support in the House from Miller as well as Ernest B. Messer and Liston B. Ramsey, ratificationists could also expect staunch opposition from Hartwell Campbell, whose delaying tactics had proved so disastrous in 1975. And Campbell had help. Indeed, Stewart, Miller, and six of the state's leading newspapers thought that ERA was five votes short.[61] Miller also knew that because of previous ratification struggles many of his fellow legislators had "had their fill of ERA." The same was true in the Senate, where antiratificationists had other friends as well. The most important was probably the new lieutenant governor and presiding officer of the Senate. Although Jimmy Green had claimed neutrality on ERA during the preceding election, few who recalled his support of Campbell in the House during the 1975 session had illusions about where he stood. As a conservative with sixteen years' experience in the House, he could be counted on to use his influence to support his own view of government, not that of Jim Hunt.

As STOP ERA petitions piled up in the Legislative Building, the coalition countered with instructions to get their own cards and letters coming in. To underscore the fact that ERA had the support of experts, NCUERA called a press conference to introduce its advisory board. It was a carefully selected group of ex-governors and business, educational, and religious leaders whom ratificationists considered to be among North Carolina's most responsible citizens. Heading the list were former Governor Robert Scott and his wife, Jessie Rae; Duke University President Terry Sanford, also a former governor; former chancellor of the University of North Carolina, Willian Aycock; U.S. Secretary of Commerce, Juanita Kreps; business and civic leader and arts patron, Gordon Hanes; North Carolina National Bank executive, Luther Hodges, Jr.; Baptist leader and editor of the *Biblical Recorder,* Marse Grant; and perhaps most appropriately, Elizabeth Koontz, former head of the Women's Bureau of the U.S. Department of Labor. This array of prominent citizens was supposed to establish the legitimacy of ERA, but antiratificationists thought the reverse was true. By "dragging in" the "biggest names in the state," proponents had exhibited their "elitism." Moreover, Jessie Rae Scott, as wife of a former governor, and Kreps, as Secretary of Commerce, together with Koontz, the former head of a federal bureau, were not really "everyday housewives."[62]

The issue, however, no longer belonged to competing groups of women but to the "Honorables" of the General Assembly. Recognizing the damage wrought by anti-ERA lobbyists in past sessions, Miller was eager to get the ratification bill onto the floor as soon as possible. Accordingly, it was introduced by its sponsors on January 18 and promptly referred to the Constitutional Amendments Committee whose chair, John R. Gamble, scheduled hearings for January 26 and 27. While both sides were lining up spokespersons, the Virginia House of Delegates voted ERA down, making North Carolina's action all the more critical.[63]

On a cold January morning, the biennial ritual got under way after passengers sporting STOP ERA buttons streamed out of chartered buses. A "revivalistic atmosphere" suffused legislative halls as Bible- and flag-carrying people warned of the apocalyptic consequences of ratification. Old and young, male and female, they packed the hearing room in defiance of fire regulations to hear seventeen speakers repeat the familiar litany of attacks upon the federal government and communism and defenses of family, Bible, and womanhood. A Durham woman explained that she had seen ERA in practice in the Soviet Union, where women actually worked in cement and brick factories. The worst thing that can happen to a woman, she concluded, "is to be treated like a man." Retired Brigadier General Andrew Gatsis agreed, and the worst thing that could happen to the country, he asserted was to recruit women soldiers. The husband of Alice Wynn Gatsis, who co-chaired NCAERA, he was familiar with what combat could do even to men who were psychologically and physically suited for it. It would be worse for women who not only lacked the stamina necessary for combat, but "let's face it," he said, "they just don't like to beat people up, particularly in hand-to-hand combat." And combat—warned Kitchin Josey, a former legislator instrumental in defeating ERA in 1973—meant death. The fusion of death and ERA came from the assumption that if "sex" could not be taken into "account" with regard to combat troops, all citizens would have ascribed to them the characteristics of the male "sex." Intoned one preacher, ERA is a "unisex amendment and we're not a unisex country yet."[64]

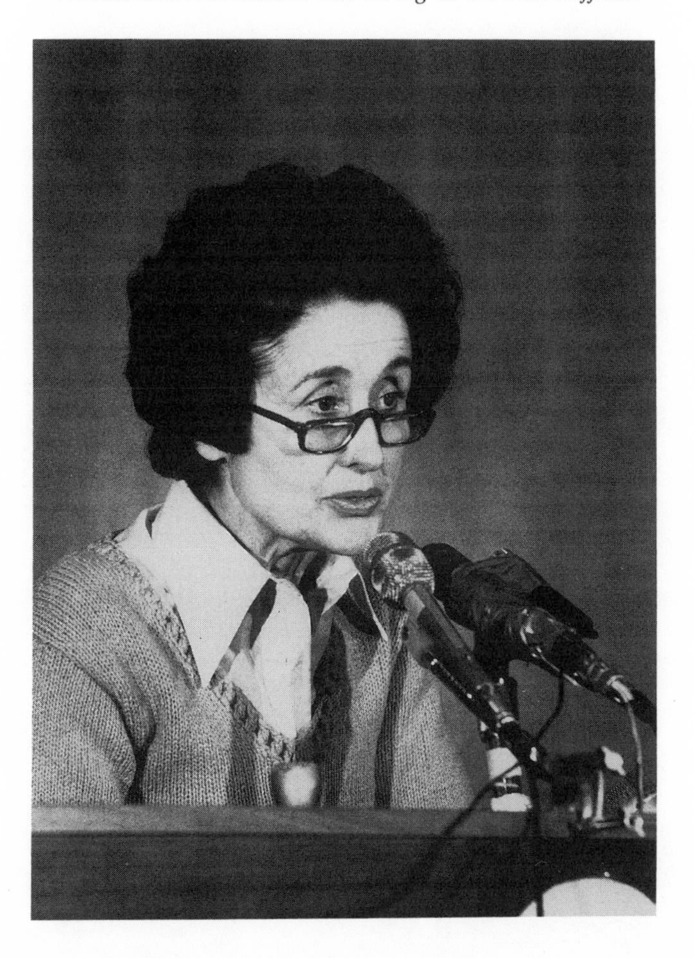

U.S. Secretary of Commerce, Juanita Kreps, was one of many prominent North Carolinians endorsing ERA. Such endorsements were dismissed by opponents who charged supporters with "elitism," pointing out that Kreps, while the mother of three, was no "everyday" housewife.

The next day, before an audience that the *Greensboro Daily News* described as younger and quieter, proponents "pulled out all the stops." They spoke with rapier thrusts of wit and down-home humor; they called up personal recollections of pain and impersonal statistics. And once again they asked William Van Alstyne for an authoritative display of legal expertise and eloquence. Collectively, they appealed to state pride and civic conscience, providing historical, legal, biblical, and anthropological evidence that would place ERA in context. Gladys Tillett placed the amendment in historical perspective. The Bishop of the North Carolina United Methodist conference and Katie Morgan, wife of former state attorney general, now U.S. Senator,

Lobbying legislators and packing legislative hearings before an array of television cameras had become a carefully orchestrated ritual by 1977.

Robert B. Morgan, talked about economic discrimination. Believers talked about religion. Recalling "Christian" defenses of slavery and segregation and confessing dismay at similar defenses of sexism, Marse Grant, a prominent editor and Baptist, challenged the General Assembly and church "to do what is right and just." So did a co-religionist whose specialty was family policy. ERA was not responsible for the problems facing North Carolina families, and it was not inimical to the biblical principles of familial faithfulness, love, and forgiveness, he said. Betty McCain, who confessed that the Democratic party over which she presided was not known for its biblical scholarship, reminded the General Assembly that both the national and state parties were committed to ERA. She did hope that Democratic legislators would keep that fact in mind.[65]

After Maria Bliss appealed to American idealism, Elizabeth Koontz told a chilling tale of sexual discrimination. The remainder of the time was set aside for Van Alstyne. Offering to make himself available to legislators after the hearings so as to answer particular questions, he focused on two issues. He was sympathetic, he said, to testimony by General Gatsis, who had raised a legitimate question as to the consequences of ERA for national security. Referring to the amendment's legislative history and the manner with which the Supreme Court had actually treated parts of the Bill of Rights in wartime, Van Alstyne said that the amendment would not be applied in combat situations. The court would not "tie the hands of the Congress . . . in waging war" but allow it instead to defer to military commanders' decisions on whether to exempt women from combat. He urged legislators to pay attention not to alarmist "speculation about the ERA" but to the way in which the Supreme Court had actually acted in the past. If they did so, they would "quickly lay to rest" the doubts of the previous day. Turning to the amendment's alleged imprecision, the eminent constitutional authority admitted that reasonable human beings could disagree on its specific objectives and "fault it," as Ervin had, "primarily because of its residual ambiguity." To rid it of ambiguity with specific exceptions, however, would be to draft an amendment "reminiscent" of the Internal Revenue Code in its excruciating detail. Van Alstyne reminded his audience that the Bill of Rights without which North Carolinians had refused to ratify the Constitution read with no greater specificity than ERA. That had been appropriate: constitutions "articulate principles not specifics."

Van Alstyne waited for questions that never came.

Decisively voting down a motion to delay, committee members sent ratification to the floor with a positive recommendation. Debate began on Monday, February 7. The final vote would occur on February 9 provided the bill passed the second reading on the previous day. On representatives' desks was a pamphlet from fourteen assemblywomen listing laws that denied homemakers economic security. Familiar to female legislators and feminists, the list was unfamiliar to vast numbers of women and the men who represented them: laws distributing property unequitably upon divorce, discriminating against women in estate taxes, penalizing them with regard to social security benefits. Outside the chamber a few carefully selected monitors patrolled the corridors gleaning information, with the help of reporters, which they could pass on to Miller. A new staff person kept letters and cards flowing. Antiratificationists had been just as active in circulating "literature" of their own, warning that ratification meant an assault on womanhood and the death of women in combat. Lest reading be done on an empty stomach, they had also sent cupcakes.[66]

Everyone expected a lengthy debate. Speaker Stewart had promised to impose no limits in the belief that North Carolinians must be assured that their representatives had considered both sides. After a few hours, the bill passed a first reading. The next day senators and large numbers of women representing both pro- and antiratificationist forces gathered once again to hear the discussion. Proponents did much of the talking. Miller emphasized that those who feared possible adverse consequences should have greater faith in government, especially when it attempted to extend rights to all its citizens. Acknowledging that religious commitments affected the way in which the amendment was perceived, he confessed that he himself could not believe that God approved of inequality among human beings. He did concede that others read their Bibles differently. Allies addressed themselves to perennial objections, especially the possibility of drafting women, for Van Alstyne's arguments had not converted everybody. Hartwell Campbell was not as concerned about the Bible or the draft as other opponents, but he did believe that the amendment was yet another way in which the federal government could assume power better reserved for the General Assembly of North Carolina. It was after all, that branch of government closest to the people, he insisted, whose responsibility it was to eliminate discrimination. The House then voted 64–52 against him; it also defeated by ten votes antiratificationists' request for a referendum on ratification. The following day representatives gave their final approval 61–55, after a short debate.[67]

Miller was delighted, as was the governor, who had made numerous calls to wavering legislators. NCUERA activists were jubilant even though they knew that the battle was only half over. Miller himself knew that some of his yes votes had come from men confident that the bill would be turned back by the Senate. Initially, however, proponents were optimistic. Chair of the Senate's Constitutional Amendments Committee was Cecil Hill, a lawyer from the little mountain town of Brevard whose wife was a staunch supporter. The committee itself, while divided on the amendment, was willing to send it to the floor. According to a survey undertaken by seven North Carolina newspapers, twenty-five senators favored the bill, fifteen opposed, and ten were undecided. The lieutenant governor was clearly opposed but would vote only in event of a tie. An Associated Press poll showed twenty-four in favor, nineteen against, and six undecided.[68]

From a national perspective the stakes could not have been higher. Ratificationists' victory in the Nevada Senate on the evening before that in the North Carolina House seemed to vindicate White House lobbying in both states. President Carter had designated a full-time ERA activist and sent telegrams to legislators in both states; he had even called a few. Ratificationists could feel the momentum as they leaned forward at least psychologically, but Nevada's Democratic House reversed its position and refused to ratify. ERAmerica strategists now looked to Florida and North Carolina. So did Phyllis Schlafly. Furious at the lobbying of the president, his family, and his aides, she let off steam to Sam Ervin. She detailed Rosalynn Carter's part in swinging the one vote that put Indiana into the ratification column and denounced the president's success in preventing Montana from rescinding. Now she was worried about the Carters' intervention in North Carolina: she had learned of telephone calls to five legislators. She wanted to know if someone could file a suit against the president enjoining him or his deputies from lobbying for the amendment. Maybe Ervin could do something. He certainly had enough "prestige as an elder statesman to admonish a President." Schlafly raged at "this grievous interference in the rights of state legislatures."[69] Ervin commiserated, but, apart from sending an eighteen-page letter to members of the North Carolina General Assembly, he did little.

Schlafly, however, apparently did more. Convinced that antiratificationists were holding on in Florida "against incredible odds" and fearful that Carolina might fall, she looked to Falls Church, Virginia, and mass mailing expert Richard Viguerie.[70] As Viguerie sent out his appeal to anti-ERA North Carolinians, antiratificationists mobilized for action. Galvanized by the House loss, they began pushing for a statewide referendum and produced an anti-ERA rally on the state fairgrounds. Ervin, Schlafly, and Anita Bryant were there to speak. The latter had recently become famous for her crusade against guaranteeing equality of rights for Miami gays and lesbians. Before a crowd of fifteen hundred, Ervin made his usual speech against "the equal wrongs amendment." Schlafly then launched into her attack claiming "there is absolutely nothing the Equal Rights Amendment will do for women." Then came the pledge! Asking the men and women who had crowded into Dorton Arena to stand and raise their right hand, she challenged them to take a solemn vow to defeat every legislator in the next election who voted to ratify ERA. With the audience clearly in the palm of her hand, she told it to let legislators know their political future if they voted for ratification! The press knew a virtuoso performance when it saw one. The occasion, a reporter wrote, suggested "a cross between a political rally and a church revival."[71]

Schlafly's protagonists acted. One pro-ERA legislator felt so threatened physically when surrounded by angry antiratificationists that her secretary hastily called her to the phone and locked the door. Another found himself literally pressed against the wall and warned that a vote for ERA was a public confession of homosexuality. Supplementing person-to-person contact was the avalanche of mail that now began to come in as the result of STOP ERA's contract with Viguerie. He had apparently sent North Carolinians eighty thousand letters containing three postcards addressed to appropriate legislators. Constituents had only to sign their names. One senator whose vote was never in question received more than eight thousand postcards; for the uncommitted the count was even higher. NCUERA leaders estimated the cost of the mailing to be $50,000–60,000—a sum equal to their entire budget.

More important was the maneuvering among opponent senators themselves as they reintroduced their by now classic ploy—the referendum. On February 25, the *Charlotte Observer* reported that a slim majority of twenty-four favored it; twenty-two were opposed; two were undecided; two refused comment. A successful referendum bill would kill ratification because the House had already voted it down. Pro-ERA leaders therefore, had their work cut out for them. Led by Bill Whichard, the House sponsor of ERA in 1973, James ("Jim") B. Garrison of Albemarle, and Lawrence Davis of Winston-Salem, they made common cause with the governor, whose legal counsel on legislative matters served as liaison with Senate leaders. Hunt himself became directly involved in the lobbying—an action that NCAERA leaders thought "unethical." Ratificationists worked furiously on the handful of senators whose votes were pivotal. "We damn near gave away the state," an insider recalled with a hyperbole that Hunt's enemies would have said was Gospel truth; but they could not prevent critical defections by key Democrats. On February 25, two senators who had voted for ERA in 1973—John T. Henley of Cumberland County and Marshall Rauch of Gastonia—announced that they would support the referendum. So did James McDuffie of Charlotte and Wesley Webster of Madison. The state attorney general's ruling that a binding referendum would be unconstitutional dissuaded no one.[72]

In an effort to change the vote of Benjamin D. Schwartz of Wilmington, ERA supporters enlisted the support of Schwartz's daughter, then living in New York State. She was a Bryn Mawr graduate with a Ph.D. from the University of Pennsylvania. Schwartz was intensely proud of her but could not join her in supporting ERA. Schwartz was a senator as well as a father, and many of his constituents were opposed, especially those who shared his values as businessmen. Rauch also refused to budge. A flamboyant ex–New Yorker who had stayed in North Carolina after a stint at Fort Bragg during World War II, he had a multi-million dollar business manufacturing Christmas tree ornaments in Gastonia near Charlotte. Active in civic affairs, he had won the B'nai B'rith award for his contribution to the improvement of race relations in the 1960s. Entering politics, he had once supported ERA. Believing that he might do so again through the good offices of an influential rabbi, NCUERA leaders traced Marse Grant, a friend of the rabbi, to his hotel room in Japan and asked him to intercede. The attempt was futile. Rauch had little sympathy for feminist "nonsense." Prayers to the "God of Our Ancestors" did not have the same ring as those to the "God of Our Fathers." More to the point, however, Rauch had never felt suitably rewarded when Jim Hunt presided over the Senate. Jimmy Green had been more astute and gave Rauch the chair of the powerful Senate Finance Committee. Rauch went with his friends.[73]

John Henley, Wesley Webster, and James McDuffie went with their constituents. A pharmacist from a small town near Fayetteville, Henley found that the people whom he saw at the country club opposed ratification. So did the crowd from every precinct who came to lobby him in Raleigh. So did two members of the bench who called him to warn about the ambiguous "ramifications" of ratification. When push came to shove, Henley had to stand with the judges and the people at home. Webster, ERA partisans had argued, could be enticed to do the right thing by appointment to a seat on the Utilities Commission. But such an offer—even if it had been made—would probably not have been enough to change Webster's vote because he, too, was under heavy pressure from constituents. One of these was the formidable Justice Susie Sharp, who was not an easy woman to deny. As for McDuffie a conservative Baptist, the "country club" fundamentalist churches in Charlotte had helped him to see the error of his ways. Hunt had tried to prevent defections. He literally looked for the elusive McDuffie in motels and offices throughout Raleigh until time for flying to a governor's conference in Washington, D.C. When Hunt finally buckled his seat belt after a harrying day of hide-and-seek and conversations with equivocating senators, he actually thought that ratificationists had the votes.

The next day in the Senate chamber ERA supporters led off in what they assumed would be a debate on the House's bill of ratification. Emphasizing the usual points, they called upon the Senate to decide the issue without a referendum. To their surprise, no one responded. "They thought they were real cute," recalled the governor's legislative liaison, Charles ("Charlie") Winberry two years later. He sighed: "I guess they were." When the roll was called, the vote was 24–26 against ratification. A disappointed Winberry left the gallery to call the governor in Washington. To his dismay, Carolyn Hunt answered the phone. Her personal investment in ERA, Winberry recalled, made his message the most difficult he had ever had to relay. "She wasn't real happy about it."[74]

Carolyn Hunt was not alone. For women who had worked so hard to run an effective campaign, defeat was devastating. This time the postmortem focused less on style, image, and opponents than on their own political expertise; it also contained a new element of bitterness. As in every defeat, there was more than enough blame to go around. Some ratificationist women, believing the governor more powerful than he actually was, complained that he had "simply not done enough." A male politician agreed; he had not lain down in front of onrushing traffic; he had not volunteered to be crucified. Less condemnatory and more appreciative of all efforts, George Miller observed that after the election of 1976 the country was becoming more conservative; the window of opportunity was fast closing. Proponent women were not so philosophical. For female legislators, ERA had become an issue that was uniquely and intensely theirs, and they were as rough on themselves as upon Hunt. If ERA were to pass, they would have to assume far greater involvement; they would, in Trish Hunt's favorite phrase, have to "start playing hardball." Politically knowledgeable women outside the General Assembly agreed but faulted the naïveté of women who had worked themselves into exhaustion. NCUERA leaders confessed their former innocence but recounted specific instances when they had turned to women in the Hunt administration and legislature, only to be told to keep up the good work. Part of the problem was that the governor had recruited many of the very women who had the political expertise the coalition so desperately needed. And once part of the "Hunt team," they were rightly preceived to have a list of priorities that extended beyond ERA. For individuals who had made passage of ERA their only priority, this irony smacked of betrayal.

And then there were problems with the mobilization of women throughout the state for ratification. Grass-roots mobilization never developed the way it had been mapped out on paper. Coalition leaders believed the problem would have been partially remedied had funds from Washington arrived early enough to hire a state coordinator to improve the lobbying effort. Had communications between the field and the Raleigh office been better, mail directed to key senators as requested by the governor might have been sent. However effective women were within the constituent organizations of the coalition, they had apparently not learned enough about politics or opposition to ERA to appreciate the urgency. Those in the field had a different perspective and complained about "Raleigh." They resented the secrecy resulting from attempts to prevent leakage of critical information at NCUERA; they wanted to know "what was going on." Unaccustomed as they were to interest group politics, they concluded that sisterhood and democracy had been surrendered to (male) hierarchy and authority.

In all this, ratificationists even discounted what they had, in fact, accomplished. They had enlarged and funded their coalition and made it more effective than was the case in most unratifying states. But the unforgiving barrage of self-criticism subsumed all that to a common theme: they thought they had not learned the rules well enough. "We must," observed an activist from Charlotte, "play the game of politics the way it is [played]." A Raleigh woman whose political education had accelerated dramatically while monitoring the legislature in 1977, put it another way: "ERA is not going to be passed because of its intrinsic merit." It would pass only when ratificationists had elected the right people to office. Some supporters had known that in 1976, but as a group they had not yet acted on the basis of that knowledge. They had not been "political" enough even though they began their campaign believing that that was precisely what they had to be and could become. Whether psychologically devastated ratificationists could pull themselves together in time to apply "the lesson of '77" to the election of 1978 was problematic.

4

"Idealistic Sisterhood Has Gotten Us Nowhere": Dressing for Political Combat

"Of course we are going to get ERA. If it takes white gloves, we will use white gloves. If it takes combat boots, we will wear combat boots; we'll use both." Uttered in 1974, the words expressed ratificationists' confidence in winning their goal by doing "whatever it took." "White gloves" elicited an image of nonthreatening gentility, what one North Carolina activist called being "clean and neat and proper—on our way to Sunday school." She added that "women in combat boots know they're in a battle." On their "way to Sunday school," where ideals and proper behavior promise good results, proponents had received generous support from President Jimmy and Rosalynn Carter, prominent citizens from Hollywood to New York, and a wide range of women's groups. "We have had," lamented an activist from the District of Columbia, "authority and celebrity on our side, but not victory." Pulling off their gloves in 1977, ratificationists were chastened and frustrated by intensifying resistance. Combat boots seemed to be definitely indicated as they contemplated defeat in Nevada, Missouri, Oklahoma, Florida, Idaho, and North Carolina; they were not consoled by the U.S. attorney general's opinion that rescission votes would be invalid. They were stung and confused by what they called "deliberate attempts to distort a simple proposition." "It is almost unbelievable, were we not witnessing it ourselves," bemoaned an ERAmerica spokesperson who castigated legislators not "man enough" to stand up to Bible-toting "housewives."[1]

It is possible, however, that neither the Bible nor the women had intimidated politicians who knew a losing proposition when they saw one. No one, observed a pro-ERA assemblyman, runs to catch a train that is not leaving the station. Public opinion surveys reported that the percentage of Americans polled favoring the amendment had dropped from 65 percent in 1976 to 56 percent in 1977. More disturbing was the fact that support, especially among whites, was disintegrating far more rapidly in unratifying states than in the nation at large. As early as 1976, pollsters could show that ERA had become linked in the public mind with feminism, abortion, civil rights, secularism, and nontraditional gender roles. Each had negative connotations.[2] That these pejorative linkages existed was confirmed by the 1977 meeting of the National Women's Conference in Houston, Texas. Created to support a feminist-defined political agenda, the conference generated ringing endorsements from three first ladies: Lady Bird Johnson, Betty Ford, and Rosalynn Carter. It also attracted a counterconference staged by Phyllis Schlafly, that, in retrospect, may have more effectively galvanized participants than the "main event."[3] To attribute declining public support solely to the political genius of the "housewife" from Alton, Illinois, however, would be to ignore the extent to which rejection of ERA was an essential part of the broad resurgence of conservatism that would eventually sweep Ronald Reagan into office in 1980.

Seemingly contrary to that trend, the U.S. Supreme Court began to address some of the most flagrant examples of sex-based discrimination. Although still refusing to make sex, like race, a suspect classification, the Court was nevertheless moving toward a stricter scrutiny. Rejecting the old criterion that required demonstration of rationally related grounds for sex-based classification, the Court seemed to be creating an analysis that demanded substantially related grounds, "substantial" being stronger than "rational" but not so rigorous a standard as "compelling." However encouraging such a shift might be to those who wanted gender-neutral laws, the fact remained that the Court's action lent support to the antiratificationist argument that ERA was superfluous. To explain that the amendment was nonetheless needed to move the Court toward a stricter standard of review was to engage in an argument that nonlawyers might regard as quibbling.[4]

In sum, developments over which ERA partisans had no control and which, at the Court, they welcomed, all pointed to the fact that the amendment was in trouble and time was running out. For it to survive, supporters had to change strategy. NOW leaders, arguing that the original seven-year timetable had been "unrealistic," launched a campaign for extension along with a boycott of unratifying states. Both measures were controversial. Opponents thought changing rules in the middle of the game unfair, a charge that congressional supporters believed had merit even as they passed an extension resolution. Less ambivalent ratificationists for whom ERA had become symbolic of a national commitment to equality argued that "human justice should not be a cliff hanger." The boycott was to be directed against five of the most-frequented convention centers in the country—Atlanta, Chicago, Las Vegas, Miami, and New Orleans—all in unratifying states. Partisans argued that the boycott could transform financial loss into political pressure. Skeptics, however, pointed out that in states such as Florida and Illinois opposition came not from Miami and Chicago but from areas unaffected by the boycott; it would punish friends instead of enemies. Supporters were not convinced because the boycott was intended to engage ratificationists throughout the country in an issue not defined by state politics. It was national.[5]

On that point—and on the critical importance of North Carolina, Florida, Illinois, and Oklahoma—Phyllis Schlafly could agree. The issue of rescission also continued to sustain the Equal Rights Amendment as a national issue. Schlafly was soliciting help in her efforts to frame a compelling argument that thirty-five states had not ratified ERA. While her husband, Fred, conferred with Philip Kurland of the University of Chicago, she asked Sam Ervin his opinion on the viability of lawsuits by Idaho, Tennessee, and Nebraska asking for a declarative judgment upholding the validity of their rescinding resolutions. Both Kurland and Ervin responded positively. Schlafly could not rest with what appeared to be "certain" victory. She did not underestimate the role of continuous vigilance in solidifying and sustaining her constituency. As a political leader she had to keep her people engaged until ratificationists had fired their last shot.[6]

Combat Boots: The 1978 Elections

Phyllis Schlafly's ally in North Carolina, Dorothy Slade, shared Schlafly's commitment and drive. She congratulated STOP ERA workers on their "magnificent achievements" in the 1977 campaign and announced plans for further action. Apologizing for waiting until July to thank her supporters, the STOP ERA activist explained that their ranks had grown so much that she now needed a staff of twenty-five people instead of the handful of friends who helped her from the start. She encouraged the faithful to register voters and work for candidates who entered the lists against pro-ERA incumbents. She added that they should also actively oppose the referendum to allow Jim Hunt to succeed himself as governor. If there had ever been an "off-season" for antiratificationists, it existed no longer. The message was work, work, work, work, work![7]

Tar Heel proponents faced the same challenge but with psychological resources sapped by the devastation of defeat.[8] Emotionally and physically exhausted, North Carolinians United for ERA leaders did what they did best: create standing committees. They also did something that was new and far more significant even though they did not understand precisely what the change meant or precisely what to do. They began to act on the wisdom expressed by a Charlotte Republican woman who—contrasting the values of ratificationists and those of legislators—angrily and cooly observed: "Idealistic sisterhood has gotten us nowhere!" She was ready to dress for the political combat of electoral politics. Unless people genuinely committed to ERA were put in office, it would make no difference how long the deadline for ratification was extended. But getting effectively involved in elections required more than sloughing off idealism; it meant trying on boots that did not always fit. Socialized in the culture of women's voluntary societies, in which political action meant lobbying for publicly beneficial issues, ratificationists viewed the "nitty-gritty" of electioneering with suspicion if not outright distaste. Members of the League of Women Voters pointed out that they were limited by charter to nonpartisanship. So was the American Association of University Women.[9]

Leadership in the new strategy, argued certain coalition activists, should come from the political arm of the women's movement: the National Women's Political Caucus. But the national caucus itself needed strengthening; its campaign chest for 1976 had been only $24,000. The North Carolina caucus was scarcely better prepared to take the field because it seemed to founder on the shoals of personal overcommitment and ideological confusion. To be sure, it had been part of the process through which women had been appointed to office, and it had indeed been founded to put women into politics. But caucus members had to work out more precisely what achieving that goal required. If it required endorsement of candidates supporting women's issues, they needed to define the meanings of support, "women's issues," and endorsement. Endorsement, they knew, did not mean what they did in 1976 when they "endorsed" candidates for governor and lieutenant governor by stamping their approval two days before the election—essentially a meaningless act. Understanding these things, NCWPC was committed to getting its house in order well before the 1978 elections. Members called upon caucus personnel in ratified states for advice and requested funds from the national body.[10]

Endorsement may have appeared to be simple enough, but it was not. Members debated how public a candidate had to be in pledging support for ERA. Some were willing to accept private assurances, whereas others wanted public declarations. They discussed oral commitments, too, and rejected them along with the national caucus because they had been burned by too many last-minute defections: no more oral commitments. This decision angered legislators. "They could keep their money," spat out one proponent; her *word* was her "bond!" Beyond the way in which a commitment was made was the issue of how all-embracing it had to be. Should the caucus try to defeat a candidate who had voted against ERA but whose record on other issues was otherwise congenial with its agenda? Fed up with the obdurant Benjamin Schwartz of Wilmington, pro-ERA women supported a weak candidate and thereby lost leverage with the angry Schwartz on every other issue. They protested that they had no alternative; he thought that they did not know much about politics. These were delicate issues of political negotiation, and ratificationists sometimes found it difficult to play by their own rules. Reserving ERA as the litmus test of political legitimacy did not apply when activists thought they could beat a man with a woman. In Winston-Salem, for example, seasoned politicians were dismayed by a challenge in the Democratic primary to incumbent Carl Totherow, who had taken abuse for supporting ERA. His opponent was Meyressa H. Schoonmaker, an intelligent and progressive lawyer who personified the kind of woman the caucus wanted in politics. To some political veterans, however, she personified betrayal.[11]

Beyond endorsements, proponent women needed a strategy to guide their own actions. By September 1977 they were asking politicians and community leaders throughout the state to help recruit pro-ERA candidates, but it was not until December that caucus, coalition, and ERAmerica representatives could resolve persisting friction enough to agree on a plan. The national and state caucus hired Betty Ann Knudsen and Lynn Gottlieb, both of Raleigh, to target key legislative races, recruit candidates to run against anti-ERA incumbents, and hold workshops on electioneering. NCUERA hired the chair of its Legislative Committee to relay information on candidates' needs to local coalition people, who in turn recruited campaign workers. With the filing date of February 6 only two months away, time was short, especially for so ambitious a goal as a statewide effort to elect friends of ERA. Handicapped by a late start, caucus strategists soon realized that, without running their own candidates, they would have to work with those who were available. Surveying the field, they sometimes felt, one soft-spoken ratificationist recalled, as if they were Nieman-Marcus shoppers in a J. C. Penney world.

Although unable to wage the campaign they wanted, ratificationists nonetheless plunged into electoral politics in ways that would have surprised them five years before. In their years of trying to mobilize the grass roots they had already learned the wide variety of legislative districts, each of which presented unique challenges. In one, a closet supporter of ERA was reluctant to invite politically experienced outsiders into his xenophobic district. In another, where textile and furniture manufacturing was interspersed with agriculture, caucus strategists later agreed that they should not have listened to the overconfident pro-ERA incumbent, who thought he could win the primary all by himself. In yet another, pro-ERA activists could not agree whether to support a young man whose position on ratification seemed suspiciously enigmatic. Maria Bliss and a few friends had more faith in the noncommital young man than did their colleagues, and they supported him. Later he justified their confidence in him by taking more abuse from opponents "than a man oughta take" recalled Charlie Winberry of the governor's office. Such nurturing was not required in districts where the caucus had only to help old friends. In districts that included the University of North Carolina at Chapel Hill, in Durham, and in Raleigh this work was relatively easy to do.

East of Raleigh and west of Asheville, recruiting for the electioneering process was more difficult. There ERA supporters were politically inexperienced and frequently passive. They could agree to mount phone banks and stake out yard signs, but no amount of coaxing could make political "combat boots" fit. Even if they wanted to fight—and many of them did—they had not been adequately socialized in electoral politics. With time and experience and agreement on more than one issue, it is possible that the coalition could have transformed itself into a political machine. But the commitment and expertise of skilled individuals could not be translated into a comparable collective effort. That ratificationists were able to do as much as they did in fact do was a tribute to them. Even when they lost a local election they might win something valuable. In the mountainous Senate District 27, the caucus had a fine candidate—Barbara Scott Price of Tryon, who had been politically active in

Illinois before moving to North Carolina. Personable and highly educated, Price generated intense enthusiasm among ratification activists. One supporter confessed not having felt such a sense of purpose since her own involvement in the antiwar movement at Berkeley. Price faced an incumbent, a local dairy farmer who was far better able than she to mobilize rural and working-class voters in mountainous counties where ratificationist networks were weak. She did not win, but she did win the respect of her opponent and he hers. In retrospect, that mutual respect may have been one of the reasons the man subsequently assured proponents that he would support ERA.[12]

Charlotte proved to be the model of successful politicking. One of the few areas of the state where a viable two-party system existed in the 1970s, the city had attracted a steady stream of middle-class newcomers from other states. With a sense of purpose that businessmen attributed to the entire city, the Charlotte caucus quickly established itself among women who were disenchanted with their marginal role in both political parties. Their networks were in an excellent position to punish Senator James McDuffie for breaking his pledge to support ERA. Leading the campaign were five NOW and caucus activists. Knowing full well that they could not defeat McDuffie if ERA were the issue, they opted for a more practical strategy. Not only had McDuffie reneged on ERA but also on a promise to support local option on liquor-by-the-drink. His critics, a group called Democrats for Integrity in Government, argued that he was not a man of his word. Begging money from Charlotte allies, national feminist groups, and sympathetic celebrities, Democrats for Integrity turned to experts for political advice and to 150 volunteers for door-to-door combat. Their strategy was to make McDuffie fifth on a slate of five from which the four highest vote getters would be nominated. Supporters of ERA included three incumbents from Mecklenburg and a lawyer from Cabarrus County who was virtually unknown in the Charlotte area, from which 80 percent of the vote would come. McDuffie, a two-term senator who had led the ticket in two previous elections, found that his newly politicized fundamentalist friends and Schlafly's STOP ERA could not save him. He won Mecklenburg in the Democratic primary but lost by about one thousand votes when Cabarrus voters seized the chance to elect one of their own for the first time in seven years.[13]

Ratificationists throughout the state were jubilant, for a few moments at least. But such celebrations were self-deceiving in the context of the 1978 election. Ratificationists' entry into electoral combat ran counter to the slow, erratic, but inexorable turn to the Right in state and nation. Although Democrats controlled both Congress and the presidency, there was no mistaking a surge of popular discontent: a profound suspicion of government, mounting resentment over taxes, a pervasive yearning for traditional values. Jesse Helms had detected the shifting currents as early as 1972, although many liberals at first thought him only "a congenial naysayer standing somewhere to the right of Barry Goldwater and Ronald Reagan." That naysaying attracted so many voters that by 1978 North Carolina Democrats could not even find a viable opponent to challenge him. In Winston-Salem ratificationists confronted that conservatism directly and lost. They had defeated a pro-ERA man, Totherow, in the Democratic primary with a woman, Schoonmaker, whom they thought much better qualified, only to lose in the November election to an opponent who was, in many respects, Schoonmaker's mirror image. A housewife and mother of five, Anne Bagnal was avowedly anti-ERA and antifeminist. A newcomer to politics, she received the endorsement of antiratificationists, fundamentalists, and a conservative antitax group. She benefited as well from a weakened Democratic party and Helms's crushing victory.[14] This blow together with stiff opposition to moderate Democrats seeking reelection to Congress and the election of anti-ERA women to the General Assembly signaled a resurgent conservatism that ERA partisans could not escape. Had the election been wine uncorked at a restaurant table, North Carolina Democrats would have sent it back.

Legislators favorable to ratifying ERA appeared to have been elected to the house by a narrow margin. Opponents emerged with a clear lead of twenty-six in the Senate, leaving ERA with only nineteen firm votes. If ratification came to a vote, it would lose once again.[15] If it did not come to a vote, there were other losses to consider. As veterans of three campaigns, ERA women had learned important political lessons. They had mobilized a coalition and politicized large numbers of women; they had introduced the rank-and-file to the world of phone banks, political action committees, candidate endorsements, and targeting. Many of them political novices, they knew that ERA partisans had much more to learn, but they were confident that they had at least learned to count, and the votes for ratification were simply not "there." Pragmatists were prepared to concede as much and move on to other issues. Movement women understood, however, that ratification politics had initiated and reinforced commitment to the women's movement. Ratification leaders felt obliged not to admit defeat until they had tried everything, even if it meant acting contrary to the political lessons they had struggled so hard to learn. Summing up the dilemma, a North Carolina ratificationist observed: "We agreed that . . . [we] couldn't afford another loss, but we couldn't afford not to go for it." They went.

Going for It against the Odds

After the election of 1978, national leaders of the pro-ERA movement understood all too well that the odds were against them. In Illinois the Equal Rights Amendment had been bottled up in Senate committee since 1975. In Florida ratificationists lost a bitter and expensive referendum campaign as well as legislative seats in the election. In Oklahoma ERA had also failed three times and faced an uphill battle in 1979. The future was not promising, but ratificationists recognized the compelling need to move North Carolina. They agreed with the Raleigh leadership that the Senate was a major problem; and they agreed to help solve it. Sarah Weddington, the Texas lawyer whom President Carter had put in charge of women's issues, scheduled lunch at the White

House with activists who might be able to persuade Jessie Rae Scott to head the lobbying effort in the General Assembly. Perhaps she could augment Governor Hunt's influence and detract from that of the Lieutenant Governor Jimmy Green. There would be no need for on-the-job training because, as the daughter-in-law and wife of former governors, Scott understood Carolina politics very well. And she might possibly be able to call upon the magic of the Scott name, which had evoked positive responses in Carolina for more than thirty years. She consented with the understanding that there would be no unsolicited calls from the president.[16]

The selection of a Senate sponsor was more complicated. Scott preferred Bill Whichard, House sponsor in 1973, whom she thought "a man of great integrity," intelligence, and eloquence.[17] Whichard was reluctant. He had alienated people in the eastern part of the state with an "environmentalist" Coastal Management Act. James Garrison, who, like Whichard, had seen ERA go down in defeat under his tutelage, was no more enthusiastic. Scott's husband's uncle, Ralph Scott, was ruled out as unlikely to win over senators close to Green. Then a Charlotte woman lawmaker suggested Craig Lawing, a supporter who had cooperated with the caucus in 1978 and pulled off a minor political miracle by engineering a law to allow local option on liquor-by-the-drink. More important, the powerful Democrat was a political ally of Green and could also work with Hunt. As the next president pro tempore of the Senate, Lawing could operate easily within that body's power structure. Lawing was offered the job. Charlotte's women legislators persuaded him to accept.

Jessie Rae Scott and Lawing then conferred with the governor. Going over the list of swing votes in the Senate, Hunt promised to get three if Lawing could get three. Putting Joe Thomas, Sam Noble, and R. C. Soles on his own list, the governor indicated that he would talk to Thomas's wife as well as to the senator, whom he had appointed to the seat left vacant by the death of a proponent. Hunt would also ask union people to talk to Joe Palmer, since his district had an atypical concentration of unionized labor. Robert Davis would receive encouragement because nothing in his twenty-six years was likely to have prepared him for the wrathful lobbying of opponents. The challenge was enormous, but it was exhilarating, too. Everyone seemed to have forgotten that they had once thought the votes were against them.

The woman with whom Hunt and Scott would work as president of NCUERA understood both the practice and symbols of politics. Beth McAllister was trained as a social worker, socialized as a mother, and apprenticed as a League of Women Voters' member. She had headed Wives and Mothers for ERA and acted as a legislative monitor in the 1977 campaign with an instinctive political sense that impressed women more experienced than she. McAllister herself had great respect for Maria Bliss who, she remarked, "walks on water as far as I am concerned." In leading NCUERA, McAllister would have to produce miracles of her own. To be sure, she had the benefit of good advice, but it was not clear that advice and commitment could compensate for having lost too many friends in the 1978 election.

As legislators arrived in Raleigh in early January 1979, reporters predicted that the session would be "low-key," "tight-fisted," and "tax-frenzied." With the exception of a crime control package that included relief for battered women, uniform sentencing, and merit selection of judges, Hunt's legislative package had been passed in the 1977 session. Fears of recession and the corresponding decline in revenues meant no major new initiatives. Indeed, with a tax revolt careening across the national political landscape, Hunt—whose liberalism never exceeded the pragmatic—had begun talking about tax relief. Against this mood of retrenchment and the numerical logic of the election, NCUERA was preparing to lobby once again. At the invitation of Trish Hunt, pro-ERA women in the General Assembly met to talk about prying open support for ratification. They agreed to oppose the appointment of anti-ERA nominees to coveted positions and to support the governor on his agenda if he "pulled out all the stops" for ERA. They also authorized a hand-delivered confidential letter to fellow lawmakers reminding them who their friends were. It was not clear how these actions would affect the hard count of no votes: reporters surmised that ERA had little chance of passing. It was not a serious issue anymore.[18]

In contrast to the dim view of the press, NCUERA, with the help of ERAmerica, invited to the state a roster of pro-ERA celebrities that included MASH star Alan Alda, housewife humorist Erma Bombeck, and feminist Elizabeth "Liz" Carpenter.[19] Jimmy Green himself introduced Bombeck at a

Actor Alan Alda headed a roster of pro-ERA celebrities whose presence revived ratificationists.

reception sponsored by the League of Women Voters. When asked if his introduction signaled a change of heart on ERA, the lieutenant governor replied that he had insisted North Carolina statutes be studied to ensure equal rights for all. As far as the amendment was concerned, he was in no position to say; he did not vote on legislation. Carpenter's prediction of a change of heart after plying him with Texas-sized doses of flattery suggested that Green was as capable as she in the art of nuanced political conversation. Any such conversion, observed a political commentator, was as realistic as expecting legislators in the tobacco state to endorse a ban on smoking.[20]

Outside the party circuit, Jim Hunt was trying to coordinate strategy. With Lawing, House Speaker Carl Stewart, George Miller, Scott, and his own legislative liaisons, he assessed the legislative situation. Stewart and Miller agreed that the vote would be close; House members were saying they would vote for ERA only after it cleared the Senate. Why not go first in the Senate, Stewart suggested. If it failed, Green would have to shoulder the blame. Hunt agreed, tossing the ball to Lawing; Scott mentally noted how quickly Hunt had taken charge of the meeting. It was a good sign! Other good signs followed by the end of January. A veteran opponent suddenly died and was replaced by the local Democratic party with a pro-ERA man. Within two days two senators announced support. The count was now twenty-two for and twenty-four against—four votes short. In early February, having gained another vote (total: theoretically twenty-three), strategists decided to gamble on a ratification bill. Lawing would file in the Senate on February 5; Miller would file in the House. Lawing promised to keep the bill moving. He was, he told reporters, as confident of ERA's passage as he had been the previous year on liquor-by-the-drink.[21] Privately ratificationists knew that picking up the additional three votes was a long shot. Counterbalancing Lawing's confidence were announcements by key Democratic legislators that they would vote no. On February 10, William D. ("Billy") Mills declared that he would remain with opponents. Joe Thomas, despite pleas from the governor, followed suit.

Republican Senator Walter C. Cockerham of Greensboro was officially undecided pending the results of an unscientific newspaper poll. To influence results, activists followed trucks of the *Greensboro Daily News* from stop to stop, scooping up copies of the paper with the mail-in ballot as soon as they hit the pavement. After "awaiting the popular verdict," the senator finally declared that he would not be swayed by the governor, the president, or "anybody else." As if to counterbalance Cockerham's poll and vote, the Raleigh *News and Observer* repeated a story on an October poll in which 51.1 percent of North Carolinians polled favored the amendment and 37.1 percent opposed. The figures were deceptive if Tar Heels replicated patterns exhibited elsewhere. Support for ERA was actually soft, but proponents could nonetheless use polling data to justify their position.[22] They believed that popular support was with them and that it would make a difference. As Hunt and prominent Democrats urged NCUERA to flood the mails with popular support for ERA, a reporter for the *Washington Post* observed that ratification in North Carolina had become an establishment rather than a feminist issue.[23]

Opponents had been saying so for some time. Styling themselves the people in revolt against establishment liberals, they flooded into Raleigh for another round of hearings on February 12. Leaders of fundamental churches had pledged to send thousands to Raleigh for hearings on ERA and state funding of abortions. Aware that five thousand people had poured into a downtown mall to protest state regulation of Christian academies, few doubted that these zealous folk would return in equal numbers to protest ERA. On a gray, chilly morning, they piled out of buses by the thousands, brandishing STOP ERA buttons, red rosebuds, placards, and Bibles. After singing and praying, they surged into the Legislative Building, engulfing the handful of ratificationists who bore pink, heart-shaped signs with the words "LOVE IS RATIFYING ERA."[24]

When the doors of the hearing room swung open, one of the first to enter was an eighty-four-year-old former suffragist who had sat outside the chamber all morning calmly knitting. Most of the audience was younger. There was a large number of children from fundamentalist Christian academies who thought ERA was a laundry soap. It was somehow ironic, wrote one editorialist, that antiratificationists, many of whom had long opposed busing of school children to achieve racial balance, now used the same method to defeat sexual equality. Above the children was a familiar face who had perennially

With hearings on ERA and state funding of abortions scheduled on successive days in 1979, members of fundamentalist churches dramatized their opposition to both by busing the faithful to Raleigh in unprecedented numbers.

opposed busing. It would not have been a real hearing without Sam Ervin. "When the good Lord created the earth," he observed, "He didn't have the advice of Bella Abzug and Gloria Steinem." A Baptist minister and civil rights activist argued that the amendment was totally consistent with the Western liberal tradition, and fundamentalists in the audience agreed. That was why they opposed it. By February 14, the capital had finally cleared of protesters.[25]

With or without public drama, ERA was being kept alive by artificial means because there had never been enough votes to sustain it. As a favor to his allies, Bill Whichard got the Senate Constitutional Amendments Committee to delay the bill for a week, but he thought his patient already "dead." After word of delay leaked out, the Senate Judiciary (I) Committee was convened by its chair, Julian Allsbrook, to change a referendum bill into a bill for ratification. He would then send it immediately to the floor for defeat. The ploy failed because the veteran Democrat had forgotten the Senate rule allowing a special committee meeting only if all members were notified. An ailing Robert Davis had not been officially informed, and he rushed to Raleigh from his sickbed. After an encounter with Davis, whom he attacked with vicious sarcasm, Allsbrook had to back down, but he vowed to kill ERA by the end of the week.[26] The governor, knowing that Allsbrook could deliver, was eager to avert a vote that could further polarize a Senate afflicted by what had become one of the most acrimonious battles in legislative history. Lawing met with opponents to count votes; they were against him. He knew that the vicious Allsbrook would "bloody" proponents if ratification came to the floor; he had no recourse but to "pull the plug." Upon his request the pro-ERA majority on the Constitutional Amendments Committee killed the bill.[27]

The action angered many. Women legislators, privy to strategy sessions throughout, were stunned when they learned that the fate of ERA had been decided without them. Furious that key women lobbyists had not been included in the decision, NCUERA activists found it incomprehensible that ERA supporters would kill their own bill in order to spare the feelings of allies. Claiming assurance from two opponents who had promised to vote yes if the bill came before the full Senate, McAllister insisted that the decision was premature. With more time, ratificationists could have persuaded a third senator to switch or at least "take a walk" on the day of the vote. But she never reckoned with Allsbrook's ability to force a vote in the meantime. An irate Liz Carpenter declared it "appalling that the destiny of women in the twentieth century is to be second-class citizens at the mercy of men playing games like little boys." Don't get mad, get even, Lawing rejoined, urging rank-and-file ratificationists to focus on the 1980 elections; but for the moment, the anger was overwhelming.[28]

Votes and Vigils

In the inevitable postmortem, Miller pointed out that the 1978 elections had been crucial. Electing twenty-six pro-ERA senators in an increasingly conservative climate had been problematic; to convert the necessary six after the election was impossible. And the problem was not merely that of North Carolinians. Ratificationists had lost Arizona, Nevada, and Virginia and faced efforts to rescind in Rhode Island, Delaware, Kansas, Montana, New Hampshire, Ohio, and Texas. A few women once again faulted the governor, arguing that it was a mistake to go ahead without the votes—which is precisely what women lobbyists had wanted to do. It was charged that Hunt—a devout Presbyterian—had failed to understand the religious complexity of the issue. Joe Thomas, a Hunt loyalist, attended a church that opposed ERA, a fact that should have prevented the governor's appointing him to replace a ratificationist.[29] Ignored was the fact that the law mandated that Hunt approve an action already taken at the local level. Jessie Rae Scott, too, shook her head in thinking about Thomas; he had managed her husband's campaign in Craven County, yet the Scott name had carried no more weight than Hunt's. It troubled her to think that the folk who had put two Scotts into the governor's chair had come to the capital with STOP ERA buttons. "Our people," she mused, have changed; but so had discourse about gender.

Lawing came in for his share of criticism, too. The decision to go with the expansive Charlotte auctioneer was endlessly replayed. He had not been effective with either undecided senators or Jimmy Green. Once confessing that they should have relied on insiders to sponsor ERA, ratificationists now thought that they had chosen someone who was too far inside. The muttering that perhaps Lawing had deceived them suggested more about their political socialization than his commitment; politicians do not enter contests in order to lose them. Lawing had faced an uphill battle, and lobbying had not moved legislative opponents. Scott recalled them as "emotional and intense. You can't get through to them." When she had tried to talk about the principle of equality, one man had responded with impassioned laments about the high divorce rate, women in the labor force, and homosexuality. Senators, she was convinced, were angry with feminists. In "getting even" with those they saw on television by voting down ERA they had acted, Scott recalled, "like peacocks strutting their feathers." She thought the male "ego" strange, echoing Liz Carpenter's indignation at being "at the mercy of men playing games like little boys."

As if to ease the pain, ratificationists staged a vigil in front of the General Assembly.[30] Scheduled to occur just before the vote, it was to have symbolized the fact that opponents had no monopoly on religion. Collins Kilburn, a spokesman for the North Carolina Council of Churches and its Committee for ERA, had promised legislators a statement from people of faith with "a different kind of style." As people gathered at the Edenton Street Methodist Church, ratificationists put on gold and blue buttons bearing the words "PEOPLE OF FAITH FOR ERA." In her opening prayer, Maria Bliss celebrated the ministry of deliverance through Moses, Aaron, *and Miriam.* To the music of a familiar hymn, the congregation invoked the image of flight from Egypt as others had before them:

Lead on, O cloud of Yahweh, the Exodus is come.
In wilderness and desert, Our tribe shall name its home.
Our slav'ry left behind us, A vision in us grows . . .

Though those who start the journey, the promise may not see.
We pray our sons and daughters may live to see that land
Where justice rules with mercy, and love is law's demand.

After a prayer of confession, Steven Shoemaker, president of the Raleigh Ministerial Council, rose. In blunt, powerful tones, he proclaimed: "We are in the middle of a battle for equality, and in a battle there is not much time for sorrow." "Where do we get our hope? From God. . . . And how do we do good to our enemies? Well, we try to convert them." Elaborating on his theme, he added: "We try to persuade them equality is a good thing. We try to appeal to their sense of fairness and justice. We try to appeal to their conscience." But, "if they harden their hearts like Pharaoh," he continued, then "we, like Moses, may have to force our enemies to stop oppressing us. . . . The Jews did not have the vote in Egypt, but we do." Removing his vestment and shifting to politics, he urged the crowd "to get commitments for ERA from candidates in 1980." After asking his audience to pray for people "afraid of change," he concluded, "We have hope. By the grace of God we have courage. And *we have two years.*"

After the benediction, the crowd marched silently through the chill night air to the Legislative Building. There marchers lit candles and stood silently as a lone security guard looked on. Breaking the silence, a woman, moved by the emotion of the moment, said "Let's sing 'Bless't Be the Tie that Binds.' " Someone demurred: by referring to "Christian love," the hymn excluded Jews. Then, with scarcely a pause, someone began to sing "We Shall Overcome." "I haven't sung that since the civil rights movement," whispered a woman in tears. As the crowd dispersed, a small group went into the Legislative Building to sit quietly in the gallery. The lieutenant governor was recognizing the work of Senate pages. Referring to "these fine young men," Green caught himself in midsentence and added, "and young women." Women in the gallery, noting the slip and quick recovery, exchanged glances. The General Assembly could stop ERA—but not the movement that embraced it.

Making the Boots Fit Better: The Election of 1980

By 1979 defeat made partisans more aware than ever of what Jill Ruckelshaus meant in 1977 when she told the National Women's Political Caucus, "We are in for a very, very long haul." In anticipation, ratificationists began hunkering down at a conference in Washington, D.C. There they heard ERAmerica representatives warn of a *new*—although North Carolinians could have told them a lot about persistence of the *old*—conservatism. They heard Eleanor Smeal, new president of NOW, recount the twenty-four-hour days required to secure extension and, like Ruckelshaus two years before, warn of the financial and emotional costs yet to be paid. It was Liz Carpenter, however, who brought delegates to their feet amid tears and laughter as she recalled highlights of the seven years of labor and the sisterhood they had generated. Thus inspired, activists digested what political consultants and veteran lobbyists had to teach them before returning to combat. To Beth McAllister and her Tar Heel colleagues two things were clear: they had a coalition rivaled by few in the nation, and they had a chance to make a difference in the 1980 elections.[31]

Ratificationists changed strategy. Smeal moved the National Organization for Women into the vanguard. NOW campus crusades would entice women from their studies to work for ratification; NOW missionaries would canvas door to door in unratifying states; NOW "message brigades" would bombard legislators with postcards. NOW rallies and marches would dramatize the issue, while NOW media projects would educate the public. Not to be outdone, the League of Women Voters formed a National Business Council to mobilize business on behalf of ERA, and BPW added its own media project to those of NOW and the league. Ratificationists also produced new material on the positive impact of state ERAs.[32]

Opponents renewed their own efforts. They began litigation to test the constitutionality of extension and the legality of the national boycott of convention sites. Having in the first ten months of 1979 defeated ratification in eight states, Schlafly had reason to feel confident. She nevertheless warned her eighth annual STOP ERA convention not to let up in the battle for "God, family, and country." She put the delegates through the usual workshops on public communication and practical politics. Escaping from the intensive work of politics, they produced skits with lyrics, sung to the tune of "Mame," roasting ratificationists with the plaint: "Who opens car doors for us? Who gives up seats on the bus for us? Who buys houses in our names? Who gives us insurance policies and alimony? No-o-o one!" But these women were not helpless. Their laughter was merely relief from the deadly serious work that lay ahead.[33]

In North Carolina proponents reorganized in a flurry of activity. NCUERA joined with NCWPC in May 1979 to form a single ERA Political Action Committee (ERA PAC). McAllister was one co-chair; veteran Republican ratificationist and caucus president Grace Rohrer was the other. They planned their next campaign to recruit candidates almost six months ahead of their 1977 schedule. They appointed political activists and experts to their steering and advisory committees and continued the political action of the previous campaign by improving support networks for pro-ERA candidates.[34] Working on a full-time basis, McAllister supervised the effort to provide targeted campaigns with whatever they needed. By the February 1980 deadline for filing candidacy, the ERA PAC had viable candidates, support systems, and money-raising projects—and an expanded coalition. NOW created its own PAC and was accumulating information to use in influencing legislators. Concentrating first on the Democratic primary, both PACs endorsed Jim Hunt, who seemed destined to win over the former governor, Robert Scott. In the race for lieutenant governor it would be House Speaker Carl Stewart over Jimmy Green, whom ratificationists believed had done so much to hurt them.

Stewart seemed to be one of the ratificationists' own. Following the lead of Scott and Hunt, he was at forty-three in the bright-young-man tradition of the state's moderate-to-liberal Democrats. After earning degrees in history and law, he had launched a career in politics that had attracted from the North Carolina Center for Policy Research an accolade as "most effective legislator" in the General Assembly. Elected to a then-unprecedented second term as Speaker of the House, he was quick to identify public need and adept at stretching the budget to meet it. He was eager not only to make North Carolina the South's most progressive state but also to move beyond regional comparisons. Green, almost twenty years older and a wealthy tobacco warehouseman, had built a legislative record as a fiscal conservative. He epitomized a conservatism that had made Southern Democrats before him more comfortable with Republicans than with the party of Franklin Delano Roosevelt.[35] Convinced that ERA's best chance lay in Green's defeat, ratificationists made themselves indispensable to Stewart's campaign, mobilizing precinct workers, staffing phone banks, and raising money.

Ratificationists also worked in legislative races, consistently endorsing pro-ERA male incumbents, contributing funds, and, in many instances, moving beyond traditional "women's chores." Confident that their increased efficacy would pay off, they were devastated when Stewart lost an agonizing cliffhanger by one tenth of one percent of the vote. Counting on a sweep of urban areas as well as a large black turnout, Stewart ultimately won only 60 percent of the black vote while narrowly losing the cities of Greensboro and Asheville. For ratificationists, it was a "real heartbreaker." For Hunt forces it was a lesser disaster. The governor's advisers, many of them personally partial to Stewart, feared that a Green defeat could mean a massive defection of conservative Democrats to the Republican column in November. They suspected that Jimmy Carter's weakening presidency would leave North Carolina vulnerable to a Republican takeover of the state; it had happened in 1972.[36]

The ratificationists' anxiety was at least partially justified. Except for the reelection of Hunt, the election of 1980 was a conservative festival. Its character was best represented in the defeat of Senator Robert B. Morgan by Jesse Helms's National Congressional Club. The club could not make the governor's challenger, I. Beverly Lake, Jr., appealing, but it could certainly make Morgan unappealing. Lacking the popularity of his predecessor, Sam Ervin, the junior senator faced a colorless ultraconservative political science professor from Eastern Carolina University who had never run for public office. Helms's allies linked the fiscally conservative Morgan with "taxing-and-spending" colleagues and excoriated with merciless repetition his votes to give away the Panama Canal and send aid to "the Marxist government of Nicaragua." As the club's misrepresentation of Morgan's record hurt him, the Republican party targeted North Carolina for the maximum support allowed by law.[37] It was not surprising, therefore, that Phyllis Schlafly should have returned to the state as one of a fifteen-member-delegation of conservative celebrities. With her came Utah's Republican senator, Orrin Hatch, Radical Right activists, and military and religious figures. ERA, abortion, and a weakened family life were linked with Democratic foreign and military policy. Salvation lay in a blend of old-fashioned religion and New Right politics.[38]

Hunt was less vulnerable to the conservative onslaught than Morgan and state legislators. For one thing, Hunt's political network reached into all one hundred counties, while the senator's organization was weak; for another, Hunt was more visible and aggressive than Morgan. When the Republican candidate and his fundamentalist allies called him a "Ted Kennedy liberal" and charged him with undermining the family, Hunt replied by sharpening the distinction between himself and Lake. He would, he said, run a campaign based on human equality and equal opportunity. A devout Christian and family man, he responded to negative ratings from the fundamentalist organization Churches for Life and Liberty by refusing to change his stand on the right of poor women to tax-funded abortions. Confident of the outcome in terms of his own reelection, Hunt could even devote energy to the Carter campaign at a time when regional pride in the president had declined.[39] He did not stump North Carolina for the Equal Rights Amendment; no one did.

A Gentleman's Agreement

Ronald Reagan's landslide of November 4, 1980, did ratification in. Antiratificationist women had—as they had in 1978—targeted, spent, and campaigned effectively in states both sides agreed were most likely to ratify. In 1978 in Illinois, STOP ERA had established a reputation for being "at least as politically adept as pro-ERA groups," if not more so. In 1980, opponents had escalated their effort with considerable assistance—in North Carolina at least—from the Religious Right. Unwilling to share the details and traumas of electioneering with researchers as proponents had done, they were perceived by some legislative candidates as simply having more clout than ratificationists, observed a consultant working North Carolina in 1980 for ERAmerica. Candidate reports to the State Board of Elections revealed that STOP ERA forces had been as engaged as proponents in contributing to the campaign. Whatever the local variations, the results in targeted states throughout the nation were the same. Women on both sides had immersed themselves in electoral politics as never before, and in three important unratifying states—Illinois, Florida, and North Carolina—antiratificationists had carried the day.[40]

ERA died on election day and would be buried in January, predicted an antifeminist Republican assemblywoman from Winston-Salem. Reagan's coattails had brought five new Republicans into the North Carolina Senate and ten into the House to add to legislative veterans' votes against ratification. Hunt's winning 62 percent of the vote could not force support in the General Assembly without a gubernatorial veto, and the lieutenant governor, who appointed committee chairs and presided over the Senate, would be opposed. Green had won office again, and he had not had a miraculous conversion. Indeed, Craig

Lawing doubted ERA would even come up for a vote. The man who had led the fight in the 1977 Senate agreed, and added to his list of casualties additional funds for abortions and a new state agency to enforce fair employment practices. At a legislative briefing sponsored by the North Carolina Council on the Status of Women, Jane Patterson, deputy secretary of the Department of Administration, acknowledged that proponents were three votes short: "The Equal Rights Amendment will not be ratified this year," she said, adding, "It may not be ratified in my lifetime."[41]

Beth McAllister disagreed. As head of NCUERA and the ERA PAC, she had been confident that ratificationists' greater political sophistication would pay off at the polls. Devastated by the election, she nonetheless refused to see it as a referendum on ERA. Moreover, she regarded the election itself rather than the liberalism of the 1960s as an "aberration." To forgo an amendment and deal with state and federal laws one at a time, as Patterson suggested, was to McAllister a "piecemeal solution." According to the latter, the American Bar Association reported eight hundred discriminatory statutes at the federal level alone. The task of attacking each seemed daunting. ERA had been the right solution. She put the electoral poll out of her mind and focused instead on a public opinion poll taken only two days before the election that showed support for the ERA holding at 56 percent in favor, 35 percent opposed. She clung to the logic that had made ERAmerica target North Carolina as the state most likely to ratify in 1981. Hunt and Speaker-elect Liston B. Ramsey supported the amendment, she argued, and so did a majority in both houses if certain "closet commitments" by unidentified legislators were kept. With NOW representatives, she pledged the resources of NCUERA and the PAC to continue working with the governor for ratification. She recalled Carrie Chapman Catt's tally of the effort required to pass woman suffrage: "Fifty-two years of ceaseless campaigns." Fifty-six campaigns for referenda; 480 campaigns for suffrage amendments; 277 campaigns to influence party conventions; 30 to address presidential party conventions; and "19 campaigns with 19 successive Congresses."[42]

Such persistence was no virtue for North Carolina legislators. They were eager to foreshorten the biennial ERA battle and get down to business. After weeks of behind-the-scenes manuevering, Senator Ollie Harris, a veteran Democratic opponent and an undertaker, announced that he would bury ERA with a referendum bill that was promptly referred to the Judiciary [I] Committee headed by a fellow opponent, Republican Minority Leader Donald Kincaid. The strategy was for the anti-ERA majority to amend the referendum into a ratification bill that could be killed in committee. Because reconsideration would have forced supporters to put together a two-thirds majority, they attempted a countermanuever that failed.[43] They did not have the votes, and everyone knew it. Uncertain as to the status of the referendum bill in Judiciary [I], opponents competed among themselves as how best to kill ERA.

Finally Senate Republicans introduced a ratification bill in order to table or postpone it in committee. Unwilling to let Republicans be credited with killing ERA, Green offered proponents a face- and peace-saving deal; he would refer the bill to the Constitutional Amendments Committee with the understanding between him and proponents there that no further action was to be taken. Lawing, whose vote count tallied with that of opponents (twenty-seven or possibly twenty-eight against and twenty-three or possibly twenty-two for) was convinced a change in the count of more than one was impossible. Unwilling to endure a vicious and emotionally debilitating floor fight leading to certain defeat, he agreed. On February 27, Senate Majority Leader Kenneth Royall announced to the press a "gentleman's agreement." Opponents would allow the ratification bill to remain dormant in committee, suppressing their desire for a "clean kill," explained the veteran Democrat. In exchange, supporters agreed that the Equal Rights Amendment not be debated or voted upon for "the remainder of the 1981–82 session of the North Carolina Senate."[44]

Response was swift. Hunt, when apprised of the situation, telephoned Royall in a vain attempt to talk the powerful Durham legislator out of the pact. The governor canceled his trip to Washington, D.C., for a meeting of the Democratic National Committee and told the press that he disagreed with the action. Female legislators were irate. Republican antiratificationist Mary Pegg, having wanted to execute ERA in public, blasted the compromise as a "mockery of government." Senator Rachel Gray a Democrat who had nothing in common with Pegg, raged at women's having been completely bypassed by the "gentlemen" in a short-circuiting of the the legislative process. Proponents outside the General Assembly argued that the issue was too important not to be discussed and challenged the binding character of the agreement. "We are just damned determined to carry on," stated the executive director of the North Carolina Council of Churches at another vigil in front of the Legislative Building. ERA was still alive, he insisted. If alive, ERA had to be freed; if dead it had to be resurrected.[45]

Getting around the Gentleman's Agreement

And it had to be done before the end of June 1982, the deadline for ratifying. Linking that imperative with a plan of action was difficult. Ratificationists from Washington explored the problem with legislators and coalition leaders in a tense April meeting. North Carolinians pointed out the implications of the gentleman's agreement: under the rules of the General Assembly, that body's short session, which met in even-numbered years, was devoted exclusively to the budget and certain kinds of bills. These included bills that passed one house the previous year or won a suspension of the rules (by a two-thirds vote) to get to the floor for consideration. ERA fit none of these categories. It seemed unrealistic to believe that North Carolina could ratify.

NCUERA leaders could not give up.[46] That the coalition was planning a May 2 march on the state Capital seemed an exercise in futility to national leaders who had been told repeatedly that such demonstrations did no good in the Tar Heel State. Local ratificationists were downhearted, but they came anyway, chanting "ERA won't go away." Lined up five abreast and whipped by a brisk breeze, they marched by the thousands along Peace and Wilmington streets, wearing banners and singing "We Shall Not Be Moved" and "Bread and Roses." There were people in wheelchairs and babies in arms, older men in three-piece business suits and younger ones in jeans and T-shirts. Democratic party officials, highly placed members of the Hunt administration, chaired professors at the University of North Carolina, ministers, teachers, housewives, and gay student activists—all wended their way to the State Capitol Building carrying banners that ranged from NOW's familiar "ERA YES" to the less-familiar "CHARLOTTE MEN FOR ERA." As small boys climbed into the limbs of trees and atop a bronze equestrian statue of Andrew Jackson, the rally opened with an address by the governor. Reading the amendment, he told the crowd: "The words . . . stand for simple justice. They are right and we will ultimately win this struggle." Hunt was followed by a woman senator who blasted the gentleman's agreement and an elderly former suffragist who vowed to fight on.[47]

It was Beth McAllister who captured the spirit of the march. "Together we have begun a movement that will not stop with ERA and 1982," she declared. "I say to those few men in the Senate who have tried to hold justice hostage that *they cannot—they will never—hold our spirit and commitment hostage.* They cannot stop us or our work. . . . Because of you," she continued in strong, impassioned tones, "and all that you are willing to do and give, equal legal rights will become a reality and our nation *will* become one nation with liberty and justice for all people." Quoting from suffragist Louise Alexander, she concluded with a ringing cry for action. " 'I call on women to raise fewer dahlias and a lot more hell. The place is here. The time is now. The opportunity is ours.' " "The time *is* now," McAllister reiterated, and we *are* the people. It is up to us to seize this historic opportunity to move the dream of legal justice one step closer to reality. Because of you . . . , 'failure *is* impossible'!"[48]

Failure was, of course, possible. The words and ritual of defiance came out of the experience of people who were accustomed to some degree of success. In their march and brave speeches they were saying that as long as they remained true to their goal and to each other they would eventually "make a difference." But they were content with neither a vague "difference" nor what they did best, which was to talk to each other. They wanted a specific goal that was really a means: ERA. In the fall of 1972 the goal had seemed such a small and logical thing to ask. In 1981 it seemed to be even more logical and much, much larger. Now they were in the awkward position of being ready to charge, but, having heard the trumpet, of not knowing where and how to move.

The governor—who had never "done enough" to suit a few women, called a secret meeting in May to chart one final assault. The session included Smeal of NOW, Jane Campbell of ERAmerica, McAllister of NCUERA, and Patterson and Martha McKay from the Hunt Administration. Also present were key Democratic legislators, such as Ruth Cook and J. Allen ("Al") Adams, who had been active ratificationists, and Russell Walker, chair of the North Carolina Democratic party. They were presented a way to introduce ERA into the short session of the 1982 legislature. Recommendations from the Legislative Research Commission could always be considered by either house or by both simultaneously and would require only a simple majority. If passage of the Equal Rights Amendment emerged as a recommendation of a Legislative Study Committee and the Legislative Research Commission, ratification could be introduced once again. The question was how to get ERA before a Legislative Study Committee. The answer was to introduce an innocuously titled bill—one that might even seem to be a sop to feminists—to set up a commission to study the economic, social, and legal problems and needs of North Carolina women. Sworn to secrecy, the group broke up. Introduced by Adams, House Joint Resolution 1238 passed well before the General Assembly recessed on July 10.[49]

The ratificationist leadership now faced a delicate task of mobilizing while concealing from the press and STOP ERA activists the tactical function of the study committee. The official position was that ratification would be considered in the short session if legislators were persuaded by constituents to do so. To that purpose NCUERA was kept intact. NOW, however, chose to go it alone, a decision foreshadowed by its forming a PAC in the 1980 elections. The national organization had already pulled out of ERAmerica and launched a major campaign in the wake of Reagan's election to double membership by 1982 and strengthen its financial base. To do this NOW had to convince women that it was the true expresssion of the women's movement and principal actor in the ratification struggle. In North Carolina, Smeal had a personal friend and disciple who could be relied upon for total cooperation with the national office. Terry Schooley, state NOW president who had first participated in a North Carolina ratification drive in 1979, was—like ratification leaders before her—convinced that the state had never been properly organized. Eager to apply NOW's new tactics, she saw the new campaign as a NOW enterprise.

In contrast, NCUERA broadened its network by aligning itself with its most politically active constituent organization, the North Carolina Association of Educators (NCAE), and the North Carolina Democratic party. The process had been initiated in 1979 when the ERA PAC added the NCAE president, John Wilson of Raleigh, to its steering committee and appointed top-level Hunt politicians to an advisory committee. Since the governor had reorganized the state party by putting his people in leadership positions, it was axiomatic that closer contact with the party would result from working with Betty McCain. She had, after all, chaired both the party and Hunt's campaigns. The linkage was supported nationally because ERA became a Democratic issue after Republicans repudiated it. One result of this cooperation was the targeting of twenty-nine county conventions, many in anti-ERA areas, in which ratification resolutions were to be introduced. They passed in twenty-one and failed in the others by close votes. These results, ERA sponsors hoped, would prompt legislators to reconsider their opposition. Reporting to a press conference in June, McAllister was joined by the vice-chair of the

Democratic National Committee, who read the committee's resolution calling for ratification. NCUERA's politicization continued when it moved its Raleigh office to NCAE headquarters, elected NCAE's political affairs specialist Jo Ann Norris as president, and held ERA Countdown Workshops run by NCAE activists. For the coalition, 1981 was not only the year of the gentleman's agreement but of "new sisters and quality men."[50]

It was also the year of the Legislative Research Commission. The Legislative Study Committee on Women's Needs was co-chaired by two staunch Democrats, both ratificationists: Representiative Ruth Easterling of Charlotte, a former president of the National Federation of Business and Professional Women, and Senator Helen Marvin of Gastonia. It included seven additional legislators—all ratificationists—and one nonlegislator, Alice Wynn Gatsis. The head of NCAERA, who had been added at Green's behest, was under no illusions as to the ultimate agenda. The committee met periodically between the two sessions of the General Assembly and held hearings on the economic status of women in the state. Before the critical final meeting on April 5, 1982, Hunt telephoned every ERA supporter to be sure that plans did not go awry even with this lopsided majority. At the close of that meeting, after a long discussion of child support, paternity cases, equitable distribution of marital property and tenancy by the entirety, Senator Wilma Woodard, a Democrat from Wake County, moved that the committee recommend ratification of the Equal Rights Amendment at the 1982 Budget Session. Passed 7–2, the resolution was also approved by the Legislative Study Committee on Aging, chaired by Ramsey, and sent forward to the Legislative Research Commission, chaired by Lawing. Confident that ERA would become eligible for consideration for the sixth time by the General Assembly, Hunt assembled his team.[51]

The Final Round: Doing Everything Right

An impressive group of players gathered in the governor's office on April 13. House Speaker Ramsey, a nine-term legislator from the mountains of western North Carolina, was a man of few words and strong principles. Chief among the latter was a devotion to party discipline and to that peculiar blend of fiscal conservatism and New Deal liberalism characteristic of many North Carolina Democrats. Described by reporters as a roughhewn, beefy man with a gravelly voice and a deep chuckle, Ramsey was widely regarded as one of the most astute men ever to hold the job of Speaker—and one of the most partisan. He was, in McAllister's words, "a man you can trust." That characterization also applied to Raleigh lawyer Al Adams, who would serve as ERA's House sponsor and a worthy successor to George Miller. Also included was the state party chair, Senator Russell Walker, an unabashed liberal prepared to make the party a full partner in the ratification drive. So was Doris Cromartie. ERA Liaison to the Democratic National Committee and an equal employment opportunity officer for Duke Power Company, Cromartie had been active in politics since the 1960s, when she was president of Democratic Women. Martha McKay was there, too. The governor's intentions were made abundantly clear when he asked Betty McCain to serve as his full-time personal lobbyist for ratification. NCUERA's Jo Ann Norris and NOW's Terry Schooley would generate constituent lobbying.[52]

The initial meeting of "the team" was, in Norris's words, "open and candid." Sensitive to the political damage created in the more remote areas of the state by the perception that ERA was "not a North Carolina issue," Ramsey wanted assurances that national supporters would stay at home, at least so far as the public was aware. The Speaker was even more adamant that there be no media campaign for or against specific legislators during the primary. Getting down to basics, the group went over lists of House and Senate members, comparing information provided by the ERA PAC, NOW, and NCAE's PAC as to where individual legislators stood. After a vote count that indicated forty-nine for, forty-five against, and twelve undecided in the House and twenty for, eighteen against, and twelve undecided in the Senate, the discussion turned to the people to be targeted in order to get the twelve votes needed in the House and the six required in the Senate. The governor then assigned specific individuals to work on each of the targeted legislators with the understanding that poll data would be available to establish constituent opinion.[53]

Never one to waste time, McCain took over an office and WATS line in NCAE headquarters, which promptly became Democratic party headquarters. A woman with boundless energy and infectious enthusiasm, McCain was the daughter of a lawyer who had sent his son to medical school and his daughter to Teacher's College at Columbia University for an M.A. in music. Marrying a physician, McCain had reared two children, taught Sunday school at the First Presbyterian Church, participated in a wide variety of volunteer organizations, and in the 1960s discovered an aptitude for politics. By 1972 she was on the Democratic National Committee and by 1982 had co-chaired two statewide campaigns for Hunt. She had also chaired the state's Democratic party and become the first woman appointed to the state's powerful Advisory Budget Commission. She understood charts, but, better yet, she understood people. Surveying information on "targeted" legislators and their "contacts," she assigned area coordinators within each home district. Most of them women and in many instances "Hunt keys," all could be expected to deliver. In the meantime, there were details to block out for a public opinion poll and media advertisements. With Patterson, Norris, Schooley, and Mollie Yard from the Washington office of NOW, McCain chose the areas to be polled—one House district and eleven Senate districts, all in rural areas of the state. Television ads proposed by NOW were screened with the understanding that there would be no mention of specific legislators. Radio spots provided by ERAmerica were also chosen with the understanding that all would carry the tag "Brought to you by North Carolinians United for the Equal Rights Amendment." Agreement on credit lines for NOW's television ads, however, was more difficult to negotiate. McCain and Norris argued for alternative wordings, believing that the NOW image was not a positive one in targeted

When Betty McCain took over as Governor Hunt's personal lobbyist for ERA in 1982, the ratification effort in N.C. acquired a new level of political expertise.

districts and that winning was more important than advertising NOW. It was a point that Martha McKay and Jane Patterson had tried to convey in a private meeting with Smeal and Yard in 1979. Ultimately McCain, with the help of the governor's former press secretary, got an agreement that read: "Sponsored by the North Carolina ERA Countdown Campaign." The clash was symptomatic; NOW officials complained that for all their resources they were accorded "little credit and even less respect."[54]

Put off by McCain's style and loyalty to Hunt, NOW leaders rejected her judgment because—as one put it—she was a politician, not a feminist. That politicians not feminists were to be converted did not affect NOW's analysis. Smeal, moreover, was convinced that concern about NOW's image was not legitimate. She later claimed in all sincerity that she was better known in North Carolina than Jimmy Green. "We are of the people," she insisted, "we believe that." McCain did not. Remembering baby pacifiers that a NOW member had once sent to opponent legislators in the wake of defeat, she was determined to run a campaign that would not leave legislators so alienated they would refuse to support women's issues in the future. Moreover, neither she nor other proponents had much sympathy with NOW's sectarian refusal to make this final push a genuine team effort. If full cooperation proved too much to hope for, McCain tried at least to ensure some degree of coordination with regular Monday morning strategy-briefing sessions.[55]

First in a series of carefully orchestrated events was the governor's announcement on May 20 of the results of a Harris poll commissioned by KNOW, Inc., a nonprofit, nonpartisan organization devoted to women's issues. The poll, a telephone survey conducted April 24–29 with a randomly selected group of 801 adults, indicated a solid pro-ERA majority throughout the state. On hand to discuss the results, pollster Louis Harris noted a "dramatic increase" in support since 1979, when the percentage of those favoring ERA was 49 percent as compared to the current 61 percent. He attributed the increase to the rise in the number of working women and the economic difficulties they encountered. The explanation was plausible inasmuch as North Carolina, the tenth most populous state in the nation, had more than 60 percent of its female population in the work force. Consistent with findings in previous national polls, support was higher among younger groups than among those over fifty. Men, as in other states, indicated greater support for the amendment than did women; the percentage of blacks supporting it exceeded that of whites. Not surprisingly Democrats were found to favor ratification in larger numbers than Republicans, just as self-styled liberals endorsed it in greater numbers than did conservatives. What was most significant, however, was that ERA was favored by a majority in all parts of the state among all age groups, both races, members of both parties and independents, and people at all points along the political spectrum.[56]

Opponents disagreed. Interviewing Alice Wynn Gatsis about the poll results, reporters found the NCAERA leader armed with figures of her own. According to an April mail poll conducted by the conservative Greensboro marketing firm of W. H. Long, Inc., a majority of North Carolinians opposed the amendment: 74.1 percent of 1,029 people in ninety-six counties. Long, questioned about his findings, explained that in his mail polls he used a pool of 16,400 registered voters from the state's one hundred counties, who were regarded as "decision makers." Although he declined to comment on Harris' findings, Harris readily commented on Long's, explaining that mail surveys tended to elicit responses from people with sharply opposing positions on an issue, leaving out the so-called silent majority. Gatsis suggested that the differing results were attributable to the fact that Harris, in posing his questions on ERA, used an abbreviated version of the text that omitted: "The Congress shall have the power to enforce, by appropriate legislation, the provisions of this article." Although Harris insisted that inclusion of the fuller text made no difference in terms of responses, Gatsis argued that the omission was significant because, consistent with her own views, most people opposed federal intervention in their lives. The bottom line, she insisted, was that "polls are polls"; they did not necessarily translate into votes.

She was right, but the governor was determined to do everything he could to see that she was wrong. Armed with individualized printouts, he called in each targeted legislator for a chat to point out that those favoring the amendment represented the majority of constituent opinion. Meanwhile, ten of eleven Democratic Party Congressional Districts came out solidly for ERA; the single holdout was the Third District—the lieutenant governor's home base. County Democratic chairs and vice-chairs and district chairs were invited to lunch at the Governor's Mansion, where Hunt appealed for help. In an effort to make the governor's lobbying a bipartisan effort, former First Lady Betty Ford flew in to talk with Republicans. The American Bar Association was enlisted as legislators discovered a personal letter from its president refuting issues raised by opponents. The North Carolina Association of Trial Lawyers added its endorsement, and Betty McCain sent copies of William B. Aycock's 1976 ratification speech to legislators along with a pro-ERA statement by conservative radio commentator Paul Harvey.[57]

Reinforcing the governor's effort at the local level, ratificationists lobbied with an intensity that had eluded them in previous campaigns. Having finally established functioning coalitions in all one hundred counties, NCUERA leaders left nothing to chance. Veteran ratificationists McAllister and Knudsen crisscrossed the state, holding training sessions and encouraging local groups. Determined to have no repeat of the situation in 1977, when local groups failed to follow through on the governor's request for letters, ratificationists developed a call-in system that ensured compliance. With a field staff of sixteen from NCAE alone, they were confident that the old dependency of campaign leaders on sometimes indifferent local organizations was a thing of the past, especially in those areas where linkages with party activists had been forged.[58]

Meanwhile NOW, in a massive mobilizational effort, had set up lobbying groups in areas of the state where ten years before ratificationists had struggled to find a single contact. Pleading their cause before groups ranging from the War Resistors' League to the American Legion, organizers set up action teams in small towns where new recruits signed up. For every victim of "ERA burnout," there were new replacements. Women who had previously assumed that the drive for equality would proceed apace without their efforts had found in the 1980 election results and the 1982 ratification deadline ample reason to volunteer. Inexperience was no handicap. NOW's carefully prepared Countdown Campaign packets explained how to recruit new workers, circulate petitions, and lobby legislators. Weekly goals for visitations and phone calls were backed up by careful reporting procedures and weekly and, if necessary, daily supervision from women from headquarters. The NOW effort more closely resembled that first put together by opponents than it did the multiorganizational coalition originally assembled by ratificationists.[59]

The "Hunt team" meanwhile debated the complicated question of when and where to introduce the ratification bill. A few representatives thought that passage first in the lower chamber would send a favorable signal to the Senate. Since that body had not responded to that signal in 1977, Hunt saw no point in putting the House through a senseless ordeal if passage in the upper body were not assured. He thought—he hoped—he could persuade enough wavering senators to put together twenty-six pro-ERA votes. The Senate it would be. Robert B. Jordan III would sponsor, backed up by a Senate team that included two former sponsors. Timing and committee assignments would be crucial. Opponents could be expected to introduce the bill with the intention of tabling it in committee. Green's action would also be decisive. Although he had said publicly that he would send any ratification bill to the Constitutional Amendments Committee, a decision to send it to Judiciary [I] would mean certain death. In light of the uncertainties, the governor agreed that Senate strategy would have to be reassessed continuously. The one sure thing was that action would be swift. Legislators were eager to get the session over with and get back home before the June 29 primary, and the absolute deadline for ratification was June 30. Ratificationists agreed, too, that with fifty-six of the sixty-one votes needed in the House, they would prefile on Wednesday, June 21, with Jordan poised for action in the Senate. In the meantime, key members of the governor's staff were lobbying furiously.[60]

On June 2, two days before the General Assembly convened, Jordan learned that Republicans were filing a ratification bill with the intention of tabling it in committee. He then filed his own sixty seconds ahead of opponents, thus assuring proponents control of their own bill. With the General Assembly scheduled to convene two days later, supporters worked furiously to keep the bill from being tabled, arguing that the amendment ought to be decided on its merits and not on procedural grounds. If action could be delayed at Friday's opening session, ratificationists would have the weekend to continue lobbying. And they would need it. All this action was focused on the same legislative body that one year before was known to have been opposed to ratification. Of the nine men targeted for the six votes needed by pro-ERA forces, only three were certain supporters. Joe Palmer, who had been a potential yes in 1979 despite his uncommitted stance, had come aboard, as had a fellow mountaineer, R. F. ("Bo") Thomas, for whom the 1982 campaign was a first. The only real convert was J. J. ("Monk") Harrington, a veteran opponent. The Bertie County legislator claimed that letters from constituents in his rural coastal district were no longer running against ERA. The Harris poll had also influenced his decision to switch. So had his secretary. Whether James Speed and Henson Barnes could be brought on board was still uncertain.[61]

Tipped off that antiratificationist legislators were holding a strategy meeting in the chapel of the Legislative Building, McCain and a colleague rushed over, checking off senators as they walked out. After everyone had apparently left, McCain walked in, switched on the lights, and spotted under a pew a political ally, whom she and the governor had been lobbying intensely. Word of her "discovery" leaked to the press. Colorful copy—the embarrassed legislator claimed he was only trying to escape reporters; it was discouraging news for proponents who had believed that Speed was "switchable." Considered a "hunt man," the lawmaker from Louisburg had been shown poll data on his district that McCain believed should have eased his conversion to ratification. Sixty-three percent of constituents polled supported ERA; 30 percent did not. But antiratificationists, increasingly skilled at targeting, were also applying pressure. One of them was his wife and another was his pastor, who were more persuasive than pollsters. Speed simply did not believe that a majority of his constituents were for ERA. He would not vote for it.[62]

As if to prove the polls wrong, five thousand antiratificationists invaded the Legislative Building the next day, clutching Bibles, brandishing familiar red STOP ERA signs, and carrying plastic sacks reading "ERA IS NOT MY BAG." Alerted that STOP ERA leaders had applied for parking permits for fifty buses, ratificationists scheduled a news conference the same morning, hoping to show that fundamentalists did not speak for all people of faith.[63] Three days later, on June 6, ratificationists demonstrated that they, too, could turn out by the thousands. Although initial plans for a NCUERA rally at the opening of the legislative session had been canceled so that funds could be spent on lobbying, McCain was dismayed to learn that a call had gone out to NOW members across the country to march in North Carolina, Florida, Illinois, and Oklahoma. Annoyed at Smeal's disregard of earlier promises by stumping the state and advertising NOW in pro-ERA TV commercials, McCain was resigned to NOW's limited commitment to a team effort. To violate assurances given the Speaker of the House was altogether another matter. Confronting Schooley, McCain was told that a march was indeed in the works and that it would be a NOW affair. Replying that if there were a ratificationist march in North Carolina, it would be under the direction of the governor, McCain called Hunt, who had put in yet another call to Smeal. Agreeing that the actions of the General Assembly clearly had national implications, he had sympathized with the frustration of ratificationists elsewhere who were understandably fed up with the recalcitrance of the state's most conservative lawmakers. But the very legislators whose votes were critical were not, as he had repeatedly explained, the sort of men who would be persuaded by delegations from New York, Illinois, or Georgia. Having reached what he thought was an agreement that the march would go forward as a team effort, Hunt reassured the skeptical McCain, who was not surprised when NOW leaders applied for a permit in NOW's name only.[64]

Aware that time would be short and procedural matters critical, ratificationists had assigned a team to cover every legislator. Anticipating tactics of opponents, they had prepared a list of countermeasures for every contingency. The bottom line, however, was numbers. When the Senate went into session on Friday, June 4, opponents also had their strategy in place—and the votes to implement it. When North Carolina's fifty Senators wended their way through the color-coded crowd to the Legislative Building, opponents in STOP ERA red and supporters in NOW green crowded outside the packed gallery to watch as the national media focused on the first of four state legislatures that would determine the fate of the Equal Rights Amendment. Immediately after the opening prayer, Jordan rose and asked that his bill, cosponsored with a Republican, be temporarily displaced and put on Monday's calendar. Green ruled against Jordan, explaining that the calendar had not been properly called yet. He then proceeded to read it, whereupon ratificationists moved adjournment in order to claim the weekend for lobbying. They were defeated. The crowd in the gallery sat silently as Marshall Rauch, the powerful chair of the Senate Finance Committee and a principal architect of opponent strategy, arose. He moved to table the bill without debate. The motion passed 27–23; proponents had failed to pick up the needed votes. Then Julian Allsbrook made a motion that would in effect have prevented further reconsideration. A ratificationist immediately moved to adjourn, but he failed by four votes. Realizing what had happened, ERA partisans in the gallery began chanting "Green, Green, Green." The lieutenant governor called security guards to clear the gallery. By another vote of 27–23, Allsbrook's motion passed.[65]

"We knew we had twenty-three votes," Jordan told reporters. "We kept hoping this morning and probably even up to a couple of minutes before the session that we would get a few more people to give us the weekend. But that just didn't work." "We lost fair and square," concluded a woman senator, "We just didn't have the votes." Governor Hunt, learning of the action, called a press conference. Expressing his disappointment, he pledged to fight for ratification as long as the General Assembly was in session. But time was running out. The only hope was to get the House to pass a referendum bill to put ERA on the June 29 ballot. If it passed, the governor could call the General Assembly back for a special session on June 30 before the amendment officially expired at midnight. It was a long shot, but he was willing to try.

As incumbents rushed home for a weekend of campaigning, ratificationists gradually began converging on Raleigh. On a surprisingly cool, breezy Sun-

Before a packed gallery and the national media, the North Carolina Senate, voting 27-23, dealt ERA a critical defeat only days before the deadline for ratification expired.

day, they marched ten thousand strong to the Capitol with a sense of élan that defied both the message delivered by the Senate two days before and the one pulled by an airplane overhead—"LIBBERS GO HOME." A sea of white with splotches of green, the crowd moved along the march route to the Capitol, linking themselves with suffrage not merely by clothing but also by a 1915 convertible carrrying a former suffragist. McCain opened the rally with a poem she had written that reflected her sense of humor and the dedication of the marchers. Norris of NCUERA and NCEA followed with a challenge to prove wrong the senators who believed that they had finally gotten rid of ERA. As long as the General Assembly is in session, she insisted, "you and I have work to do. We cannot and we will not stop now." Hunt, wearing a bright green blazer, underscored the sentiments in his own address, as did NOW's Schooley. But while NOW leaders could take pride in the fact that North Carolinians had rallied in numbers comparable to those simultaneously demonstrating in the capitals of Florida, Illinois, and Oklahoma, further action by the General Assembly was impossible.[66]

A referendum was not feasible. Even if passed by the House, the director of the State Board of Elections pointed out, the legislature would have to suspend so broad a range of technical requirements that action was problematic. Moreover, approval would also have to be secured from the Civil Rights Division of the U.S. Department of Justice, a procedure that usually required sixty days. Legality, not practicality, was the first concern of ratificationists who assembled in the governor's office on Monday morning. Desperate for anything that might work, they went over the legal issues with counsel. Satisfied on that score, they faced the next hurdle: how to get the signatures of seventy to eighty members of the House. That night thirty representatives, meeting in a Raleigh law office, pledged to sign and recruit others, but they wanted assurances that if House members went out on a limb before the primary the Senate would back them up. As sponsor of the now-tabled Senate ratification bill, Jordan approached Green the following day only to discover that opponents for the first time in the ERA struggle would not support a referendum. So informed, its backers conceded defeat.[67]

Rank-and-file ratificationists, frustrated by their impotence, could do little in response. A few expressed their anger by sending small boxes of chicken droppings to anti-ERA senators. Others protested Green's role by turning their backs on him at the state Democratic Convention a week later after vociferously applauding the governor. Still others chose to emphasize retaliation at the polls by sending anti-ERA legislators a white glove to make the point that "the ladies" were prepared to exchange their lobbying gloves for the heavy boots of electoral politics. In the days remaining before the primary, many ratificationists did just that, transforming their lobbying effort into intense campaign activity. The results, however, were mixed. Ratificationists had no alternatives to the twenty-seven no votes, every one of whom ran unopposed. They were, however, able to protect three senators who had moved into the ranks of ERA supporters—Harrington, Thomas,

ERA supporters converged on the State Capital in the wake of legislative defeat, reinforcing their sense of moral victory and belief that they spoke for the majority in North Carolina and the nation. (UPI/Bettmann Newsphotos)

and Palmer—and once again to trounce James McDuffie. But they lost challenges in three other districts.[68]

Proponents also learned that politics is not always about friends and enemies; sometimes it is ambiguous. In Senate District 1, in the northeastern tidewater, Winifred ("Winnie") Woods, a staunch partisan for ERA with good political connections, had challenged incumbent Melvin Daniels. Daniels was one of the soft no votes whom Hunt had persuaded to vote for ERA in case of a tie vote; but there was a condition. Hunt would have to appear at a major campaign event on his behalf. To gain a possible yes, the governor gave his word; he felt obliged to keep it even though Daniels himself was never put to the test; it was a matter of honor. For ratificationists—and for Woods—who knew nothing about the agreement, Hunt's show of support for Daniels was inexcusable. In self-righteous indignation ERAmerica retaliated by pulling out funds earmarked for Hunt's Campaign Committee. This time, Hunt had done too much.

Despite internal tensions, ratificationists believed that in 1982 the critical ingredients for a win had fallen into place. A grass-roots organization, expanded and invigorated significantly by NOW's organizational efforts and an infusion of skilled field personnel, had provided sustained and vigorous constituent pressure at the local level in tandem with the Democratic party—at least that portion responsive to the governor. An effectively coordinated campaign had been placed in the hands of individuals, for the most part politically skilled, who represented not just pioneers in the ratification effort but key women legislators, the top female leadership in both the Hunt administration and the Democratic party. Included, too, were representatives of national ratificationist organizations whose commitment of resources was essential to an effective drive. Critical also was the leadership provided not only by key legislators committed to ERA but by a governor prepared to mobilize in public view all the resources at his command. It was, in sum, a combination of essential players and one long in the making. Within the constraints of time and the prevailing political environment, ratificationists could say with considerable justification that they had at last done everything right. It was a textbook campaign, an ERAmerica activist would recall from her own experience as a state legislator in Ohio. In terms of grass-roots organization, the North Carolina coalition, she continued, was the best in the nation—the product of hard, creative work on the part of "some pretty remarkable women."

That said, however, there were the inevitable recriminations that followed in the wake of defeat. The governor, the women who worked for him, and the Democratic party had not been "committed enough," and they came aboard too late. Others commented caustically on NOW's arrogant refusal to participate in a team effort. Far more relevant were realities that ratificationists were frequently loath to acknowledge and over which they had no control. The sustained and effective lobbying of antiratificationists was once again a significant factor. So was the fact that the lieutenant governor's political ambitions rested on a conservative network that "keyed" in part on defeat of ERA. Other factors could not be discounted: intraparty resentment at the growing reputation of a governor whose national prestige would be enhanced by ratification, apprehension about the political repercussions of a pro-ERA vote a few weeks before the primary, and opposition on the part of a small but growing Republican contingent in the General Assembly. There were other factors not unique to North Carolina that helped to explain the amendment's failure in the other key states in those critical weeks before the expiration of the June 30 deadline: the original softness of support for ERA, the backlash against the political and cultural liberalism of the 1960s, the emergence of a technologically sophisticated and aggressive Right energized by the Reagan landslide. As Green himself pointed out, there was the real difficulty of winning over the votes of legislators elected in a campaign in which most had taken a stand on the amendment. The lieutenant governor had gone to the heart of the matter with his observation that "ERA was killed Nov. 4, 1980 when these fifty senators were elected."[69] The positions opponents had taken, moreover, were often affected by profoundly personal concerns that resisted logic based on poll data. Even the most perfectly executed lobbying campaign could not succeed against the logic of the ballot and the nature of the issue itself.

Telling the narrative of these campaigns from the vantage of ratificationists illuminates the issue from one side only. We must now consider the way in which both partisan and opponent women understood the amendment and the institution that voted it down. Ratificationists—including the governor of North Carolina—did everything they "could" do; the power to mold the outcome of events does not always lie with those who initiate them. The experience of confronting in however limited a way the meaning of the twentieth-century gender revolution has touched most Americans. In contemplating the Equal Rights Amendment some of them found ways of responding to that revolution that are not to be evaluated by explaining what bright, aggressive, talented, middle-class women "could have done differently." Wanting to know only the answer to that question distorts their experience as well as that of those who opposed them. More interesting is what the ratification experience meant to them, to the opponent women who were alarmed by them, to the institution in which both *seemed* to place their destiny. What does it mean to address an issue that seems so simple and is yet so complex because of its ability to make us think—with the words "equality" and "rights" and "sex"—about what it means to be our selves.

5

"We Are Called and We Must Not Be Found Wanting": ERA and the Women's Movement

Commitment to the Equal Rights Amendment was mobilized in coalition. Feminists were only one constituency among the many who worked for ratification. Traditional women's groups such as the League of Women Voters and Church Women United joined newer organizations such as the National Organization for Women, the Women's Equity Action League, and the National Women's Political Caucus to achieve "equality." The goal was neither concrete nor specific but abstract and general, evoking the language of eighteenth-century revolution and twentieth-century idealism. "Equality" to the coalition—whatever the private or sectarian gloss—meant a repudiation of tradition, stereotype, prejudice, norm, that is, all fixed categories attached to the word "sex." "Equality" meant establishing as a principle in the highest law of the land that when women were to be evaluated, judged, or protected in the law, they would receive the same treatment as men. In symbolizing this principle, the amendment mobilized women as no other issue since suffrage; and it is in the forging of a collective experience that its significance lies.

Women who led the fight for ratification looked upon the world as something they themselves could make. For all their diversity, they differed in critical ways—and primarily in their social activism—from their female adversaries and fellow citizens. North Carolina proponents were, for example—like their counterparts in Massachusetts and Texas—for the most part white, middle-class, well-educated, married women. Although many in North Carolina had rural or small town backgrounds, they were nonetheless cosmopolitan in outlook, liberal in politics, and, if religious, affiliated with churches or synagogues either liberal or mainstream. More likely to be employed outside the home than antiratificationist women, they were also more likely to have held professional or managerial jobs than salaried anti-ERA women. Socialized in homes where parents tended to be college educated and politically active, they continued their own political education in professional associations, the American Civil Liberties Union, Common Cause, Planned Parenthood, or women's organizations. In North Carolina, many had also joined forces to fight racism, and the Vietnam War, free political prisoners, and oppose capital punishment. The ideology implied by this kind of activism distinguished proponents from women who lobbied against ERA.[1]

Enlisting in the Women's Movement

Given this incipient activism, it is not surprising that ratificationist women should have thought of themselves as serving the best interests of the Republic. They engaged in debate on a wide range of issues, followed the rules of acceptable behavior, and acknowledged values that they believed had made their nation exceptional—equality, justice, liberty, individual rights. Like Alexis de Tocqueville's ideal citizens, they had not limited their loyalties and horizon to a circle of friends and relatives involved only in a private world of immediate and personal issues. They had found the familiar abstractions of American civic life important not merely as symbols (although symbols were important, too) but as promises. Especially important were "equality" and "individual rights." There was a goes-without-saying commitment to these ideals, although the commitment was not developed into a specific social theory or a conscious plan of action. Equality and rights were among the good things of American life, or at least they should and could be. The "normality" of such ideals was as questionable as that of women whose civic world was a stage for reaching the goals of their moral idealism.

At the same time, proponent women challenged normality. Their ideas about individual rights and equality were developing in such a way as to challenge normality where gender was concerned. Women came upon the Equal Rights Amendment through the creative chaos of social change in the "women's movement"—a term that was less specific than evocative. Within the broad sweep of the movement, ratificationists became committed to freeing women of traditional gender-based constraints through individual choices and collective action. The relation of anticipated change to the past was clearer than the nature of their goals. Most were not prepared to surrender mothering or heterosexual intercourse, for example, but they did believe that women should not be bound by traditional gender definitions and they were in the process of discovering the subversive implications of that belief. They were thus collectively—and sometimes individually—ambivalent about the meaning of the Equal Rights Amendment. Most did not believe that the amendment itself could effect a cultural revolution. Some were skeptical because they understood how deeply entrenched were the foundations of what they called "patriarchy." Others believed that equality was merely an extension of rights conceded at enfranchisement; these had been denied by vestigial remnants of the common law and stereotyping in law. Purging the law of "sexism" was therefore "normal" because it would be done according to American norms of justice.

This point they labored to prove by producing information on the unrevolutionary but nonetheless praiseworthy effect of equal rights legislation in other jurisdictions. In such exercises they were separating themselves from the cultural radicalism identified with women's liberation and embracing instead values sanctioned by time and American mythology. Thus it seemed accurate to insist that the amendment would affect only state universities (not private ones), governmental actions (not familial ones), legal disabilities (not cultural patterns), public policy (not personal relationships), and the structure of law (not the structure of the family).[2] In making such distinctions ratificationists were not pushing an unexceptionable principle (equality) in order to provide the basis for a cultural revolution (androgyny). The distinctions between public and private, legal and personal were meant not only to reassure the wary but also to underscore the significance of ratification for women regardless of political orientation. Alerted to the need to improve the condition of women, ratificationists were driven by the rules of political combat and male hegemony to make distinctions that seemed natural, pragmatic, and acceptable. In doing so, however, they frequently distanced themselves from feminism and misrepresented even to themselves the significance of what they were doing.

The distinctions cited above between state and private, legal and cultural, public and personal held as long as proponents addressed legislators, hostile men, and wary women. In their own minds, however, as ratificationists thought about the expectations of the women's movement, distinctions between private and public could not be sustained because the movement had so consciously fused them. The language of movement was the language of challenge and change; and it promised an open future for women willing to repudiate stereotypes embedded in law and custom. Thus the North Carolina Women's Political Caucus newsletter called women of the Tar Heel State to arms in January 1972: "IF YOU WANT EQUAL ACCESS TO EDUCATION, TRAINING, JOBS, PAY, CREDIT, INSURANCE, PROPERTY OWNERSHIP; IF YOU WANT TO LOOSEN THE BONDS WHICH MAKE WOMEN SECOND CLASS CITIZENS; IF YOU WANT TO PROVIDE OUR COUNTRY WITH A VAST STORE OF RICH HUMAN RESOURCES, SORELY NEEDED BUT UNTAPPED, RESPOND! WE ARE CALLED AND WE MUST NOT BE FOUND WANTING." The words were addressed to women who already considered themselves part of "the movement" and suggest how the amendment was to be understood. These were words of change meant to evoke action: "if you want equal access"; "if you want to loosen the bonds"; "if you want to provide . . . , RESPOND!" The implication was that ERA would *help* women break down barriers, improve their status, and contribute to their own welfare. The obligations of women to each other demanded that they support ERA. "We are called and we must not be found wanting."[3]

The "call" to action—understood in conjunction with the incipient expectancy of the *Yale Law Journal* article—contrasted dramatically with the language of reassurance. The ambivalence had plagued suffragists as well; indeed, movements directed at "objective" changes—not mere "salvation"—but which cannot wrest concessions through selective pressure upon vulnerable areas of the economic or political systems must appeal to the public in such a way as to create a majority in legislative bodies. To do so, public discourse must appeal to ideals or interests of which an incipient majority approves. Thus suffragists had envisioned changes perceived to be radical and felt by them to be refreshingly innovative, but they elicited support by appealing to evocative ideals that were presumably normative for all Americans. That legislators thought more of the votes to be won than the merit to be accrued from enfranchisement does not detract from suffragists' rhetorical strategy. In one sense they were appealing for change in the name of tradition.

So were partisans for the Equal Rights Amendment. Looking back to 1923 when ERA had first been introduced, they thought it so natural to have moved from suffrage to equality that they wondered why it had taken so long and why it had been so difficult. Bewildered by extravagant accusations hurled against them as destroyers of the family, they asked in wounded amazement if suffragists had encountered the same arguments. It was not reassuring to be told that they had. This brief lesson did not entice ERA activists to think about their history as woman so much as about the inevitability of victory. Suffragists, after all, had won. For ratificationists ERA was part of the inexorable realization of equality in American history. They chose the former suffragist, Gladys Tillett, to preside over the campaign of 1975 because she personified that process. Her earlier activism, age, achievements, and position could make ratification as respectable as enfranchisement. Moreover, recent gains by blacks in the civil rights movement were thought to have enhanced the egalitarian ideal in such a way as to provide further legitimacy to the fight for women's rights. Taking charge of their own destiny characterized the ratificationists' historical consciousness. That consciousness was not knowledge of how women had lived in the preindustrial past or of how women's lives had been changed by industrialization, urbanization, education, fertility control, and improved health care. Few ratificationists understood the development of women's networks in the nineteenth century and their own role in the movement of women from private to public spheres. Nor were they familiar with past heroines; most had probably never heard of Alice Paul, Carrie Chapman Catt, or Cornelia Jerman.

Historical consciousness is not simply past narrative; it is also an attitude toward the scope and meaning of activities beyond the personal realm of relationships—friends, kin, and neighbors. This framework of action could, of course, be called "public" to distinguish it from the "private." Such a distinction would indicate a level of action engaging people sharing a common culture, language, and government in matters of interest to all or at least to a majority. The idea of "public," however, does not capture the *dynamic* of historical change—the sense in which a self-conscious group interprets public actions as part of its destiny stretching back into the (imperfectly) remembered past and forward into the (positively) anticipated future. The assumption that history justified women's collective action in bringing about a better future came to characterize ratificationists' self-understanding. Their past and future allowed them to transcend the present as a mere stage in the inexorable realization of equality. The struggle of women, Afro-Americans, Chicanos, and native Americans for their rights seemed to make those actors more "real" or at least more "American" than those who had made the struggle necessary.

Unlike their opponents who were angered by the changes of the 1960s, proponents were pleased to have benefited from them. Unlike those who invested personal identity in the ahistorical categories of gender and family, proponents understood themselves as part of a process through which women worked to guarantee that sex and family would no longer necessarily define them. Opponents seemed to think of history as tradition to be received; ratificationists believed that an essential part of that tradition was making history. Although not sentimental romantics gushing at meaningless clichés, even the most tough-minded among them—and that would have to have been Martha McKay—believed that the Equal Rights Amendment was part of the historical and moral logic of American history.

This conviction was evident in letters of North Carolina women to Senator Sam Ervin. Well-informed and often angry with Ervin's determination to keep them in what one woman called "protective custody," these women demanded change. Many of them referred to personal experiences of discrimination: lost income, lost promotions, lost benefits, lost opportunities, humiliating treatment, and condescending bosses. We women, wrote a librarian, "have never known anything except discrimination." They had been denied jobs, credit, mortgages, and other forms of economic independence. They had been assigned the "legal status of a child," and they were more than a little annoyed. They resented being dismissed as "women's libbers" and being sent condescending " 'calm-the-irate-women' form letters"; but they were indeed angry at the protective tyranny imposed by stereotype. For more than a few women, Ervin was not the beloved "Senator Sam," but a mischievous and irritating old man. His attempts to enscribe his views into the Constitution with amendments to ERA were particularly offensive. "We do not need any special protection, we do not need social concessions or social supervision, we do not need any law that treats us differently from any other citizens." Such laws had resulted in the "political, economic, social, psychological and cultural subjugation of our sex"; and they were insulting. "We are not," wrote one wife, "the weak minded, weak-bodied, weak-spirited individuals you take us for." And it was as individuals they wished to be treated. This was as true for the student who hoped ERA would prevent principals from suspending girls for wearing slacks as it was for an older woman who insisted that "even if *most* women were delicate creatures" government had no business "restricting the opportunities of all women."[4]

The student and the older woman had something in common that neither age nor class nor race could hide. They both knew—as did the woman who wrote of the "psychological and cultural subjugation of our sex"—that the ways in which women and girls defined themselves had been dictated by someone other than themselves. And they, along with countless others, wanted to be free to determine for themselves what sex—that is, gender—meant. They naturally and easily fell into the sanctioned language of American opportunity, appealing in one form letter for "equal pay for equal work and equal consideration of our capabilities." The phrase, emphasizing as it did equal work, pay, and opportunity, had the respectability of a cliché, but it actually belonged with objections decidedly not respectable. Women offended by exaggerations attributed to feminism and wary of being labeled a "far-out woman liberation type" would still brave this stigma in order to serve notice that they were fed up. No one ever said that ERA would end the laughter of hostile men, but personal hurt festering under such laughter made women feel acutely the need for change. No one ever said that "subjugation" would end with ERA, but ratification was a place to begin. Merely trying to achieve "equal consideration of our capabilities" had aroused too much opposition from men like Ervin to make ratificationists that naïve. His attempts to amend ERA were thought to indicate that he did not understand what the past twenty years had been all about. Referring to Ervin's most hackneyed arguments, a fifty-one-year-old mother of two sons wrote: "I could care less about the dower right, the alimony business, the child custody bit. . . . I have already done my military service and am an honorably discharged (1946) veteran of the Army of the United States . . . I have even been a Harvard student. . . . Stop putting down North Carolina women!"[5]

It was clear from these letters that their authors felt themselves at a historic moment when women could take charge of the future. This anticipation created problems that were never fully appreciated except perhaps in hindsight. For example, during the frantic lobbying efforts of both the 1977 and 1979 campaigns, North Carolinians United for ERA headquarters in Raleigh failed to keep women in the outlying sections of the state informed as to what was going on in the General Assembly. The lack of communication hurt and angered women who had worked so hard in previous campaigns. To some observers the response seemed to be politically naïve; "Raleigh" was too busy to keep activists informed even if confidentiality had not been a major concern. What was at issue for those left out, however, was their continuing participation in action that they had helped to initiate. And the leadership understood this, for one of the reasons it had proceeded with the chancy lobbying effort in 1979 was to sustain the historical continuity needed for future action. To take a kindred example with regard to the maneuvering of ratification bills through the General Assembly: activist women thought of such bills as literally "theirs." They sometimes did so with a degree of possessiveness that made them as suspicious of pro-ERA legislators as of their avowed opponents.

That ERA should have been understood as a way to enlist in change is not surprising when we consider the way in which it was presented. Ratification, activists were encouraged to say in their speeches, would create a greater range of opportunities for women. It would halt federal and state discrimination on the basis of sex. It would strengthen extant laws against discrimination and guarantee fairer practices in obtaining credit. It would mean the abolition of laws restricting women's control of their own property. It would mean equal access to benefits for women in the armed services and equal access to jobs and education. To explain what equal access could mean, the League of Women Voters told its speakers to consider this example: ERA would mean that stereotypes of what men and women should do with their lives would be dissolved. One can imagine a stir in the audience at this point. This was to have been a practical example. Men make more money than women—it was a fact of life and one that was unfair. One way to deal with this fact would be to train girls and women for jobs traditionally reserved for men. (One can imagine another stir, especially from the women who were offended by "girls" working as "flagmen" on road repair gangs.) With ratification, sex stereotyping in education would be suspect. The result would be for boys to be allowed to take cooking—chefs were after all men—and for girls to be allowed to take mechanical and technical courses.

Even if the league's examples were not often used, they nonetheless symbolized recent changes affecting women. Merely moving to the lecturn identified speakers with those changes. Even if they did not actually refer to sex stereotyping, speakers could not have hidden their pride in expanding opportunities or the fact that they themselves personified those very opportunities. The professors, lawyers, and other independent women like Nancy Drum who spoke for ratification allowed a broad spectrum of inferences from their activism over which they had no control. Although arguments for ratification always emphasized that ERA had nothing to do with personal life and would require no change in family roles, they were made by women who—wives and mothers or not—had clearly identified with a public issue raised by women who did not define themselves simply as wives and mothers. To the extent that any speaker in the flush of anticipation could "reassure" women that ERA was a natural part of changes that were eroding the sex stereotyping of the past, she had increased the anxiety of women who interpreted what she called stereotyping as the normal action of women. Even if boys did become chefs and girls could become mechanics, the assumptions that made it possible for women to break the stereotype could seem to be too close to the "lib" movement for comfort.

From Better Homes and Gardens to Activism

The world of flux and change from which ratificationists came was not merely that of self-discovery and empowerment, but of knowledge. And it was knowledge toned by empathy with millions of women put at risk by tradition, restricted by laws on credit and property, and demeaned by male attitudes and values. It was knowlege about how these attitudes, frequently expressed in the honeyed modulation of presumably complimentary phrases, provided mannered reinforcement of women's secondary status. It was also knowledge about women as an aggregate and as a class—information about women in corrections, business, government, the professions, the wage labor force, and nontraditional jobs. To ratificationists it was significant and shaming that female heads of household earned 48 percent of what men earned and that college-educated women earned less than men with a high school education. Such statistics demanded action in the view of women who had, as one recruit remembered in admiration, "very full lives . . . within their communities, within their families, within their organizations. And have got all their antennae out, and are touching base on a lot of things that are important to them."

These women had, for example, tried to help homemakers suddenly bereft of husbands by desertion, divorce, or death. Such women all too often lacked the experience, skills, and credentials needed to commend themselves to employers. They needed help, and women of the women's movement saw that they got it. For women caught up in the nuts and bolts of helping displaced homemakers, ratification was a natural commitment. For those committed to helping women change their lives, ERA was both symbolic and practical. Women who would be accused of using ERA to decriminalize rape had already organized rape crisis centers and distributed rape evidence kits to hospitals in the state; they had already helped police departments become more sensitive to rapists' victims, counseling them while opponents of ERA remained immobilized in disbelieving silence. Ratificationists helped to rewrite laws on sex offenses in order to make conviction of rape more certain. Consistent with these actions they publicized widespread evidence of domestic violence and lobbied for laws to protect women and children. For such people, the amendment was one of many changes required to make the world as safe for women as it was for men.

Change was an essential part of ratification activists' lives. To be sure change meant no self-consciously radical break with the past; few could accept the revolutionary implications of Marxist or radical feminism or could embrace what Phyllis Schlafly in derision and contempt called the "unisex" society. But ratificationist women were—even if members of Wives and Mothers for ERA—participating in a process that was separating women from house and home. Proponents said that ERA meant freedom of choice for women, and they believed it. The said that women could remain within familiar domestic precincts, and they believed that, too. But they also believed in expanding opportunities for women, and those opportunities were outside the home. They promised an end to legal discrimination to women who had never known it (even if they had suffered it), and they promised equal access to credit for women who either did not want it (yet) or who did not understand what it meant. The chasm between proponents and opponents on these matters reflected different orientations, the former to public issues, the latter to personal experiences identified with the home. In promising equal access to benefits in education, jobs, or the armed forces, ratificationists once again seemed to be saying that ERA was for women outside the home. Try as they did to explain the importance of ratification for women whose world was caught up in the family, proponents were always inhibited by a mental framework that identified them with the movement of women from personal into public consciousness. Activists who could honestly say that they were housewives and mothers (without adding perhaps that they were management consultants) were, through working for ratification, also historical actors.

Activists faced a dual challenge based on the assumptions of interest group politics. First, they had to learn to lobby. Their model was the way in which women had won congressional support. Women's organizations had been mobilized to flood congressional offices with mail, produce experts to testify, and create in the view of their elected representatives a group with interests to be met in much the same way as any other lobbying group. Organizations were accepted as authentic mediators of what women wanted; their telegrams, letters, postcards, and telephone calls were accepted as quantitative evidence that such organizations spoke for a constituency of women; and popular reference to the "women's movement" gave further legitimacy to the belief cultivated among the members of Congress by activist women that they spoke for *all* women.

Beyond formulating a lobbying strategy, ratificationists had to create grass-roots support to legitimate it. By the rules of interest group politics, constituent organizations are supposed to have helped develop a collective identity to prepare members to work for their interests. Mobilization of support for a position on impending legislation, therefore, would be an educational process directed to a constituency already alerted to its interests and prepared to act on its behalf. The unity of an interest group is based upon common acknowledgment of the need for collective action to achieve agreed-upon goals and an understanding that one may have to suspend cherished personal beliefs and suppress minority agendas for the good of the whole. The claim of an interest group to speak for all or most persons identified as actual, potential, or inferential members of the group must be sustained if it is to convince lawmakers of the wisdom of its proposals. Mobilization of North Carolina women was complicated by the fact that groups contributing to the pro-ERA coalition had not prepared their constituencies to think of themselves as well served by the Equal Rights Amendment. Activists had to supply that conviction.

Ratificationists began by emphasizing that "the Constitution of the United States is not applied to women, who remain special categories of persons under the Fourteenth Amendment. I don't care about symbols; I want equality under the law." In thus explaining what it meant to "write women into the Constitution," Martha McKay condensed in two short sentences hundreds of pamphlets, handouts, articles, speeches, and lectures on legal equality. Defining ERA as a legal issue had the inherent virtue of placing authoritative explanations of ERA in the hands of experts. Ready to provide information on and interpretation of matters too complicated, complex, technical, or otherwise inaccessible to any but specially trained people, experts have become so important to modern society that their testimony—if not challenged by other experts—can raise certain questions above controversy. Since the amendment was about the law, the person in charge of the first ratification campaign had been an expert, a lawyer, "someone," recalled Elisabeth Petersen herself, "who would explain this kind of thing to legislators who wanted to understand it. We were terribly innocent about that." Proponents had hoped that by restricting discussion of ERA to legal issues, where they thought it legitimately belonged, they could avoid entangling it in other, more emotional issues.

Just the Facts, Ma'am

But avoidance of emotional conflict was only one aspect of emphasis upon the law. More significant was the way in which this emphasis in the minds of proponents seemed to give their cause objective validation. Thus, whereas opponents justified themselves by appeals to the self-evident and "unamendable" categories of nature and womanhood, proponents tried to do the same thing by appeal to the presumed objectivity of the law and the compilation of data: facts. This combination of style and substance, so essential to modern society, took on a special meaning to proponents because they believed that they were being done in by lies, misrepresentations, and ignorance. Themselves neither radicals nor dupes, they resented the implication that they were one or the other. They resented, as well, the terrible reduction of ratification to a feminist conspiracy against womanhood. Angered by alarmist rhetoric that shrieked "RAPE" at women hypersensitive to the power of men, they appealed for a calm deliberation of what was at stake. Angry accusations of betrayal from anti-ERA women made appeal to "the facts" a way of distinguishing proponents from the opposition in their own minds. In facts lay an objectivity that would lead to ratification. Accordingly, lecturers were warned against getting involved in arguments "to which there is no objective answer."

Hard fact versus emotional misinformation, therefore, was the dialect of debate as far as ratificationists were concerned. They issued "fact sheets," "cool facts for the hot-headed opposition," and sober explanations of the constitutional background and legal implications of ERA.[6] One of the major and perennial responsibilities of the Raleigh office was to provide facts and information to legislators. Fact sheets of six pages suggested relevant cases, instances, examples, arguments, and corrections needed to vindicate positions, solidify commitment, and correct false impressions. It was a fact, for example, that the Supreme Court could under the Fourteenth Amendment still hold sex as a "valid basis of classification" in law—contrary to what the opposition said. It was a fact that ERA would "shift the burden of proof to the state" to show why discrimination based on sex was "necessary." Following would be a list of examples suggesting the impact of ratification in technical aspects of the law, followed in turn by a list of facts designed to correct or place in proper perspective (a significant difference) the "nonfactual" objections of opponents. Information, too, was frequently needed to reassure skeptics at an elementary level. Ratification, explained Senator Katherine Sebo, would mean that a woman prostitute could be guaranteed a female probation officer just as a man would be guaranteed a male, and a woman patient a female nurse.

Such technical "facts" represented one generic range of response. They were statements supporting the contention that ERA would not replace unreasonable or arbitrary standards based on sex stereotyping with other unreasonable or arbitrary standards. Thus, "facts" meant that the ERA would not result in sexually integrated public toilets. To proponents the "potty ploy" symbolized the inability of opponents to face facts, and it seemed appropriate if annoying and frustrating to have to answer such silly and groundless fears with "facts." But what was relatively easy in one instance was relatively difficult in others. Facts on the sex of professionals and clients were not so undeniably real as Senator Sebo believed them to be. The issue was whether ERA would impose silly, unreasonable, or arbitrary standards. Sebo thought not, based on the intent of Congress, authoritative interpretations, and legal precedents. Accepting her reassurances required *not* belief in the reliability and authenticity of her sources but in her own "objectivity." And this would have been impossible to acknowledge for someone who believed that, by championing ERA, Sebo had surrendered any claim to authoritativeness.

The same could be said for attempts to cast the cool light of facticity upon Sam Ervin's arguments. Ervin had charged ERA with depriving divorced women of support. Proponents flatly denied this charge and argued that support systems would be established on the merits of each individual case and not on the basis of sex stereotyping. ERA might require spouses to "contribute equally to the support of the children within their means." For insecure traditionalist women wary of the pervasive power of men, this requirement was not necessarily the reassuring observation that it was meant to be. Looking back at the argument and contrasting it with the change in divorce and support laws since that time—changes in line with the stated goal of ERA—we may find the "facts" less clear than they were then. Women in these circumstances have not necessarily benefited from "equality." Facts also seemed to belie Ervin's accusation that women would lose something by the abrogation of protective legislation. The fact was—or seemed to be—that it had been used to discriminate against women. The Equal Employment Opportunity Commission, the courts, and the state legislatures had seen that fact; Ervin's inability to do so seemed perverse.[7]

Despite the "facts," Ervin's accusations seemed to be etched in stone. The proponents' indefatigible research, careful presentation, and encyclopedic knowledge appeared to be of little use, their command of the facts never acknowledged for the impressive feat that it was. Proponents made the discovery of relevant information a household industry. Led by and appealing to relatively highly educated women, ratificationists became both intensively and extensively well informed. They learned the relevant Supreme Court cases and why the Fourteenth Amendment could not be used to achieve their

The resistance of opponents such as Ervin to "facts" about what ERA would or would not do was a frustrating reality for ratificationists.

goals. They learned the legal status of women in North Carolina and some of the laws that still treated married women as legal incompetents in the use of their own property. Responding to Ervin's apocalyptic warning with regard to divorce and marriage, they found that North Carolina men had been allowed to receive alimony since 1967. In its General Statutes 50-16.1–9 North Carolina was already in compliance with ERA, a fact that proponents believed should have reassured tremulous opponents. This and other facts should have made a difference, activists believed, in evaluating unfair and misleading attacks on ERA. They should have made the difference in deciding whether to trust other statements of the opposition. Such facts were not earth shattering, to be sure, but together with others they helped to frame a mental context within which ratificationists could sustain their argument. At least that was the theory.

One of the most frustrating and infuriating facts with which ERA partisans contended was the perceived discrepancy between demonstrable popular support for sexual equality and the failure to ratify. The "fact" of support was deduced from public opinion data reporting that a majority of persons polled favored passage of ERA. The *Washington Post*/ABC News, Gallup, Louis Harris, and National Broadcasting Company polls suggested that a majority of Americans were with the proponents. Moreover, when the language of the amendment was read to those interviewed without reference to ERA, an even greater number of people approved, sometimes by a ratio of 2:1. Majority support for ERA held at about the same level from 1975 to 1982, with one exception. A Louis Harris poll taken in the spring of 1982 reported a surge of approval in the previous three months. It was this upswing that North Carolina ratificationists used to justify their eleventh-hour efforts. Without referring to anti-ERA arguments in formulating questions, pollsters received a response decidedly favorable to ratification. Of course, the fact that thirty-five states had ratified the amendment supported ratificationists' contention that they spoke for a majority. The facts were with them.

The "fact" of popular support, however, was ambiguous. This ambiguity derived not from antiratificationists' polls, which reported substantial opposition to ERA, but from the nature of support for ERA, which was "soft," as a recent study indicates. Two-thirds of the people who had an opinion on ERA said that their position was not "strongly held." Furthermore, as the political scientist Jane J. Mansbridge points out, after 1970 attitudes on gender were in flux and often contradictory: polls reported 84 percent approval of married women working and 76 percent disapproval of mothers' working outside the home unless "financially necessary." A majority of men favored an "equal marriage" but opposed changes that would require them to do more housework. In addition, Mansbridge reminds us that Americans are well known to favor "rights" in the abstract but to balk at their application in specific instances.[8] Finally, popular support in opinion polls could not easily be transformed into political capital. Only as a last resort—in 1982—would proponents try to gamble on an advisory referendum. They had previously feared that the apocalypticism of the opposition would appeal successfully to the very stereotypes that the amendment was designed to abrogate.

The dangers posed by the referendum suggest that facts are often not as objective as people like to believe. The opposition did indeed make declaratory statements that were untrue; but as statements of *meaning* they were undoubtedly true. That is, it was simply not true that rape would have been decriminalized by ERA even though opponents might have believed that abolishing classification by sex was in effect symbolic rape. Proponents were quite right in pointing out that antiratificationists ignored the legislative history of the amendment, the intent of its authors, the importance of sex-neutral legal language, the discriminatory aspect of protective legislation, the predicament of women in the military—a whole range of "facts" which were not acknowledged as just that. Proponents discovered that there was no way to make opponent women and legislators play by the rules assumed and practiced by ratificationists. There was no way to force them to acknowledge those facts that had led ratificationists to ERA. Even when opposing groups could agree upon facts—for example, that ERA would mandate the inclusion of women in selective service—they could never agree upon what the facts meant. In the beginning it had all seemed so simple. Armed with the facts, "all we had to do," remembered one activist with self-deprecating humor, "was to provide simple, rational, and logical explanations."

Fact and logic were sometimes confused, especially in discussion about the implications of ratification. Although not facts in the same category as statistics, consensually acknowledged events, or natural catastrophes, legal implications were thought to have a concrete and objective status, especially when contrasted with the perceived hysteria of the opposition. Proponents shared a basic assumption that equality of rights was a positive principle. This interpretive orientation meant to them that attack on irrational and arbitrary classification by sex would not in turn become irrational or arbitrary. The "facts" of ratification could be inferred or deduced from the "facts" inherent in the legislative history, clear intent, and authoritative interpretation. In their impressive study and presentation of facts, women were relying on their own competence and abilities. Because of their faith in expertise and relevant data, ratificationists approached the General Assembly forthrightly, seeming to say, even if they did not: "These are the facts, and this is what you should do." Antiratificationists never presumed to tell assemblymen what to do or say or how to argue.

Proponents could not be so reticent. They projected themselves into the legislative fights with the firm belief, in the early campaigns at least, that what every legislator needed was a full complement of facts. The response of politicians was often unenthusiastic. Some women understood the coolness as an expression of sexism. They had become authoritative informants through careful study of the facts; they knew all that was needed to deflect criticism and make a good case. Like professors lecturing to uncomprehending undergraduates, they were frustrated by a studied lack of enthusiasm. One woman recalled the obvious distaste of a male legislator for her information on Supreme Court cases. Men just could not take authoritative information from a woman, she recalled. And she may have been correct— but it never occurred to her that the man might have had all the legal facts he wanted. If she had been able to tell him how to embarrass, shame, blackmail, trade, or force a no vote to become a yes, he might have been more attentive, her sex notwithstanding. She may have had the wrong set of facts.

Women Warriors: Facts and Principle

In one important matter appeal to facts was not an appeal to information over ignorance or truth over falsehood but to dispassionate "perspective" over alarmist "distortion." The issue was women in the military. For proponents the issue symbolized the anomaly that opponents attributed to them, the alarmist distortion of a sensitive issue, and the extent to which they themselves would go to insist on sexual equality. They thought the issue a scare tactic—"Watch out," screamed an antiratificationist pamphlet, "They're going to draft your daughter!" Consequently, ratificationists frequently found it difficult to take the issue seriously, dismissing the women who introduced it as unworthy of their sex in fleeing civic responsibility. Such flightiness represented the ultrafeminine softness that men used to preen their egos and deny women their rights. Thus it was not surprising that a woman college student, when faced with an objection to ERA because "I don't want to be drafted!" should have hissed, "You bitch!" Her independence, self-confidence, and commitment to a career in medicine simply could not abide this simpering frailty. If women had to serve as part of the adjustment demanded by equality, they should be willing to do it. The woman's vivid expressiveness seemed almost out of place when responding to a reasonable enough plaint—who *wants* to be drafted? But her outburst represented more than the fleeting anger of a moment; it also expressed disgust for the traditional dependence of women and the advantages it had reserved for men.

In addressing the matter of women in the military, activists rarely if ever allowed themselves the luxury of such epithets. They were too preoccupied with getting the facts straight.[9] ERA would not require *all* women to serve; the fact was that most men had never had to do so. Mothers would not be torn from their families; the fact was that there would be exemptions for parents expressed in gender-neutral language. Facts about women already in the military also came to mind as soon as opponents mentioned the anomaly of the draft. Facts of women's ability to compete effectively with men in an ever-broadening range of activities were all evoked by the issue. To be consistent, proponents of ERA could not very well concede special exemptions for women from the military because of the pervasive discrimination against women already serving. Moreover, the achievements of female personnel were an essential fact of the women's movement. Claiming military exemption would have meant repudiating those achievements and the women who had made them; it would have meant hiding behind the very stereotype against which ratificationists were so doggedly fighting.

Beyond compulsory military service, women in combat was an even more troubling issue. In public hearings, NCUERA stood with its eloquent champion, William Van Alstyne of Duke University, who insisted that under ERA the Supreme Court would probably not send women into combat.[10] Some proponents conceded that that would seem reasonable as a judgment of what the Supreme Court was likely to do—although they might have been persuaded by lawyers of the National Organization for Women to challenge this interpretation in different circumstances. But for many partisans the difference between Van Alstyne and NOW seemed irrelevant. They followed the lead of Martha McKay, who in 1973 dismissed the issue as a way of diverting attention from real needs. Her view and that expressed by Trish Hunt during debate in 1975 seemed to speak more directly to the position of pro-ERA women. They insisted that women could fight as well as men if computers and machines did the damage, and they could—as a few did—point out that Joan of Arc and Deborah had not been exactly incompetent; but in the final analysis such "facts" were really beside the point. The indelible fact that Hunt did not want a majority of her male colleagues to ignore was that women were just as patriotic and valiant as men and their rights should not be determined by whether men were more willing to sacrifice sons than daughters in the "incredible stupidity" of war.[11]

In their commitment to facts, however, ratificationists were prepared to accept the possibility of combat and death. Here facts were a function of principle. On a commonsense level there is no compelling reason to say that equality of rights—even for men—dictated committing women to battle. Certainly the differences between Van Alstyne and the Supreme Court on one hand and the Yale authors and NOW's lawyers on the other demonstrate that these are matters about which reasonable persons disagree. Yet proponents took the harder course, as if to "outtough" the men who seemed to use the threat of death to intimidate them. Their essential argument was based on a historical fact that was assumed therefore to be a norm: women were already an important part of the armed forces. And this fact was supposed to deprive the symbol, women warriors, of its dangerous connotations. Thus sterilized, the "fact" was to establish a moral claim upon women to unite in forcing government to use sex-neutral rules and regulations for service personnel. "Facts" were thought capable of softening the harshness of symbolic anomaly so skillfully freighted with danger by the opposition. Facts were believed capable of shielding women from hysteria and engaging them in discourse that could confirm ERA as an opportunity for women. The task was formidable.

Suffragists, too, had been accused of meaning to put women in the army, and they had deflected the issue by arguing that military service was not a prerequisite for full citizenship. Moreover, they defended themselves in part by appealing to social role defined by gender. Supporters of ERA could not do this. They were more sensitive than most suffragists to the ways in which stereotypes linked with gender had deprived women of equality, and they were not inclined to make concessions to canons that men could use to restrict them. It was a matter of principle, and principle dictated the way in which proponents interpreted their critics. To proponents, the issue of women warriors represented archaic male dominance more than actual combat, suffering, and death. Since *all* men were not subject to the draft or combat any more than women, the issue seemed to be framed in hypocrisy, self-interest, and bombast. The pious and loudly proclaimed determination to keep "girls" out of foxholes seemed to be mawkish nonsense unless one understood it as a way of condemning feminists as unsexed and aggressive. Some women actually suggested that it was a sign of masculine insecurity. Such male strutting would have been humorous to the women—and some of course did laugh—if it had not had such serious implications for women's rights. Opponent men's refusal to deal with the fact that women were already in the armed forces represented the traditional male disregard for women out of place. Given their understanding of the meaning of the draft and combat to the opposition, proponents were not prepared to concede anything that diminished their commitment to principle.

That commitment, it has been argued, may have damaged chances of ratification. At least this is the thesis of a recent study of "why we lost the ERA." In it, Mansbridge points out that Common Cause, the National Federation of Business and Professional Women, Senator Birch Bayh, and indeed most senatorial proponents of ERA took the position articulated by Van Alstyne that the amendment would not have mandated the use of women in combat. A small group of feminist and academic lawyers, however, emphasized the need for consistency and successfully committed pro-ERA women to the "egalitarian" interpretation. Mansbridge believes that greater clarity and the grounding of proponent arguments in a "military exemption" might have saved ERA, avoiding the radical nimbus of the course actually taken.[12] But the general fate of proponents' facts in confronting the opposition suggests that this view attributes more power to precise definition and careful reasoning than knowledge of the ERA conflict justifies. Since North Carolina legislators ignored Van Alstyne's argument, one would suspect it would not have made any significant difference elsewhere. Certainly none of the anti-ERA legislators who heard Van Alstyne were converted. Moreover it is possible that the line of reasoning would have been as damaging as the egalitarian argument for, when confronting it himself, Sam Ervin had replied that it had thus conceded "sex" to be a reasonable classification. Absent the egalitarian interpretation as a feminist position and the opposition would have charged a naïve inability to face the realities of "equality" that, they would have insisted, was symbolized best by women in combat. It is true that the dissonance between traditional womanhood and woman warriors made the draft-combat issue the most damaging of those raised by the opposition. But this result does not mean that managed differently the point in dispute could have been neutralized. The anomaly of women warriors was too powerful for the opponents to surrender no matter what the facts. As for proponents, the gender revolution made it virtually impossible for them to have hidden from the logic of their position even if they had wanted to do so.

The Celebration of Equality

In the principle symbolized by women warriors, it is easy to see that fact was a function of ideology. And if women in the military symbolized the principle of equality, concrete examples of actual "inequality" could symbolize the "fact" that ERA was, as handouts insisted, a "bread-and-butter issue." It was not difficult for ratificationists to list laws, practices, and situations in which women were distinctly "unequal" to men. These facts were meant to enlist women in a cause as practical as woman suffrage had been. An encyclopedic range of facts established the lack of equitable and fair treatment for women. The list was an indictment: sex stereotyping beyond the law, disproportionate numbers of men in supervisory positions over women, the inflated importance of male-held over female-held positions, the relative absence of women in policy-making positions, the low status of homemakers. The implication was that ERA would address these and similar issues.

But devotion to the principle of equality made facts carry greater burdens than their patterned design permitted. ERA would not have addressed the above indictment. Nor could it have affected other "facts." For example, ERA would not place women on the Supreme Court or in elective office, put them on corporate boards or change the sexual (im)balance of power, or call a halt to sex stereotyping by high school guidance counselors. And yet these flaws were all used to justify ratification. When ERA failed ratification, women did not lose access to state universities and women students did not lose support for their athletic programs; women did not lose access to professions or the rule of equal pay for equal work. The complementary fit of a detailed sexism on the one hand and the implied solution from passage of ERA on the other was simply not very snug.

This situation is not surprising. Proponents saw such details as fragments of a larger whole, flaws leading from the major fault that demanded something more than the glue of statutes and incidental litigation. They contended for the principle of sexual equality. Technically the principle was not precise—principles rarely are. Sometimes it meant making sex as well as race a prohibited classification in law. Sometimes it meant making sex a suspect classification, but whether suspect or prohibited, classification by sex was to be placed under critical scrutiny. Proponents were willing to enter an ambiguous future under the umbrella of ERA because they believed the principle itself was so appropriate. Even in 1982, six years after the Supreme Court in *Craig* v. *Boren* (1976) seemed to extend the equal protection clauses of the Constitution to void sex discrimination, the proponents' legal champions were still insisting that the principle needed a more secure place in the Constitution through amendment.

"Equality" to proponents was an essential part of their nation's symbolic capital. In the concatenation of themes that surged into the harmony of the movement, the celebration of equality was the dominant motif. Ratificationists believed it to be the one basic value validated by struggle in the 1960s. They would all have applauded Beth McAllister, who presided over the coalition from 1979 to 1981, when she spoke of equality as something that could only "enhance" her life. She had seen it as a driving force when, as a high school student in Charlotte, she had been so impressed by the courage of a black girl who braved a threatening mob for equality. It had led her as a college student to the sit-ins in Greensboro and as a young mother to head Wives and Mothers for ERA. She was not mystical about it, but *Equality was,* she believed—as the coalition slogan had it—*for everyone!* Indeed, in explaining the impact of ERA, proponents reassured their audiences that ratification would extend benefits conferred by law—and privileges and opportunities, too—to both sexes. This extensive function had a broader implication, however. The amendment would place the issue of sexual equality in a position where the U.S. Congress, the federal government, the federal and state judiciaries, and state legislatures would be forced to come to terms with it. The careful, understated style of the 1973 and 1975 campaigns in North Carolina did not convey what proponent women really felt. They labored not for an issue of restricted range but for a vision.

The vision could not be contained. Although proponents had tried to reassure the wary that ERA would not affect personal relations, they nonetheless understood the amendment to be part of a process through which those relations would indeed be transformed. Angered by the apocalyptic foreboding of opponents, they insisted that the amendment would not rescind good manners, affect "private action," or reshape "social" relationships. And yet they sometimes seemed to be saying the exact opposite. In public testimony in 1979, for example, Jessie Rae Scott lightly but pointedly observed that the old sexual politesse of opening doors for women was an appropriate symbol of what women were up against. They did not need or ask for courtesies that set them apart from men, she said, but for the more meaningful courtesy of equality symbolized in walking with men together through doors they both had opened.[13] The audience applauded. Her homely example was natural in the context of time and place, but it also underscored the proponents' difficulty in keeping discussion of ERA free of debate over behavior and values. It is not surprising that opponents should have hooted at ratificationists' disclaimers of any cultural impact for ERA. No matter how much they denied any connection between ERA and cultural change, proponents could not restrain even themselves—much less the opposition—from making it.

Again, it was natural for partisans to promise that ratification would mean little more than legally "defining all duties neutrally in terms of function and needs of the people involved, rather than in terms of sex." But this presumably modest change was neither simple nor "merely" technical; it had implications that allowed proponents to argue that ERA would "lend new dignity" to the roles of wife and mother, an argument surfacing to meet objections that ERA had declared war on wives and mothers. Indeed, according to the Republican National Committee in the 1970s, ratification would "strengthen the family by establishing a true partnership that would be beneficial to husband, wife, and children."[14] "True partnership" in marriage represented a considerably broader range of effect than that intended by the amendment's sponsors, indeed broader than the amendment could actually have produced. And yet both the idealistic goal and the technical adjustment of sex-neutral language did indeed belong together in symbolic conjunction. The latter was designed to free women and men from stereotypes associated with their sex in order to grant them greater opportunity to set personal goals within as well as outside marriage. The ideal of true partnership meant that the two persons involved would—on the basis of ability, preference, and reciprocal concessions—negotiate a relationship that was equally rewarding to both. True partners would not have to act according to what husbands or wives "were supposed to do" but according to what the two persons involved agreed was appropriate for themselves. The connection between sex-neutral language mandated in law by the Equal Rights Amendment and sex-neutral partnership in families is obviously indirect, tenuous, and problematic; and most proponents did not see it as a cause-and-effect relationship. That the argument was made in their pamphlets does not mean that they embraced such a view of change but that they valued sex-neutral language.

Such language, symbolized in the broader women's movement by the word "person," represented the rights of women as individuals. North Carolinians quoted the Republican National Committee in presenting ERA as a way for women to "reach their potential through equal opportunity, freedom of choice, and upward mobility." The words did not seize the radical imagination, perhaps, but they did reflect the sentiments of North Carolinians as well as Yale lawyers. In interviews and speeches, private discussions and public debate, Tar Heel women spoke of ERA in the same breath as "freedom of choice," women's "significant contribution," "individual attributes," "full citizenship," and "a fair opportunity to develop their abilities and to use them in the avenues of their choice." The words implied impressive psychological benefits derived from meaningful labor. To be sure, proponents wrote and spoke of economic penalties against working women, and they also understood that economic independence did not always result from jobs that facilitated "reaching one's potential." Nonetheless, "reaching one's potential," "individual self-fulfillment," and similar rhetorical flourishes seemed to be inherent in the shattering of stereotypes associated with ratification. Once these stereotypes had been challenged in constitutional law, a broader effect would slowly take place, and the cultural ramifications could not fail to benefit all women: that was the theory.[15]

Sex-neutral language, like the draft, became a symbol of what proponents wanted and what opponents feared. Both agreed that sex was the issue. For ratificationists, sex had too long been defined by men; it had become a way of reserving women as the appendage, accoutrement, or showpiece of men. In desultory discussion of their nemeses in the state Senate after the disappointments of 1979, one activist reminded her colleagues of the crusty senator from eastern North Carolina who boasted of marrying his high school sweetheart and raising her to be the woman he wanted her to be. In response, someone observed that he probably got a blue ribbon every time he showed her; but no one laughed. The attitude behind the boast was too pathetic, pervasive, and powerful to be taken lightly. From the proponents' point of view, male resistance to ERA represented contempt for women expressed by the senator's "macho" cleverness and evidenced men's inability to deal equitably with women. Sexual affectation seemed to have benumbed male legislators who were annoyed by women's entry into politics. When in a testy exchange a male senator commented on the uncommon attractiveness of a woman colleague, everyone in the room knew that he had insulted her in the form of a traditional "compliment" by telling her in effect that she was out of place. Little had changed since U.S. Senator Lee Overman had used the same language almost sixty years earlier. As the adaptation of tasteless coctail party banter to political discourse, it represented a deeply ingrained tendency to think of women only in terms of sexual allurement. Sex-neutral language therefore was one way to give women the power to define what gender meant.

The Three Faces of ERA

The principle of equality expressed in sex-neutral language helped recruit, mobilize, and transform women. In working for ratification, they learned what equality meant and how elusive it was. They learned political reality.

Feminists who had already been taught by experience and theory about the nature of power did not share the optimism of those who saw the women's movement on the verge of victory. But they could see that the ratification process was consciousness raising. The ratification process was a painful lesson in who held power in North Carolina (and the United States). It was painful because defeat of ERA became a personal defeat. One woman mused that she would never feel completely "validated as a human being" until ERA was adopted. How many proponents felt that way is impossible to say. The woman who said it probably would have acknowledged that even with adoption there would have been no guarantee that her humanity would have been validated. Her confession was a statement about minimal expectation—at least this much! At the same time, losing the struggle for something so self-evidentally right taught her important things about herself and other women.

The process was a renaissance represented by Betty Ann Knudsen's butterflies. Everyone who knew the remarkable Wake County commissioner had seen them attached to sliding glass doors, shower guards, refrigerators, and less obvious places. They were symbolic of the Christian doctrine of Resurrection, she had said—coming out of the cocoon to be transformed into a new being. The women's movement was like that, she had also said, breaking women out of the cocoon of dependence and fear and birthing them anew. (Knudsen herself was like that, Beth McAllister had pointed out. She, like her butterflies, had "broken out of the cocoon of selfishness and taken the risk of reaching out and touching others.") "The only way to assure success in achieving goals is to be willing to risk," Knudsen said. "To lose the security of the cocoon. No one can ever go back to it. So if we haven't learned anything else, we have learned to risk."[16] There were many women like Knudsen in the struggle for ratification, an impressive number of activists who defined themselves not by their enemies and their lost votes but by the bonding of the movement. They spoke unselfconsciously of how they had discovered a community of women who had made the movement concrete. They had also made it something that opened up a new world for all women. For *all* women? What about the opponents of ratification? Maria Bliss had responded to such a question by pointing out that "there are, in fact, on either side good people [who] . . . with the best of motives are trying to do what they think is the best thing." Two years later Betty Ann Knudsen could still say, "We have to make room for these women because they have just as much right to express their viewpoint as we have to express ours. And that's ha-a-a-ard to swallow. But they are women, and we need to love them, too."

If ratificationists thought of themselves as working for all women, they had learned early in their first campaign that they could not assume female solidarity and that was, as Knudsen said, "ha-a-a-ard to swallow." Yet at the same time one of the achievements of the ratification process was that so many different kinds of women had found a way to work together for a goal. Housewives and college professors, lawyers, medical students, realtors, nurses, secretaries, church volunteers, and social workers were held together in a loose coalition of feminists and self-styled nonfeminists. Within this broadly inclusive movement there was a creative tension between three different ways of working and thinking about the Equal Rights Amendment. These three ways were not definite factions even though they sometimes represented different loyalties and styles; nor were they exclusive to the extent that women could not be a part of all three either at different times or simultaneously. These three expressions of mood and logic are defined by their primary orientation: coalition, movement, and politics.

Coalition represented traditional women's groups—those already in existence before the coalescence of the feminist movement in the 1960s. They therefore represented the historical continuity of the Equal Rights Amendment with women's long-standing attempts to achieve equality. Coalition symbolized the respectability of ERA, as the League of Women Voters pointed out to its lecturers, something anti-ERA women perceived when they commented sarcastically on "country-club feminism." The logic of coalition was that women's groups such as local chapters of the League of Women Voters, BPW, and church women's societies, when taken together as a whole, represented, by implication most of the women of North Carolina. United Methodist Women, the YWCA, the league, and BPW were thought to embody not merely their own organizational goals but also those of women in general as such groups conceived them. Ratification strategists assumed that the local membership of each state society agreed with the pro-ERA position of state officials and representative delegations. Working through these familiar organizations, ratificationists could avoid the stigma of radicalism and rightfully underscore the relative antiquity of ERA; it was, after all, fifty years old in 1973. They could also exploit familiar and trusted channels of communication; members would be inclined to support an issue explained in terms of accepted goals and traditional self-definition. Ratification strategists also thought that women's organizations, as structures within which women had been active for some time, had taught women how to get things done. The coalition ideal stipulated that officers of each constituent organization would enlist members in the ratification process.

In representing traditional women's organizations, coalition also represented their assumptions and behavior. Such groups could do things among people like themselves but had not yet learned how to get things done outside the boundaries of sorority. The League of Women Voters, for example, was self-consciously nonpartisan, which meant that it remained aloof from activity that would reveal its powerlessness in affecting people who held power. The league's endorsement was always valuable, to be sure, but it was more in the form of a prayer by the pope over crusaders than in the provision of knights and retainers. As Martha McKay pointed out, women's groups often had elaborate organizations with a broad range of offices that gave members the "illusion" of "doing something." Activity outside this organizational matrix was not designed to get people outside the group to act but rather to provide a forum within which people could act in a controlled and restricted environment. Emphasis would be on how members benefited in terms of self-development. Thus in each ratification campaign, she noted a persistent need among activists to congratulate themselves on what they "did right." They had, she pointed out, substituted the standard of "learning" for the standard of "winning."

Movement meant change rather than continuity. It meant dramatizing a new assertiveness among women just as the civil rights movement had erupted from the assertiveness of blacks. Caution was not so important as clarity and forcefulness of expression. As an activist said of the 1973 hearing when proponent women laughed and groaned at testimony, "We were rowdy . . . and we acted out too much." It was not prudent behavior. Prudence was perceived as deference to male domination. The emotion expended through laughing at outmoded gentility was therapeutic; in it, women could feel social solidarity and moral superiority. Moreover, the "struggle"—that is what they called it—was renovating in itself. It created a new world of women within which to hammer out tactics within strategic frameworks and achieve significant results in a variety of tasks—all reinforcing a sense of belonging and meaning. "I did not realize," beamed one woman, "that there were that many dynamic, talented, committed, and interesting people that I could trust, and who would really just go full tilt!" They could go "full tilt" without winning. The rhetoric of public display that accompanied the struggle demanded marches, rallies, and demonstrations to reinforce solidarity. In them, the most important things were confrontation and clarity. Even in parts of the state where legislators adamantly opposed ratification, defiant women worked indefatigibly to make clear that they, at least, were for ERA, wearing their defeat as a proud badge of honor. One of these women recalled that after the last defeat in 1982 there seemed to her to be more pride among proponents than victors. Whether she was correct, the fact that she thought she was reveals the nature of the movement. Ratificationists could feel triumphant without winning; in losing they were not "losers."

Politics was not necessarily antithetical to either movement or coalition, but it did have a different logic. On the trivial level it was the difference between women who donned "shades" and boots to set themselves apart from other women and the sensibility of Senator Helen Marvin. She laughed as she recalled how on rainy days during the legislative session she would leave her rain boots in the car because her male colleagues were so unaccountably skittish about a woman's wearing boots. Maybe Nancy Sinatra's singing about "Boots"—"gonna walk all over you"—had touched a male nerve. Who could understand phobias? Marvin's was a political decision, avoiding the taint of movement and thus avoiding unnecessary aversion to her politics. Representative Margaret ("Maggie") Keesee always attended sessions of the House with her long hair carefully pinned up—and for the same reason that Marvin got her feet wet. Politics sometimes required different relations with men than those dictated by the logic of movement.

Beyond personal appearance and its symbolic effect, the logic of movement shaped tactics that the logic of politics would have avoided. The decision to have Charles Deane and Bill Whichard introduce bills of ratification into the General Assembly was partially personal and partially ideological. The men were relatively liberal in their politics and therefore comfortable with a commitment to human rights. Moreover, Whichard was one of the primary managers of a liberalized abortion bill, so it seemed ultimately sensible for him to sponsor ERA. Women legislators also wanted to introduce bills of ratification, but liberal men had already been assigned that role. The range of options considered therefore was defined by the logic of movement rather than politics. It seemed natural to proceed with people who by inclination, age, and ideology could be identified with the women's movement, something new and different—new ideas, new commitments, and relatively new legislators. A decision dictated by politics would have recruited one of the older, experienced, and powerful "uncles" of the legislature to lend age, reputation, influence, and craft to ratification. Since the logic of movement almost won the day anyway, it is a fair inference that a more political decision might have carried the vote. But no one at the time thought to make it.

Although probably not to a fatal degree, movement drama and political pragmatism collided throughout the ratification struggle. Women could be quite put out at the absence of marches and rallies and the presence of "dainty strategies" dictating a "low profile on human equality" and silence in response to "the opposition's big mouth." Others were angry with politicians who agreed to vote for ERA without making the amendment a major issue in electoral campaigns: "I find that an unacceptable stand!" exclaimed a western activist. Others resented what they thought was the defection of women politicians who seemed to have placed the political future of Governor Jim Hunt ahead of all-or-nothing efforts to win ratification. Political calculation beyond the issue of ratification was not part of movement. Yet the logic of politics said that having lost on this one issue, we can expect to win on others by using public declarations of those who defeated ERA as leverage to gain help from them on addressing domestic violence, adopting tenancy by the entirety, and commending studies of sex discrimination.

Politics assumed a future relationship with people who disagreed with you, one that could be used for eventual political advantage. But movement logic disdained compromise, caution, and the tendency to minimize points of difference and friction. After the terrible loss of 1977, strategists agreed with the colleague who observed that "idealistic sisterhood has gotten us nowhere. We must play the game of politics the way it is [played]." What she meant, as we have seen, was that coalition, movement, and political women would have to enter electoral campaigns and support the right candidates or become candidates themselves. And—as we have also seen, the strategy did not work because, by the time it went into effect, ratificationists faced a broad conservative resurgence that used ERA as a symbol of dangerous liberalism. Although ERA supporters believed that no one was ever defeated for favoring ratification, they found that pro-ERA people lost enough seats for other reasons to prevent ratification. The game of politics had rules and logic that promised victory no more than had lobbying. Moreover, movement women did not always play the game by political rules, as a pro-ERA male senator Carl Totherow from Winston-Salem discovered in 1978.

Conflicts between movement and politics were identified with NOW in the 1982 campaign. Refusing to subordinate agenda and ideology to politically experienced women judged to be insufficiently feminist, NOW remained aloof from national and state coalitions. At the eleventh hour, however, the organization agreed to cooperate, but in a manner that underscored its determination to call attention to the women's movement as a dynamic force and to its presumably preeminent role within it. The result in North Carolina was that there were three parties in the ratification camp: politicians led by Governor Hunt, Speaker Liston B. Ramsey, and Senator Robert B. Jordan III; the coalition represented by Jo Ann Norris and Betty McCain; and the National Organization for Women under Terry Schooley and the national president, Eleanor Smeal. From the beginning, NOW's style and goals conflicted with those of the other groups. Not surprisingly, there were disagreements about the political wisdom of using pro-ERA materials carrying the NOW label in conservative areas of the state. Moreover, North Carolina Democrats disapproved NOW's targeting specific politicians in primary elections before the vote on ratification; nor did they approve Smeal's stumping the state. Her call to enlist out-of-state women for a march on Raleigh was, they believed, contrary to promises NOW had made. Fed up with NOW's political obtuseness in doing things she thought NOW leaders had specifically promised *not* to do, a coalition leader observed, "NOW sure knows how to plan a march!" It was not a compliment.

The drama of public display against entrenched patriarchy was exhilarating for marchers, however. Such display was also cherished by NOW in the General Assembly. Schooley wanted a showdown vote on ERA even though she knew it would lose. "To have a record," she said. The "record" thus enscribed symbolized the values of movement, but it was difficult for politicians to see how the record could serve women's interests. Like pacifiers sent anti-ERA legislators in 1973 and those boxes of chicken droppings sent in 1982, focusing on a record of defeat was a gesture made from weakness. It was intended to have been a way to hold legislators accountable to women; it implied a promise of revenge but from people lacking the power to inflict it. The suggestion to "have a record" was made without thinking of the damage that acrimonious use of the record would have on women's issues in the General Assembly. Symbolic action was posed against concrete, if minimal, achievement—the posture was at odds with the action of women who had already gained something through electoral and bureaucratic processes. NOW's actions were not necessarily illogical, irrational, or naïve, for they were the logic of movement. Symbolic protest might have been good for movement participants, but it did nothing for ratification. Indeed, many activists believed that by 1982 ERA had become important to NOW as a way to recruit new members. Afterward Schooley could boast 1,600 NOW members for the state—fewer than a fair-sized Baptist church in Charlotte.

NOW's logic was, however, compelling in its insistence on mobilizing women outside coalition, and it did so superbly in the months before the 1982 vote. The movement, too, by remaining as free as possible from the logic of male-dominated politics, could make an important contribution by making women confront their powerlessness. At first it was difficult to define precisely what their powerlessness was; it seemed to be "ignorance" of tactics. Ratificationists assumed that if they had known exactly *what* to do, they could have learned *how* to do it, and—acting according to the *rules*—they would win. They had come so close in 1973, 1975, and 1977 that their powerlessness seemed to be only a matter of a few votes and time, a fleeting moment, surely. As time passed, they learned the scope and range of their powerlessness. The lesson angered them. The anger was with themselves for being too middle class, too white, too naïve, too idealistic, too ready to believe that "because a thing was right, we could win." They became angry, too, with their friends and allies: the governor, political women, volunteer women, careerist women, *other* women, and allied men. They acted according to a natural "law" of political psychology: In disappointing political losses where there is a high degree of personal commitment and where allies have not suffered corresponding defeats, there is a tendency to ignore the enemy and blame your friends.

Undaunted, at least as a group, ratificationists had attempted to gain a measure of power through planning, hard work, and ingenuity. By 1978 they had created a network throughout the state that they divided into manageable units for the effective flow of information and coordination of action. They had an impressive and (on paper at least) efficient machine. It was not, unfortunately for them, a *political* machine. It reflected the powerlessness of women because it worked more effectively within the boundaries of women's networks than between the machine and electoral politics. In commenting on this situation, one politically experienced woman observed that the proponents were "overorganized"; they thought that activity within the machine was equivalent to political activity outside it. She might have remembered the breakdown of communication in 1977 when Governor Hunt had thought the "machine" could produce needed pro-ERA letters—immediately!! NCUERA had gotten promises to "send a memo" instead. With more than a trace of bitterness, she complained of "little church ladies" and "little BPW ladies" who simply did not understand politics and grass-roots action. As a stinging metaphor of politically unsophisticated gentility, "little ladies" was both appropriate and unfair. It was appropriate as an insight into the way in which women's organizations had been stunted politically by being developed within women's "separate sphere." Condescending though she was, she understood how women had been locked into their own peculiar political culture and thus immobilized.

The slighting dismissal of "little ladies" was also unfair. The woman who made it in frustration confessed as much. A dedicated, tough-minded, and activist feminist, she knew that women were venturing into unknown territory. The ratification process was a way of learning the extent of women's powerlessness and what to do about it. Ratificationists were trying to learn how *not* to be "little ladies," and they were frustrated by their own ignorance and their teachers' ineffectiveness in dealing with it. Coalition women looked to political women to teach them how to do what they had told them that they had to do. To uninitiated women there seemed to be mysteries withheld; from the political perspective there seemed to be an incomprehensible inability to act. This lack did not afflict women so long as they were dealing with others within movement and coalition who shared their goals; but when it came to working with political veterans and unfriendly legislators, there seemed to be a failure of sorority. Political judgments were sometimes evaluated as lack of commitment rather than sound political analysis. Women in the administration and General Assembly told "us," remembered one discomfited woman, what "you" should do, not what "we" together should do. Even when politicos tried to bridge the gap, "something was missing." In one such exchange one person remembered, "I felt like a little girl." She did not want to feel that way again. Acknowledging that she had once felt that way was not only a comment on political mentors but also on how far she had come. Once she knew that politically she had been a little girl, she could grow up. The movement had recruited women she admired and envied; envy can be an effective if harsh teacher and had brought her to a critical point. "All my life," another woman confessed, "I have waited for someone else to do it for me. I realized [in the movement] that I had to do it on my own. I found I *could* do it." The next step was to move beyond such personal victories to the public arena.

Looking Ahead

Looking to the public arena would require a change in ratificationists' understanding of their powerlessness. They had to quit blaming themselves and each other and learn political realities. The first reality was that they could not gain a majority in both houses of the legislature; an even grimmer one was that they would probably not gain one in the near future. As they learned more about North Carolina politics, they attributed their powerlessness to the fact that they were women. It was not that they did not know that fact already, for they had learned through each disappointing campaign the extent to which it was true. They saw the personal side of it in the "macho" strutting of opponent men after each defeat for having put women in their place. Opponent men seemed to have been "threatened," the women said: it was one of the clichés of the movement, but the threat was difficult to comprehend. There had to be some explanation other than the psycho-sexual problems of aging, middle-class men. The leadership cadre of the coalition did understand that there were systematic problems. Women did not have the clout or credibility of realtors, insurance companies, and savings and loan associations. When they were divided as they were on issues such as ratification, their influence was diminished even further. The way in which opponents had defined ERA made ignorance and perversity appear to be the enemies of women rather than the patriarchal structure of politics. The diversionary images of women warriors seemed to be so irrelevant to the world of facts that had brought them to ratification that they began to look for the rational framework behind the irrational.

In a desultory conversation among representatives of the coalition after the 1978–79 campaign it became clear that defeat, like jalapeño peppers, had a way of clearing the head. What ratificationist women had threatened was the distribution of power in North Carolina. All that clamor about rest rooms and draft, the Bible and family, was a "smokescreen" one woman said in anger—a much better word than "mystification." She repeated it several times, angry with people who did not see that, angry perhaps with herself for being unable to fan the smoke away, certainly angry with the people who had laid it down so expertly. "It's money and power we're talking about, and not about church people" and their values, she insisted to people whom she thought particularly obtuse on this point. "Money and power" did not mean "payoffs" but the intricate fusion of vested interests, private ambition, and a dollars-and-cents mentality that posed public issues in terms of what it would cost powerful people if they had to be called to account for the inequities of the sex-gender system. Powerful people opposed ERA and equal participation by women in the market (and legislature) for the same reason Furnifold Simmons and Cam Morrison had opposed woman suffrage. They would lose something important if they yielded. It was one thing to see that, however, and an entirely different thing to do something about it.

Ratificationists are still, in 1990, wrestling with problems of powerlessness. Because they lost ERA, they may have been too ready to concede that they lost more. But the historical impact of the fight for ERA has not yet been played out. As people moving in and out of the historical process, ratificationists have yet to run the course of their intellectual, social, and political intersection with the powerful. Failure to ratify did not void the insights of the movement, although it may have shattered the fragile networks of organized women that might have provided the matrix from which to move into politics. As students of social science, they still believe that statistics can reveal the discrepancies between ideals and action; they still believe that knowing the fact that most poor people are women and children can compel political action. Doubtless, they still believe in equality because they could not surrender faith in the values for which they fought if defeated by what they believe was ignorance, fear, and self-interest. If the Equal Rights Amendment was not the main thrust of the women's movement, it was nonetheless its focus of hope and symbol for a while. Its history is not yet complete because history has a future as well as a past. Ratificationists taught women something about themselves and power. If they can use the strength of the former to challenge the latter, it could be said that they had indeed answered "the call" of January 1972 and not been "found wanting."

6

"We Don't Want To Be Men!": Women against the Women's Movement

Antiratificationists skillfully wove the symbols of ERA, "women's lib," and "family" into a web of apprehension, defensiveness, and anger. The result was to enmesh ERA in a psychic trap that separated liberation from obligation: wanting to be free meant no longer wanting to be female. To those who had used the "privileges" of womanhood as a protective shield against the consequences of their weakness, ratificationists seemed to be saying that women must become men. Feminists' attempt to strip away the protective coloration of exploitation, especially that in marriage and family life, was particularly troubling to women who saw solutions to the problems they conceded in male-female relationships to reside in reinforcing traditional patterns of behavior rather than in abandoning them. Such women had lived by the rules that defined the responsibilities of their sex by gender—even if men had not. They believed that changing those rules would harm women by making male-defined standards the measure of their life and work. "I am a widow," one woman wrote, "have three children, and work to make ends meet. I am still against ERA. I am a woman—and I want to be treated as a woman—not a man!"[1]

Ratificationists thought that there was a world of difference between voiding legal classification by sex and masculinizing women. But many opponent women distrusted change sponsored by feminists who seemed to think marriage, love, babies, and families restrictive and burdensome. The view seemed to reflect "male" values, and if attributing it to all feminists was unfair, it was nonetheless a fair interpretation of what feminists had in fact said. Feminism seemed to demean what women traditionally had valued and, through the

Equal Rights Amendment, seemed about to be written into the fundamental law of the land. For antiratificationists, it was as if the perceived chaos of feminists' own unhappy lives was about to bring chaos to their own.

Women's opposition to ERA presented ratificationists with a disturbing paradox that they tended to avoid in explaining defeat. The standard interpretation was that ERA had failed because men had killed it—which of course they had. Conservative businessmen had indeed financed an opposition that enjoyed the prayers of male religious leaders and the votes of male legislators. The latter had justified themselves, however, by the obvious if unpleasant fact that women opposed the amendment. The perception of feminists was that antifeminist women had broken female solidarity by snuggling within the arms of patriarchal hegemony. In this very understandable feeling, feminists failed to appreciate that it was they whom opposition women believed had broken female solidarity by acting like men. This perception suggests a situation far more complex than that implied by the assumption that anti-ERA women had been duped and manipulated by men into sustaining their own victimization.

Trying to understand anti-ERA women does not address head on the question of why ERA failed so much as how large numbers of women responded to feminism. And if the "why" may be implicit in the "how," the focus is on listening to people who received a message feminists did not intend to send. Probing the complex impact of feminism means not dismissing nonfeminist women as nonpersons with no will or imaginative life of their own. As unbelievers resisting conversion to an understanding of what critics believed the world was really like, such women have something to say that is dismissed as irrational or false only at the risk of surrendering critical perspective altogether.

Women who opposed ERA were not easily identified by group or class according to traditional ways of defining "collective consciousness." As we have seen, opponents as a group were politically and religiously much more conservative than ratificationists; they were a little older, somewhat less educated, and more likely to be housewives. If employed outside the home, they were less likely to hold professional and managerial jobs than their pro-ERA counterparts. They were also less likely to have grown up in politically active families and more likely to be oriented toward local than cosmopolitan values and experiences. Ideologically at home with the Radical Right, anti-ERA leaders were solidly identified with ultraconservative organizations such as the John Birch Society, the Eagle Forum, Women for Constitutional Government, and Daughters of the American Revolution. Locally, opponent women were likely to have been leaders in church groups or in organizations traditionally concerned with children such as the Girl Scouts. Prolife groups also provided a base for recruitment in Massachusetts, although not in the non-Catholic South.

This profile reveals less about opponents than about those among them who provided a text through which to understand the cultural bases of their actions. If it made little or no sense to ratificationists, the text made sense to those who created it and suggests a pattern of meaning based on the elusive subjective experience that anti-ERA women shared with each other. This experience, what the theorist Charles Taylor calls "intersubjective meanings"[2]—those meanings experienced in social practices that lie behind the flawed and incomplete expressions of them in public debate. The personal and existential response of anti-ERA women to the amendment suggests objections rooted deeply in intersubjective meanings inseparable from identity defined and expressed in traditional gender roles. These roles were the basis of status, moral worth, and self-awareness linked together in a system of "interworking meanings." The importance of the phrase "interworking meanings" is suggested by the anthropologist Clifford Geertz, who—while acknowledging the importance of class interest and social "strain" in shaping ideology—insists that belief systems reveal a dimension of human experience beyond interest. That dimension is the meaning which people attribute to symbols and through which they represent social reality. The insistence that ideology be understood as metaphor and symbol warns us to probe the meaning behind presumably "irrational" statements. Such statements, Geertz argues, are part of the "head-on clash of literal meanings" that provide "novel symbolic frames against which to match the myriad unfamiliar somethings that are produced by the transformation of life." The transformation identified with a resurgent feminism—and the socioeconomic developments that made it possible—constituted a crisis that elicited and stimulated two popular ideologies, each of which "named the structures of situations in such a way that the attitude contained toward them [was] one of commitment."[3] This was true of opponents no less than ratificationists.

ERA and Phyllis Schlafly's Apocalypse

The place to begin with anti-ERA women is their belief that the amendment was, as Sam Ervin had warned, an attack on "homemakers and mothers." Although not fluent in feminist theory, anti-ERA women had been jarred by its metaphors and challenge to tradition. Consequently they accepted the credibility of an apocalyptic future that Phyllis Schlafly sketched out for them if ERA were to be ratified. She warned that women who had never been economically independent would be forced—regardless of personal preference—into the job market, compelling the state in turn to establish day care centers. There, psychological distancing that had already endangered relationships between parents and children in public school education would be reinforced. Nurses and teachers, as surrogate parents, would be freed to remove the last vestiges of family identity in favor of loyalty to the community (state). The remaining remnants of family life would be further transformed by a revolution in gender symbolized by an implicit sanction of homosexual marriage. To this sign of *abomination* was added the warning of *danger* in the mandate that ratification would give to abortion.[4] It was not so much an argument as a profusion of images.

To ratificationists, this hypothetical history symbolized the emotional volatility of their opposition. They took opponent objections apart piece by piece, as if separating elements of the whole could demonstrate that women had nothing to fear from ERA, but the exercise was misleading. It reinforced belief in the irrationality, ignorance, and false consciousness of opponents and diverted attention from why the whole paradigm would be believed. Equally misleading was analysis that attributed to opponents a victimization that demeaned them to the status of marginality. Rather more to the point was what Schlafly's apocalyptic "history" meant. The assumption of meaning even in presumably "false" statements is the first assumption of those attempting to understand the rhetoric of conflict. And meaning is understood best by divining patterns of symbol and word and analyzing them according to their evocative power rather than their accuracy. If the ideology of proponents dictated that they treat anti-ERA rhetoric in the analytical fashion that they in fact adopted—denying or reinterpreting "facts"—we are not restricted by their political goal.

Since political discourse always employs symbols, analysis should not be thought of as a way to avoid substantive issues in favor of those that are "merely" symbolic. Symbols are not "mere" things. They capture the imagination, stimulate commitment, and represent the most basic values of contending parties. Where specific issues are imprecisely stated or deal with what Geertz calls "myriad unfamiliar somethings,"[5] the symbolic frame can be called upon to represent the meaning of change. And for many opponents of the Equal Rights Amendment there were too many "unfamiliar somethings" in the future should the amendment be ratified. As Sam Ervin and other conservative opponents pointed out, the definition of "rights" in ERA was not clear. Other constitutional amendments had specified certain rights; the Nineteenth Amendment, for example, had established the right of women to vote—it was a specific grant. A broad, unspecified grant of rights would have to be interpreted by a distrusted Supreme Court that would presumably interpret the amendment with the aid of legislative history and briefs permeated by feminism. Opponents were determined to prevent this revolution.

Family-Oriented Women

Anti-ERA women's fears about ratification were a projection of their collective assumption that women's liberation was a dangerous virus cultured in the pathology of American life. The epidemic, erupting from the liberation politics of the 1960s, was facilitated by deterioration in the health of American society and by increases in the number of households in which traditional family patterns were absent, of child custody and support cases, and of female-headed households at the poverty level. Each change suggested ever-greater demands upon women as they suffered personal isolation and the loss of familial bonding and economic support. Such cold statistics seemed to underscore the debilitating effects of the feminist contagion. Its virulence seemed to be masculine. Certainly in its criticism of what women did it characterized male attitudes. The natural response was defensiveness among women who had obeyed all the traditional rules of feminine behavior. They had clung to their families through who knows what kind of private pain only to feel feminists' criticism as a denunciation of them.

References to themselves as "family-oriented women" could very often trigger fierce reaction. As one antiratificationist angrily spat out her words, these were women who were "scared to death of what's happening to them." Familiar magazines once linking them to fashion and style in the world beyond their comfortable homes now seemed to assault them with values they despised. Schools once supporting parental authority were now teaching things that they did not understand or approve. Worst of all, the love once characterizing relationships between them and their children had been imperiled. Any woman thus attacked, she concluded, would be "like a mother lion fighting for her cubs." The cubs were already in danger of being trapped outside the den. ERA symbolized commitments by schools and universities to wean sons and daughters away from their parents. It was reminiscent of atrocities against families whose children had been seized by the Nazis and Soviets. The metaphor was applied to government-supported child centers that under ERA would take children away from their families just as busing had literally taken children away from their families in the process of desegregation. Having already seen family life put at risk by policies identified with racial equality, many women were apprehensive about what a federal commitment to sexual equality would mean.

Further threat lay in widespread attempts to purge textbooks and children's literature of bias in favor of traditional gender roles. The images, behaviors, and goals that had helped traditionalist women understand what being female meant were being proscribed. In Chapel Hill, for example, feminists were writing nonsexist books for their children, hoping to prevent losing them to a sexist society. Anti-ERA women wanted to lose their children no more than did feminists. The sense of threat ran deep. Even in private conversation the apprehension could erupt into fury, the words tumbling out so fast that the argument was difficult to follow. "If feminists had their way, your children would come to hate you." The reason was not very clear. Either they would hate you for furthering the feminist goal of androgyny, or for refusing to do so, and thus shaming them under the unforgiving surveillance of their peers. The rationale was not so significant as the word "hate." The family was the basic unit of society, the woman had said earlier, where you learn how to think about yourself and your goals for life. The world of the future, growing out of conflict between parents and children, would be a terrible, forbidding place, evoking apocalyptic anxieties.

Because the metaphor of losing one's children was so evocative, anti-ERA partisans used it in conjunction with issues unrelated to ratification. In 1979 and 1980 they practiced their tactics on Governor Jim Hunt. Upon taking office in 1977, Hunt had learned that only three states had higher infant mortality rates than North Carolina. Almost half of all children between one and five had inadequate diets; about seven thousand babies had been born to girls under the age of seventeen in 1977; 40 percent of infant deaths that year were related to genetic diseases; and hundreds of mothers had given birth with no prenatal care whatsoever. One solution to such problems had already been designed before Hunt's election. The National Health Planning and Resources Development Act of 1974 had set in motion a process welcomed by North Carolina health officials. Private, public, and academic health authorities cooperated in developing a Child Health Plan to help counties identify and solve problems. The plan was bound with a blue cover and imprinted with the seals of the American Academy of Pediatrics and the state of North Carolina. Hunt wrote a letter of endorsement.

The governor had also been active. As chair of the Southern Growth Policies Board in December 1978, he had urged other governors to join forces to save children throughout the South. In his own state he asked Florence ("Florry") Glasser of the North Carolina Department of Administration to write a bill bringing the resources of each county to bear upon the problem. He received the support of county commissioners, and in January 1979 he presented a "children's budget" to the General Assembly. Based on expansion money, the budget was really a way of publicizing a public commitment to children. Consistent with this proposal Hunt persuaded the General Assembly to enact Glasser's New Generation Bill. It set up an interagency committee at the state level to coordinate all activities that served children and suggested that counties could—if they wanted to do so—set up their own interagency councils. Passing the House 98–8 and the Senate 35–9, it was, as Glasser said, an ideal act: it favored children, promoted efficiency, hired no new bureaucrats, and cost nothing.

As ERA supporters had learned much earlier, good intentions are not politically negotiable. Hunt's foes denounced the act as an attack on the family. Since the act was so simple and the Child Health Plan so complex, opposition could be raised only by identifying the plan with the act, selecting offensive phrases from the plan and defining them as goals of the act. Conservative Republicans, fundamentalist preachers, and the ultraconservative National Congressional Club expertly managed this intricate political vivisection. Two of the most visible opponents were Representative Mary Pegg and Senator Anne Bagnal, both from Winston-Salem, both conservative Republicans, and both outspoken opponents of the Equal Rights Amendment. The act, like ERA, was part of the state's warfare against the family. The act (actually the plan), they charged, had said that the "family alone" could not meet the essential needs of children in "health care and education," a truism that allowed the authors to get from one sentence to the next. Pegg and Bagnal warned that the "health care home" (the doctor's office) would "virtually" supplant "parental authority." Further ramifications were suggested when opponents learned that the governor's wife had presented a copy of the Child Health Plan to the wife of the U.S. ambassador to the United Nations. Everything became clear when one discovered that on the binding of the plan was the logo of the International Year of the Child—it was an internationalist plot.

Anne Bagnal had earlier explained her position on such matters in a private interview. She thought of herself as speaking for women who were housewives and mothers—but, and she would make this point more than once—she was not chained to the kitchen. Life was too short! She understood her public responsibility to be a representative of women who had neither opportunity nor time to make themselves heard. It was important for women to see their "sisters" active in public life. She recalled how at Winthrop College in South Carolina she had been so affected by Mary Elizabeth Massey, a highly respected historian of the South. She also mused about her mother's friends who met regularly in the living room as a local chapter of Business and Professional Women and who vented their intense anger at the deep-seated discrimination against them. But times had changed. "I think," she emphasized, "that debate over the Equal Rights Amendment has brought to the forefront matters which have benefited women. Now," she insisted, "discrimination against women is as illegal as it is going to be!"

But ERA was disturbing. It was not merely that her husband was a businessman, although she mentioned the pressure of affirmative action. Her wariness of ERA was only part of her commitment to conservative Republicanism; and she admitted to ambivalence on certain issues. But it was her "femininity" that won out over memories of her mother's friends. She recalled with satisfaction that at a party a man had contrasted her with another woman running for public office. "You're feminine," he had said, "Meyressa is feminist!" The word triggered her revulsion at confrontation—black-white, child-parent, female-male. "I hate these things that continuously divide us—I abhor them." She felt as if "they" were trying to change the meaning of the family—schools, counselors, teachers, social workers, college professors. She could fight back by fighting the Child Health Plan and the Equal Rights Amendment.

The mother lion fighting for her cubs was also interested in what the cubs were doing while she kept the hunters at bay. The woman who used the simile had been generously articulate about the way in which ERA had been birthed in the 1960s. The amendment was born of feminists' narcissistic consciousness raising, trying to create new selves under the mistaken belief that they could achieve goals that would fulfill them in ways impossible through the bearing and raising of children. Rejecting a normative motherhood, these pampered rebels had trivialized sexual intercourse as hedonistic celebration. Such lack of self-discipline would eventually weaken the internal cohesiveness of society by failing to pass on the repressed sexuality that made civilization possible. Abortion in this context meant a way of avoiding the consequences of their actions. "Control over their own bodies?" She was skeptical. "If they had had control over their own bodies in the first place, they would not have been pregnant."

Reference to abortion seemed to be natural for opponents. Although ratificationists initially tried to separate the issues of ERA and abortion, and although not everyone who worked for or supported ERA favored abortion as public policy, the two were fused in the minds of opponents. Before ERA had been introduced in North Carolina, abortion had been a difficult political issue despite the passage of a liberalized abortion law in 1967. Then, legislators had yielded to reformers within the medical profession and the energetic lobbying of a retired banker and freshman legislator from Charlotte. Attempts to liberalize the law even further failed by only four votes in the 1971 session of the General Assembly, but there was little if any antifeminist connotation in the letters, petitions, and public testimony against further reform. To be sure, there were references to a "new morality of promiscuity" and the movement to guarantee the middle class "a life of ease,"[6] but these things were not connected to feminism—yet.

How ERA and abortion came to be so natural an association among antiratificationists is unclear. The *Yale Law Journal* article accepted by opponents as definitive never in its exhaustive discussion mentions abortion. Certainly the connection between the two issues was not made by all opponents of abortion; but to many it was almost self-evident: "Equality is the right of every person," wrote one constituent to a legislator in 1979, "but this ERA is a bad bill; no one has the right to murder babies!!"[7] This was three years before the Supreme Court case *Harris* v. *McRae* (1980), which the political scientist Gilbert Y. Steiner believes made the linkage between the two reasonable. To some it was reasonable from the very beginning of the ratification process. Both issues had been identified with women's rights. ERA was, after all, a woman's issue and a demand for "equality of rights"; and abortion was, some feminists insisted, a woman's "right." ERA was an attempt to remove sex as a classification in law, a way of separating individual women from their sex. Abortion was a way for women to avoid the natural process associated with their sexuality. Thus both undermined the family by separating familial responsibilities from women. Both ERA and abortion, therefore, were seen as ways through which women could be released from traditional roles and responsibilities; possibility was perceived as prescription. ERA and abortion could become two aspects of the same threat to women whose identity was wrapped up in motherhood.[8]

Abortion was also linked with uncontrolled sexuality, itself symbolic of a society on the verge of disintegration. Women's physical body was transformed into a symbol of the social body; all attempts to release women from traditional roles were fraught with danger, predicting a pervasive loss of rigor, order, and purity. Abortion was persistently linked with pleasure, promiscuity, and irresponsibility—all having to do with loss of social, not merely personal, control. Here as elsewhere, antiabortion rhetoric revealed a moral imagination of given rules, strict boundaries, and expected conformity—an almost mechanical universe. In abortion and ERA women could abandon their familial identity. The feminist effort to entice the cubs away from mother lion would make them into undisciplined and compliant pussy cats. "I've sat here at home and I've read the magazines, and read the newspaper, and I've watched the television news, and I never did anything," explained one women in 1977. "And all of a sudden, this was it. I could just see ERA was going to deepen the mire, to do much more if I didn't stop it. . . . There's a freedom and permissiveness that I think is wrong. But these young people who are living together, having abortions, wandering aimlessly without purpose, someday they will have to answer. There are no boundaries," she continued, and ran down a list of things that used to be abnormal: homosexuality, women truck drivers, teen-age sex, and abortion. "I don't care if she takes the pill for twenty years and never has an abortion, someday she is going to have to answer—deep down deep inside herself. . . . A free permissive life is wrong. I've had it with this permissiveness and I'm going to fight it."

Others had had it, too. The permissive mood covered a broad range of action and value. It was probably the fault of the schools and colleges when they began to reject *in loco parentis,* one woman observed. It had begun in the 1960s when teachers got "down" on the same level as the students—someone remembered a college professor who had lectured students on growing "down" instead of "up." There were other symptoms: ministers being called by their first name; a pervasive failure to provide proper guidance; children growing up too fast. All this had been done in the name of equality and rights. The words seemed to nauseate one woman, the values they represented having absolutely nothing to do with normal family life. In a normal family you did not learn equality, she said, but deference to a loving authority. Not rights—imagine the absurdity of children's rights—but obligation, duty, responsibility, and self-discipline. The appeal of ERA to equality of rights, therefore, was by definition subversive of the normal family.

"Scared to Death of What's Happening to Them"

Disciplining and protecting children, important as they were for understanding the ERA-feminist threat, were not the only focus of debate. The simile of the mother lion had been part of an emotionally packed statement introduced by reference to women who were "scared to death" of what was happening to themselves. The women's movement had placed anti-ERA women under siege; the issue had been framed by Sam Ervin almost from the very beginning of debate. His theme became one of the essential articles of faith in STOP ERA. As an anti-ERA activist explained to pupils in Northeast Junior High School in Greensboro, "libbers" had launched a "total assault on the role of American women as wife and mother."[9]

The speaker was not alone in feeling the threat of women's liberation. To have felt bonds loosening and been on the verge of losing cover could mean nakedness and vulnerability instead of rebirth. And even rebirth could demand too much. If a self-generating newness carries with it the opportunity to write a new script for our lives, it virtually forces us to do so. The result could be an awful freedom. Thinking about it, one woman expressed the matter in the homely terms of private life and family sensibility. She did not want to be liberated from her husband, she wrote. "He works for me, takes care of me and our three children, doesn't make me do things that are hard for me (drive in town), loves me and doesn't smoke, drink, gamble, run around or do anything that would upset me. I do what he tells me to do. I like this arrangement, it's the only way I know how to live."[10]

For other women, anxiety about equal rights reflected a less-sentimentalized political judgment. Their problem was they they were a little too old to learn new roles. If they had known forty years earlier that one did not have to be a full-time, homemaking mother, they might very well have opted for another job. But at age sixty-two it was hard to retool, and—apprehensive about the solvency of social security anyway—they were not inclined to stand up and shout for an equality that would penalize them for doing what they had done as well as anybody, and probably better than most. One pro-ERA speaker, Grace Rohrer, who was secretary of cultural resources under Governor James E. Holshouser recalled women coming to hear her laden with anti-ERA material, asking questions, deserving answers. She felt no hostility from them, and she found herself actually bonded with them. She felt their fear that having been given so little by society, they needed to clutch ever so tightly what they knew they had. They dared not risk it all on the assurance of a woman, who was obviously capable of taking care of herself, that sexual equality would not harm them.

Opposition came from women who were economically dependent because of their "life experience." Essential to that experience was their definition of self as housewife and mother and their belief that ERA challenged their privileged status in the mythology of American life—as Jane Mansbridge points out. When one woman blurted out her anger with ERA, the first thing she could think of was that *Cosmopolitan* magazine had changed radically—its values were no longer her values and she did not like it. She was joined by many more as one can infer from survey data and the response of antifeminist women to feminist rhetoric. But the women Secretary Rohrer encountered were not worrying about prestige. They, like the women the sociologist Barbara Ehrenreich discusses, were concerned about support, social security, their rights as housewives. They understood their economic dependency very well indeed. In both cases and arguments are women who have lost or who fear that they will lose something of themselves as the result of change: comfort, self-esteem, economic security.[11]

To argue, however, that resistance to sexual equality, feminism, or ERA was essentially a revolt of housewives and mothers *concerned about the family* is to ignore certain evidence. The millions of housewives and mothers who supported ERA—indeed the survey data that reports full-time housewives were no less likely to support ERA than employed women—suggests another interpretation. But the interpretation cannot ignore the sense among housewives and mothers who opposed ERA that they did so in order to defend the family. They identified themselves with a selfless ideal to fend off what they perceived as attacks upon them by feminists. That selfless ideal was motherhood, service, and punctilious obedience to the rules of domestic responsibility. There is no reason to believe that the words "housewife" or "mother" necessarily dictate inflexible roles within a family in order for the family to be normal, however. Although much anti-ERA literature seemed to convey the idea that flexible family options were both politically and personally threatening, many women opposing ratification had themselves created various and flexible roles within the families they headed. The family as a unit was not the issue. That the family would have been weakened more by placing mothers rather than fathers at risk in battle or the labor market was really not the issue any more than the weaning and rebellion of children, although each was an important part of antifeminist ideology. If the family had been the major issue, the quality of family life would have been the major focus, but it was not. There was something else of which children, discipline, homemaking, and family were simply subcategories: and that "something else" was classification by sex. Or—as Phyllis Schlafly put it, "unisex." Or—as others might express it: the definition of social role by gender.

Admittedly the phrases refer to different things: law, life-style, socialization. Classification by sex in law had been the initial focus of ratificationists. To be sure, the issue as to when classification by sex becomes overclassification is one about which reasonable persons may disagree, but in saying that "physiological and functional differences" between men and women would be nullified by law, Ervin was wrong. It was precisely those acknowledged differences that under ERA could not have been used to deny men or women of their rights. The issue was "rights" for ratificationists. But at the same time it was also "sex" as in "on account of sex." Like suffragists before them, ERA activists resented traditional definitions of what "sex" meant. Classification by sex could not be contained within the scope of law because history and culture had put it there. As part of the new feminism, ratificationists had rejected the idea of sex as a basis for defining social roles. If Schlafly caricatured the historical process, she did agree with feminists as to the importance of gender. She elevated anatomical differences between men and women from the folk wisdom of everyday discourse into a doctrine that could, she hoped, entice women into the auxiliaries of the Radical Right. They may or may not have acknowledged her self-annointment as spokeswoman, but they agreed with her that feminism, embodied in the Equal Rights Amendment, was an assault on womanhood.

How many women felt that way is obviously unknown; but what proponent women called "scare tactics" suggested that many of them did. At least they sensed in feminism something more devastating than another rise in the cost of living. If public policy were to be based on feminism, as antifeminists claimed with reference to ratification, it was about to subvert behavior that expressed for some people the very sources of their personality. For millions of people the sexual distinction was so basic a part of selfhood, and specific gender roles so firmly implanted by the family of origin and reinforced so well by daily interaction, that changes anticipated by ratification could be frightening. "Today," a mother wrote after ERA had passed the Senate, "I am ashamed and terrified at what the future holds for my three little girls. Will my shy, sweet Tommy be drafted in six years? So modest I can't even see her undress. Oh God!" she was writing to Sam Ervin, "I can't stand it. I just can't bear it."[12] To deny the relevance of sex in law was to deny life and hope. "Sex" meant the intimate privacy of shy little girls and equality meant ravaging them, stripping away the protection of innocence and thrusting them into battle. She had seen the obscene results on television—all Americans had: ambush, rape, and slaughter. It was not a linear analysis of cause and effect that Tommy's mother expressed, but the overlay of death and innocence, images that haunted rather than arguments that persuaded.

The family had very little to do with this demonic vision. Tommy was still a child, and it was her sex and what it would mean for her as she grew up that was her mother's first thought. Distinctions based on sex would allow her to ease into adulthood, unless of course she were one day confronted by a rapist—something her mother foresaw only if she became an anomaly, that is, a woman warrior. Shy as Tommy was, she seemed to personify innocence and vulnerability. ERA would destroy that innocence. "You must pass a law," her mother wrote Ervin, "allowing parents to have girl children sterilized. This would be the only solution. Then women would not have to worry about or prevent pregnancy just like men." The idea was preposterous—she knew it was. "Everyone would be truly equal." It was absurd! "Dear God in heaven—help us women." It was a surprisingly intense cry of anguish and protest. The fusion of ERA and sterilization created a powerful symbol that shrieked danger. Tommy's mother was not being rational.

Under the circumstances, Tommy's mother would probably have doubted the rationality of those putting her daughter at such risk. Legislators and ratificationists agreed that the presence of women in the draft and combat was probably the most mischievous criticism of ERA. That women were already members of the armed services and had already served in combat zones were not sufficient assurances. Tommy's mother wrote of shame, strangely enough, and terror, and the awful assault upon her child's sexuality. She faced an absolute danger—not insecurity or carelessness or simple risk—but elemental, demonic, ultimate obliteration of self. And not just the death of personal self but of those values and beliefs that had shaped her Tommy in the first place. No more Tommy; no more Tommys. She confessed her helplessness to an awful and ultimate power: "Oh, God. . . . I just can't bear it." Moral failure and existential terror fused in the eruption of her feelings. They were too strong to be expressed in person, perhaps, but very much like those shared with others who spoke of ERA, sex, and self in ways suggesting that what was at issue was a basic understanding of what life was all about. The frameworks of significance structuring antifeminist beliefs revealed that certitude about the role of sex in shaping personal identity, private obligation, family life, and social responsibility was, for many women opposing ERA, essentially a religious conviction. Tommy's mother did not say that ERA was un-Biblical or un-Christian but that it would destroy little girls. Her addressing God did not mean that she was especially religious, although she may have been, but that she had assigned her daughter's gender an absolute quality.

Gender, in other words, was sacred. It was a given—a biologically, physically, spiritually defined thing. It was—or it was until sports scientists began counting chromosomes—an unambiguous, clear, definite division of humanity into two. In an uncertain world of revolution, counterrevolution, economic recession, climatic volatility, political corruption, military disaster, riots, and

OPEC, gender was a basic, personal reminder of the orderliness of nature. Or it should have been! Feminists, however, had denied that gender was quite the given fact that so many people assumed it to be; at least it did not necessarily dictate the way in which males or females ought to behave. The leap from "sex" to "ought to behave" was not, feminists had said, "natural" but cultural. The meaning attached to sex was actually made by humans, not God or nature, and therefore could be changed. The absurdity attributed to this distinction between "sex" and "gender" did not affect debate on public policy until the Equal Rights Amendment passed Congress and provided focus for a diffuse anger at the breaking down of boundaries, rules, and "givens" that had characterized the 1960s. Identifying ERA with feminism and the cultural and historical relativism of gender and gender roles did not result from widespread familiarity with feminist theory so much as the actions of women who were stepping out of traditional roles into the worlds of men. Ratification could become a referendum on all of these actions that challenged the concreteness and immutability of obligations identified with sex. It is not surprising that one woman should have identified NOW with the anti-Christ before blurting out, "This thing is spiritual warfare!"

This symbolic reference freighted every aspect of debate over ERA and created an emotional "jamming" of messages that ratificationists sent to politicians and the public. The moral meaning of feminism for antiratificationists so overwhelmed careful statements about the anticipated consequences of ratification that ERA partisans were almost left gasping at what seemed to be hysterical apocalypticism. The conclusion was natural if one were insensitive to antiratificationists' assessment of women's weakness in a man's world. This assessment, and the assumption that feminists were masculinized women, allowed them a rhetorical leeway that made ratificationists accuse them of "scare tactics." This charge seemed to be especially true when arguing that ERA would repeal *all* laws protecting women (rape and sex crimes) and mandate the integration of public rest rooms. This accusation made absolutely no sense to feminists because they had done so much to support rape victims and to stir public discussion about the crime—indeed, much more than had traditionalist women. And yet charges about the decriminalization of rape were a persistent part of anti-ERA rhetoric. They fit into a framework of plausibility that emphasized the vulnerability of women because it was assumed that distinctions in law relating to sex actually protected women. Remove those distinctions and the result would be to leave women open to the power of men—RAPE!

Rape was a more than serviceable metaphor: it was also a cry of outrage. The existential danger and nauseating trauma inherent in the reality represented by the word should have isolated it as a major metaphor of antifeminism. If those who accused ERA of decriminalizing rape and traditional sex crimes had actually believed that the amendment would literally do what such accusations claimed, they would have made RAPE! the primary symbol of antiratificationist rhetoric. But they failed to do so; it was only one of many charges and in private interviews almost totally ignored. Women talked instead about their own personal conviction that social role and personal identity were dictated by their biological differences from men. They spoke, as Tommy's mother had written—of the vulnerability of little girls and women and how they feared that ERA, in "writing sex out of American law," would leave them defenseless. And they spoke as if feminists had violated them. Not content with changing themselves, feminists wanted to *force* other women into new roles whether wanted or not. As antiratificationists talked and explained, the meaning of rape became clear. To be sure, it was a cry of danger, but it was also an elemental expletive expressing feelings of violation.

With these intimations of danger came also the image that seemed frivolous, overly precious, and daintily fastidious—the sexual integration of public rest rooms. It is what proponents called the "potty issue." The "potty parable" imagined that under ERA someone would try to enter a washroom of the opposite sex, be denied, and appeal through the legal system to the Supreme Court, which would issue an order to integrate. The idea of integrated toilets had obvious connotations of integrated classrooms. The potty parable described the pattern of civil rights litigation in the 1950s and 1960s. It was not the first time in American history that filth, sex, and blackness had been fused in white folk metaphor. It was not the first time that equality had been made to appear absurd. For many whites, bringing white and black youngsters together in a classroom had been absurd because—in a perverse misstatement of the issue, proximity of blacks to whites would not make them any smarter. This perceived absurdity was compounded by the busing of children to facilitate integration and achieve "racial balance" in the schools. The charge of integrated toilets served, therefore, to reduce the idea of equality to an absurdity. But this is only one reading of the potty parable.

Another is suggested by a letter from a white millworker to her state senator urging him to vote against ratification. "Mr Whichard, you do not and cannot know what the (blue collar) working woman has to put up with." She explained her situation by recounting what she considered to be a warning. "We will have to use the same restrooms as the men both black and white. . . . I was on a break on the afternoon of December 14, 1976 and since I had really felt bad all day, I was lying down on the bench in the ladies' washroom." She remembered it well. A black man opened the door a crack to thrust his arm in so that if anyone were there, she could tell him to go away. She did. He had not dared to come in; but, she insisted, he would have done so if ERA had been in force.[13] The integrated rest room represented something beyond reason—it had become the symbol of a threat deeply felt because of her own sexual vulnerability. It was a black arm that had come into the ladies' room. ERA would bring the rest of him in. It was a mixture of racism and fear, although in the mouths of white men it could become racism and bravado.

But there was more. The ladies' room was a haven and sanctuary. The men's room was a latrine and urinal. To charge that ERA would integrate toilets was to say that women's washrooms would degenerate into men's rooms. The reality of women's sexual distinctiveness, relative weakness, and sense of vulnerability seemed to be starkly highlighted by the unforgiving brutality of a flourescent feminism. The movement had denied the relevance of sexual distinctiveness despite the reminder of that fact in daily acts of purifying and expurgation. It was an outrage to deny sexual differences; to treat women as if they were the same as men! The feeling enveloped the biological fact, acts of purification, the danger and stench of the men's room until exploding in a scream of outrage at defilement. Feminists had defiled women by attempting to force equality upon them whether wanted or not. Antifeminist women were saying in essence, "You have done men's work. Fine, if you can do it. You have invaded men's clubs, although it hardly seems worth the effort. And you have become, in effect, men. Since you have already achieved what you want without ERA, there must be some secret agenda that we women don't know about. But we can guess. The only thing left is to change the traditional norms of behavior so that you can become the model of what women ought to do and be. You mean to make us like you, that is, like men. You have defiled us!"

"Dear Sir," came a letter from Burgaw, North Carolina. "I am enclosing a petition against ERA. Please vote against this terrible law. It is the most immoral mess I ever heard of to be called a law. It is filthy!"[14]

Defiled, antifeminist women transformed resistance to ERA into a rite of purification. We were, one woman recalled, "fed up with the women lib thing. Many said they were fed up, and they associated the Equal Rights [Amendment] with women's lib. And they were just fed up with that. And that they were women, and pleased to be women!" "Pleased to be women" was a constant theme, a statement about themselves as well as confirmation of the feminist insight into the power of socialization in linking sex and social role. Challenging the womanly ideal meant calling into question the personal victories of a lifetime. The transformation identified with the consciousness raising of liberationists had a counterpart among those women who were jarred into action by the opportunity to fight feminism. The process could be personally satisfying: "Well," one opponent confessed as she looked at the tape recorder with a sense of accomplishment, "I really had a lot to say for a little girl who never raised her hand in class." She had been stirred to action as well as words in working out what she believed about being a woman and in mobilizing collective efforts. "Pleased to be women" meant pleasure in self. It was the measure of normality against the chaotic disintegration of sexual identity associated with the women's movement. "There is no way you can combat abnormality today," another woman insisted, "except," she could have added, "in fighting ERA."

Filth, abnormality, and purification were constant themes among antiratificationists whose "normality" was so threatened by feminism. The word "sex" when associated with filth and abnormality made ERA symbolic of a broad range of rhetorical possibilities, including the certification of homophile and lesbian life-styles. Proponents were perplexed at this kind of free association because ERA had nothing at all to do with homosexuality. But the role played by gay people in the politics of feminism had by 1973 identified them with ratification. This connection was explained in part through vulgarizing the language of the amendment—"on account of sex." There are two kinds of sex, explained a minister, "the kind that you are and the kind that you do." "On account of sex" therefore was interpreted to mean "on account of the manner in which and the partner with whom you have sexual relations." A college professor anticipated the possibility of a class action suit brought by gays to guarantee access to jobs, to force teaching homosexuality in schools as a legitimate alternative to heterosexuality, and to establish affirmative action guidelines. If ERA were to pass, he warned, there could be "vast sexual chaos."[15]

Feminists as Women-Who-Want-To-Be-Men

Identified with anomalous men were women who hated their traditional role and bodies so much that they had assumed masculine roles. The roles traditionally dictated by sex involved a complementary and theoretically reciprocal relationship with men who by reason of physical strength and sex were obligated to provide women with the framework within which they could behave as they were supposed to. Since many antifeminist women believed that they were being attacked for their own approximation of the feminine ideal, they imagined that those responsible for the attack were women-who-refused-to-be-women. Refusing to be a woman implied that there was only one other sex role to adopt. The perceived flight from a sacred role certified the feminist as an anomaly—a person who rejected her true "nature," an individual who refused to conform to her sexual and physical class, a disquieting contradiction of the expected, the fit, the healthy.[16] Only a few antiratificationists, however, apparently identified this anomalous role with lesbianism.

More common was the belief that ratificationists had repudiated feminine for masculine ideals, behavior, and jobs. It was behavior and attitude that elicited this response from a woman who had had a career in television: "I have found that most of the women who were pro [ERA] are competitive women. They are women who instinctively, and it's not a criticism, have an element in them that makes them by nature, and it's just a reflex, competitive." She pointed out that she was not exactly "passive" about "whether I win or lose. That's not at all what I'm saying." But feminine and masculine roles were basically different, not so much in terms of precise social position but in terms of attitude. "Well, it's more than attitude. It's just something that's instinctive, and a woman that smothers it becomes frustrated." She was unable to explain precisely what she meant—it was elusive—but her discomfort was real and she did express the resentment of women who believed that feminists were trying to masculinize them. Male competitiveness seemed to represent something less civilized than the social sensibility of women. Whether in or out of the home, a woman should "ease into what is peaceful and productive, and competing with a man" was neither. A deputy sheriff pointed out that since she could not do many of the things that men could do, the sheriff had allowed her to develop her position to the limits she herself had set. Women could manage work problems on an individual basis, she insisted, valuing personal relationships and avoiding confrontation. She did not want to compete with men; she "wasn't a man."

The belief that feminists were women-who-wanted-to-be-men was based upon how feminists themselves had actually behaved by being aggressive, demanding, and competitive. Antiratificationists believed that being a woman demanded female solidarity lest males and their values threaten, punish, or corrupt women within their own world. Female solidarity depended on women's being feminine. Women who endangered this sisterhood and behaved like men were strikebreakers of the worst sort. By organizing marches, demonstrations, and rallies, feminists had entered man's world and beat him at his own game by being fiercely competitive, that is, masculine. "Masculine" should be understood as a symbolic construction abstracted from collective observation or experience. Abstractions reflecting "force" and "power" could not be extricated by antifeminists from the biological category male, so that when these were exerted in and by the feminist movement they were identified as unnatural to women. Indeed, the masculine affect touched all feminists. To their opponents they were angry, foul mouthed, lesbian (masculine), mean, arrogant, elitist, self-centered, or aggressive—to select a few recurring adjectives.

These personal characteristics could also be read as competence, self-confidence, assertiveness, and dedication to a goal, that is, things valued by feminists. Such qualities could also be read as characteristics commonly identified with the masculine ideal, and that is the way many women read them. Antifeminist women could applaud achievements of women in competition with men, but since these women could already be successful in a man's world because of their extraordinary ability, their examples and achievements were perceived as a new norm soon to be imposed upon all women—somehow. Perhaps the Equal Rights Amendment would, like civil rights acts, impose a standard of behavior that contradicted tradition. It seemed plausible. "You Olympic swimmers," nonfeminists seemed to be saying, "should not rock the boat for the rest of us who cannot swim—especially after you have stolen our life jackets!"

The Penalty of Equal Opportunity

Antifeminists came to the conclusion that feminists (whom they equated with ratificationists) wanted—and wanted *them*—to be men partially because of their understanding of equality and individualism. It is a truism that the ideal of equality allows a broad range of interpretation. For some it is the essential but unrealized goal that justifies what they call the American dream. For others it is a utopian leveling that undervalues the able people who have contributed most to American society. Between the two poles, equality has probably been able to command greatest acceptance by being identified with work—equality of opportunity. The ideology of opportunity theoretically celebrated equality for all but begged the question as to whether equal access to a field of combat was authentic equality. It has also been discreetly vague about the rules of competition. Access has frequently been so restricted that equality of opportunity has meant very little to those who could not take advantage of it. Nonetheless, it was still the predominant form of equality widely acknowledged as normative when the Equal Rights Amendment was submitted for ratification.

Equality and work were closely identified in resurgent feminism. Despite the churning cultural experimentation of the 1960s and its romantic, self-dramatizing challenge to middle-class values, the liberation of women became linked with options for "individual self-fulfillment" expressed as work outside the home. Indeed, it is possible that the identification of work and equality in American ideology made women—both feminists and nonfeminists—believe that equality demanded paid work. Equality of opportunity for women, which had once meant equal access to the pool of good men, now meant equal access to good jobs. Moving out of the home into fulfilling work became a metaphor for feminism. Thus when feminism, ERA, work, and equality were all fused, they seemed to form a divine imperative—Go ye into the work force! Unprepared and ill disposed to compete in a world they identified with men, many women were understandably suspicious of a legal innovation that, as part of a social trend, would eventually make them, or more likely women like them, extinct. Feminists could deny that equality was sameness, but, when identified with work as self-fulfillment or as economic independence, it seemed to make demands upon women that many of them thought masculine.

Antiratificationsts resented proponents' refusal to admit that they defined themselves by the work they did outside the home. "Few of *them,*" sniffed an angry opponent, "are the wives and mothers they profess to be." She was referring to public statements by ratificationists emphasizing that they—as wives and mothers—supported ERA. Since these women were, as matters of fact, "wives and mothers," the accusation seemed to be an inability to face facts. But it was actually a political statement. If pro-ERA women were legal wives and biological mothers, they had not assumed a social role identified with this fact but had symbolically become husbands and fathers. And as one opponent pointed out, families could not have two fathers and survive. In trying to leave marriage and motherhood behind even as they professed not to do so, proponents seemed to be not only two faced but irresponsible. Their attempt to flee their female responsibility seemed to be the triumph of ideology over nature. Feminists seemed to be complaining about things that could not be changed—gender interpreted as sex—whereas being an adult was learning how to assume personal responsibility in less than perfect situations. It was the same kind of argument that Sam Ervin had used. Instead of shouting "RIGHTS!" women should assume their appropriate responsibilities and quit running to government every time they failed to get what they wanted.

Appeal to "being a woman" was not an insipid reliance on the pedestal. Many antifeminist women were clearly self-reliant but suspicious of changes that they feared would make it difficult for them to fend for themselves. As realists, they understood that their place in the work force differed from that of men; it was more difficult, less well rewarded, restricted to a special status, but shielded nonetheless from unfair male competition by protective legislation. Only assumptions about differences between men and women made those laws possible and prevented bosses from forcing women to do the same work as men. On the loading dock and assembly line, and in the secretarial chair and patrol car, they had sensed in interaction with men and bosses an ethos in which coffee breaks and the ladies' lounge symbolized the necessity of treating women differently from men. It was not necessarily what social scientists called lack of self-esteem that made them wary of competing with men but a hard-nosed evaluation of the political situation of their working lives. When, therefore, middle-class women urged them to fight for equality, they doubted that such people could really understand the implications of what they were saying because they did not know what happened in the "real" workplace. Even if many opponents of ERA had heard the argument that the amendment would extend protective legislation to men, they would not have believed it because of their own experience. "Look, Honey," one anti-ERA woman observed, "these women don't need ERA. They need a union."

For women who worked and who saw in the ideal of equality a guarantee of equal competition with men who had a natural physical advantage, ERA would infuse into the workplace a struggle that would damage their capacity to hold a position. To many of them, it must have seemed as if the life of work outside the home was a volleyball game in danger of becoming a scrimmage with the San Francisco Forty-niners. That they should have stated their position in pride of womanhood, objecting that they did not want to be men, reveals the importance they attached to their sexual-social identity and suggests something of their attitudes toward men. They had learned that the reciprocal relationship envisioned in the ideal marriage was a "pipe dream." "I am a working mother," a woman wrote to North Carolina legislators, "with two children. I am also divorced." She had been offended by the way in which "outstanding North Carolinians" had recently endorsed ratification. Governors, senators, and their wives could not "know what it means to live from payday to payday and pay the bills for Mother, Daughter and Son in a home with one parent with the cost of living going up almost daily." She began a new paragraph, "I am proud to be a woman and I feel grateful that there are already laws on the books in our state to protect women like me and I DO NOT want these laws to change."[17]

Others wrote in the same accent: "If it were possible for you to stroll through [my] plant unseen, you would find more women working harder most of the time than you did men. You would also find that the women do not have the respect of the men that they once had. It is not because of the fact that morality is any less, but because of equality. . . . If we become equal (to men), we will have to load 44 lb cases of cigarettes into box cars and also receive these same cases from conveyor belts and stack them on pallets higher than our heads. There are also many other jobs that women will have to do in a place like ours because of straight line seniority and equality." She had little faith in laws; "the laws in the 1960s did not force my ex-husband to support me and my four minor children." Here was the legitimacy of personal hardship, struggle, and accomplishment. People who worked with their hands resented middle-class women's attempts to improve their own position while putting others at risk. If women's position in the factories was not so good as it ought to be, "that is to be expected as women have to be out of work because of families." "We had better stick with these 'antiquated facts,' as Representative [Martha] Griffiths calls them than go to the other extreme." Laws that tinkered with what was to be expected of women and men were dangerous. A "professional secretary" thought that "the situation which the Women's Liberation Movement advocates would create is chaos."[18]

In urging women to assume masculine roles, feminists seemed to be "letting men off the hook"—it was a favored metaphor. In a general and diffuse way the metaphor represented a general relaxing of male responsibility within which the Equal Rights Amendment was merely one instance. The pattern seemed clear. In opposing the war in Vietnam men had run away from their responsibilities as citizens. In divorce they had run away from their responsibilities as husbands and fathers. In the *Playboy* philosophy—opponents seemed to know that the magazine supported ERA—they had sought sex without responsibility and now they sought the Equal Rights Amendment as a way of further sloughing off social and personal responsibility. Indeed, many men had actually refused to "be men," which meant both actually and symbolically that they were homosexuals; gays seemed to represent the ultimate male retreat from responsibility. The stream of consciousness seemed to be quite natural. Through ERA feminists were trying "to weaken our boys," explained one mother, "and I don't like it. Homosexuality is unnatural!" Maybe gays should be allowed to do whatever it was that they did, "but don't recognize it as right by the state!" She concluded that "it's time for people like me to stand up and yell, 'No!' "

"Letting men off the hook" usually meant releasing them from the obligation to support their families. It did proponents no good to deny charges about loss of support or explain away perceived distortions because the issue for opponents was not about results. It was about moral obligation. One student of right-wing women thinks their opposition to ERA was anger at feminists for the latter's economic independence, which undermined their own fragile privileges. In resisting ERA they were trying, she writes, to "bind men more tightly to them[selves]" in the face of what they thought were men's weak commitment to marriage and challenges to their "cultural legitimacy."[19] (Many antifeminists would observe that this focus on "economics" rather than obligation is a typical feminist obsession.) These three matters—binding, commitment, and legitimacy—are moral issues. Each is about the nature of obligation and how people ought to act with regard to their commitments to each other. And many anti-ERA women who saw so much shirking of male responsibility thought that ERA was legalizing as well as legitimating the trend. Feminists seemed to be volunteering all women—not just themselves—to take on the responsibilities of both sexes. Feminists seemed to be pushing all women to accept the draft, and jobs, and careers; to assume the support of their children; to keep their own names and to do *everything that men refused to do*. Antifeminist women understood men as well as feminists did, and they could appreciate the (moral) superiority of women; but if forced to do everything—if forced to be men—they could not very well be women, too! They would—like men—become so locked into their work that they would place their own self-fulfillment ahead of the welfare of their families—as feminists were already doing.

Forced Equality: The Power of the State

Equality, of course, meant something other than work. It was an essential part of the the permanent revolution that was America. But as the murder and assault of civil rights workers demonstrated, equality and rights were not high on everyone's list of values. Moreover, they had to be applied in specific laws, defining specific behaviors and outlining specific courses of action to achieve specific goals to create a more just society. When translated from amorphous generality into policy, "equality" and "rights" lost some of their compelling power as people began to resist—as they always had—the specific meaning of the two words. Resistance to governmental policies designed to achieve equality spread throughout the United States as school desegregation was ordered for Michigan as well as South Carolina, Los Angeles as well as Charlotte, and South Boston, Massachusetts, as well as South Boston, Virginia.

Most opponents interviewed insisted that they favored the expansion of civil rights and acknowledged the justice of racial equality. But they were visibly suspicious of the word "rights" and angry with the federal government for undertaking policies to integrate public schools and establish affirmative action guidelines. Commenting on ERA during a heated campaign in North Carolina, one person wrote, "Let HEW run the nation and be done with it!" The minister who argued that his church's tax-exempt status would be endangered if he would not perform "homosexual marriages" under ERA was thinking of the Internal Revenue Service's policy of denying such status to all-white "Christian" academies. Those who objected to the "integration" of public rest rooms were obviously spinning a parody of the way in which blacks had used the courts to "integrate" public classrooms. But probably the one activity mandated by federal authorities that most affected opponents' thoughts about equality and government was the "forced busing" of children to achieve "racial balance." In discussing ERA, opponents had insisted that government had already begun to undermine the family in the name of equality through busing, and they feared the government would come further into the family under cover of sexual equality. It is not surprising, therefore, that the audience's response should have been so positive when, in legislative hearings, a woman had pleaded with the General Assembly not to "desexegrate" us.

In a world where both the school and workplace had come under federal scrutiny in unexpected ways, no one could really guarantee that the institution in which gender roles were defined and passed on—the family—would not come under similar scrutiny. In a world where social workers, teachers, public health workers, and school bus drivers had already weakened the family, no one could really guarantee that government would not intensify social engineering under the Equal Rights Amendment. Equality had come to symbolize government manipulation of the weak, a pretext for increasing the surveillance of an interventionist state. One woman summarized the argument in one brief statement: "*Forced* busing, *forced* mixing, *forced* hiring. Now *forced* women. No thank you."[20]

The implication was clear. The conjunction of equality, government, and sex in the wording of the amendment, read in the context of feminist activism, suggested imposition by the state of values shared only by a small, well-organized minority. "You are the retainer," one assemblyman was told, "of a few ambitious women who'd like to have help from the Federal Government everytime they want a job or promotion. Typically they confuse their own unhappiness or boredom with injustice. Having learned in college and in the media the possibilities for leading a happy, intellectually happy life, they think (or at least they claim) that they are maltreated if prestige, power, and bliss are not handed them on a platter. Life isn't like that. But I suppose you're not the man to tell them so."[21] Ratificationists thus became the very opposite of what they thought themselves to be. Rather than a dedicated vanguard fighting for their sisters, they were seen as a privileged few trying to use the state to serve their own selfish ends.

In contrast, antiratificationists thought of themselves as "the people" in arms. Who better than a movement of "humble little housewives," concerned citizens, feminine women, and ordinary folk, would have a realistic sense of what equality meant? Having risen above disappointment, misfortune, and limited resources to make good lives by following a commonsense understanding of equality, they seemed to be threatened by elitist social engineering. In response, and without being ordered or manipulated, they had risen en masse to defend themselves. The moral conviction of an enraged populace was a powerful image, especially when contrasted to ERA proponents' system of paid lobbyists and antidemocratic experts. It was a contrast that could be skillfully manipulated. Phyllis Schlafly told fifteen hundred eager partisans at Raleigh's Dorton Arena that Duke University Professor William Van Alstyne had opposed a popular referendum on ERA because it would refer the issue to the "lowest common denominator." "And you, brothers and sisters," Schlafly smiled, "you know who the lowest common denominator *is!*"[22]

The positive response was a vivid expression of opponents' anger. Schlafly may have been remarkably skillful, but even she could not manufacture the response to her carefully chosen words. Nor could the politeness and gracious forebearance of opponents in interaction with intruding interviewers mask the opponents' sense of themselves as "the people" under attack. Liberals and feminists seemed to think that conservatives and nonfeminists were people who could be dispensed with, not in the same manner as Kulaks, Girondists, and Loyalists perhaps, but no less absolutely. If ratification was just one event in a long historical process, they insisted that it was one event that they could do something about. They would not be passively stuffed into the dustbin of history by elitists who could not possibly understand what "most" women wanted.

The idea of "forced women" linked with the ideal of equality, therefore, was more than a rhetorical flourish. It carried the implication of making women into something they literally and physically could not be—men. The sexual division of labor and its obligations had so framed the experience of women that it had made sexual equality appear to be dangerous. To make this point, opponents gravitated naturally to the issue of women warriors and the alarmist accusation that ratification would mean abolishing sex crimes. Treating women equally seemed to ignore the fact that women needed protection from men and was in effect to violate them. In making equality for women dangerous, the opposition diminished the ideal of equality itself. To be sure, by the mid 1970s the ideal had—despite violent and pervasive resistance—become acknowledged more or less in public discourse as the American ideal of race relations. But the fate of the Equal Rights Amendment and the rhetoric of its opposition suggest that the attack on sexual equality was also an attack on racial equality and governmental policies undertaken to guarantee it. As one lobbyist pointed out, ERA was the next issue for conservatives after they had lost on segregation. An astute politician with no liberal reputation to protect, he had worked with enough conservatives to understand the anger and frustration of men and women who resented change in race relations. "They don't talk about HEW and busing all the time," a liberal legislator pointed out, but they are "seething inside" nonetheless, and it is easy for the anger to erupt at "equality of rights" in the form of ERA.

Anger was but one manifestation of a profound resentment at being "forced" to apply the diffuse abstraction of equality. It came out in "Freudian slips," non sequiturs, and the identification of forced women with "forced" mixing—all referring to desegregation. It came out in the clever attack on "desexegration" and in the transparent images of the potty paradigm. White America had entwined sex, race, and power so long that the politics of one could be used to play the politics of the other. Private conversations suggested the confusion. One woman legislator overheard a male colleague snarl: "I ain't going to have my wife be in the bathroom with some big, black, buck!" The comment symbolized the vast difference between the world views in conflict. On one side language paid homage to equality and justice, and, on the other, it warned of personal threat within the context of folk wisdom. The fusion of sex, social role, family, and equality as issues of ultimate concern had shocked ratificationists. Instead of discussing a legal issue, opponents had wanted to talk about personal identity and sexual threat. After feminism made gender an issue in the cultural politics of the 1960s, conservatives accepted the challenge in the cultural politics of the 1970s.

Cultural Politics, Cultural Fundamentalism

The result was an explosion that gave public debate the surprising intensity of religious conflict. Opponents sometimes suggested that proponents were irreligious, which was not true, although there was indeed a religious difference between the two sides. This difference lay in the manner of religious perception. Studies of antiratificationsts in other states as well as in North Carolina reported that one of the few obvious contrasts between the two sides was that opponents were more likely to be conservative in their religious beliefs. In Massachusetts, for example, the beliefs were conservative Roman Catholicism, but in the largely Protestant South they were what critics have called "fundamentalism." The term, although used widely by scholars of American religious life and opponents of "fundamentalists," is ambiguous. It refers to a series of debates sparked in response to modernist forms of Biblical and historical criticism and the transformation of science during the nineteenth century in which certain absolutes of a faith based upon traditionalist reading of the Bible were called into question by faithful Christians as well as skeptics. Many ministers who opposed ERA identified themselves as "fundamental" Christians to identify themselves as "true" as opposed to "liberal," that is, "false" Christians.

As women's liberation struggled for form and substance, fundamentalist Protestantism was becoming increasingly popular. As mainline or liberal denominations lost more than 10 percent of their membership, conservatives calling themselves "fundamental" Christians or evangelicals or charismatics gained.[23] Churches belonging to the Southern Baptist Convention and the Assemblies of God, for example, experienced an increase of 18 and 37 percent, respectively. Conservative television ministries with hundreds of subscribing stations boasted budgets of more than $50 million per year and, more important perhaps, the number of children enrolled in Christian academies increased more than 200 percent during the decade. There, children found a strict and prayerful regimen. Condemning the 1960s' radical chic, the academies emphasized patriotism, propriety, and deference to authority. Against liberation was posed the discipline of what some scholars have called "ascetic Christianity." Against the attack on sex roles was posed the reaffirmation of femininity and traditional family relationships.

North Carolina's fundamentalist churches had been increasingly active in opposing the Equal Rights Amendment since 1973. They sent letters, petitions, and busloads of women to lobby the General Assembly. Their ministers preached: "According to God, a woman's duty is to be the keeper of a house and to bear fruit." Their women transformed lobbying into a morality play and expressed a moral economy within which they had realized a "liberation" that made feminism seem idolatrous. One of these women, an intelligent political activist and schoolteacher, tried to explain how her life had received meaning and love. In Christian marriage based on reciprocity and love, husbands and wives complemented each other, she insisted. Neither was superior or inferior. She recalled the frustration and anger she herself had felt at a particularly difficult time of her life and projected it onto feminists. She thought she could understand their anguish; but, having left despair behind, she felt more meaningfully liberated as a believer than she could ever have been as a feminist. Freed to realize God's plan for her life, she used her well-stocked library and nimble mind to teach schoolchildren, raise her daughter, and fight feminism. She disapproved of placing personal pleasure (equal rights) above obligation, responsibility, and love. Feminism represented frivolity. She laughed: "Here we've spent all these centuries trying to convince men we're not stupid and then some idiots out there burn their underclothes."

In the 1979 ratification campaign this woman, her friends, and her pastor lobbied in the best tradition of interest group politics. They were joined by churches throughout the state. ERA was a second issue; the first was governmental intervention in Christian academies. Some—but by no means all—of the academies had been established as havens from the racial desegregation of public schools. Others had been the institutional expression of a growing fundamentalist collective and political self-consciousness. Racism may have been implicit in these actions, but from the viewpoint of the men and women involved, the primary goal was that their children replicate their values. Accordingly, academies adopted curricula that emphasized "basics" and obedience, repudiating the moral relativism of progressive educators, social scientists, and "secular humanists." If not exactly Christian boot camps, academies were nonetheless guarantors of a way of life and, their founders argued, essential to their ministries and therefore protected by the First Amendment to the U.S. Constitution.[24] For the Internal Revenue Service (IRS) the issue was ambiguous; since 1970 it had been prepared to withhold tax-exempt status to schools founded to avoid court-ordered integration of public schools. But IRS went further in August 1978 when, under pressure from civil rights groups, it challenged the contention that academies were authentic expressions of religion. From Wisconsin and New York to North Carolina, "academicians" were shocked into political activity through the National Christian Action Coalition and other groups.

Academies in North Carolina were also having difficulties with a government that wanted to enforce minimal educational standards. By 1974 Christian academies had begun to complain of state regulation, fearing an attempt to impose "progressive" education and Darwinian biology—among other things—upon them. Tar Heel principals were alarmed when they were required to administer standard achievement tests, and in the fall of 1977 they refused to file reports with the state Department of Education. After the Board of Education took the schools to court, they counterattacked with a legislative package to deregulate private schools. It passed in the spring of 1979, after weeks of intense lobbying. This activism by people assumed to be apolitical had already begun to attract attention throughout the United States. Soon the Religious Right would capture—and in some cases inflame—the imagination of the press. The *Wall Street Journal* telephoned the Reverend Kent Kelly in Southern Pines to find out what was going on. Kelly later explained how he had made common cause with those lobbying against the

Equal Rights Amendment. It was natural to become involved with Mormons, Right to Lifers, and North Carolinians against ERA, Kelly explained, in a society on the verge of becoming Sodom and Gomorrah. It was natural to form coalitions against values symbolized by ERA, to resist governmental intrusion into the family as well as IRS intrusion into Christian schools.

How fundamentalists and conservatives explained their opposition provides a clue to the larger cultural dimensions thought to be involved. Theirs was a moral imagination that in its simplest form rejected the flexibility of cultural forms and resisted the moral implications of ambiguity and paradox. Opponents of ERA seemed to be bonded in a *cultural* fundamentalism. Richard Hofstadter had used the term in 1965 to explain the conjunction of religious fundamentalism and domestic anticommunism in the 1950s and 1960s. There was a "congeniality of thought," Hofstadter thought, between a fundamentalist way of thinking and right-wing politics, especially their common apocalypticism and view of a Manichaean universe. Fundamentalism was not only a historical religious movement but also a way of approaching issues in ideology and religion. Politically, Hofstadter thought fundamentalism was "expressive" rather than "programmatic" because it was interested more in symbolic victories than in "concrete" ones. But he granted one exception in which symbol and program merged: prohibition.[25] Had he been alive in 1982, he would have added to the list: abortion, praying in public schools, and the Equal Rights Amendment. In thus enlarging the list, we see that these issues, like prohibition—and, in the context of the ERA, women warriors—are more than expressive. They are also programmatic. To be sure, they are broadly (or diffusely) symbolic, but the symbolism calls upon people to act or not act in certain ways and for concrete laws.

For many opponents the Equal Rights Amendment was about behavior, not the behavior that proponents discussed in their lobbying—that by government—but the behavior of individuals. The words "on account of sex" indicated personal behavior—no matter how much proponents denied it—because so many opponents believed that biological sex dictated "normal" behavior. Like other issues on the above list, ERA *meant* personal action; if the issue was symbolic of a broad range of values, it was also programmatic. The conjunction underscores the fact that a major component of political conflict is a personal investment in programs and symbols that evoke loyalty and expectation. This investment cannot be limited simply to "interest" or "symbol," as was revealed so dramatically in the 1960s when "illegitimate" demands of authority made smoking marijuana a political act. The existence of poverty amid plenty made the clothing of poor people a political uniform. The needs of the body were politicized in the racial integration of drinking fountains, lunch counters, restaurants, hotels, and rest rooms. Finally, sexual relationships were politicized.

The Equal Rights Amendment was one of those issues in which program, symbol, and personal behavior were so nicely fused as to raise serious questions about the usefulness of understanding politics as defined only by economic interest or status. The expressiveness of fundamentalism as a movement in the early twentieth century was not merely that it purged tortured psyches of inner tension—which it may have done—but that in expressing self and community in conflict it created symbolic forms to interpret the meaning of the conflict. The expressiveness that Hofstadter found so apocalyptic may not have been purgative rage alone but also a statement about the meaning of conflictive issues. Such symbolic conflict is not easily resolved because, as the anthropologist Abner Cohen writes, "Symbolic action is almost by definition action involving the totality of the person." The apocalyptic is not the edge of sanity but evidence of the tremendous personal investment in certain symbols. Through symbols of "interpersonal relationships" interwoven with those representing the "perennial problems of human existence" (illness, pain, misfortune) we create our world. Symbolic formulation creates a "special style of life, or a special combination of a variety of symbolic formations that distinguish it from the rest of society."[26] Opponents of ERA may have been mistaken from the proponents' viewpoint, but their response was an authentic social action that helped them define themselves personally and collectively.

Once fundamentalists intruded into politics, interaction with them suggested a new political ethos, or at least the possibility of creating one. The nature of that ethos is not understood if all we can say about the people who find it meaningful is that they "interpreted the Bible literally."[27] Rather it was how they thought about the Bible, history, God, obligation, and culture that defined them—the implicit meaning behind their symbols and doctrines. The two most obvious characteristics of religious fundamentalism are radical supernaturalism and the rejection of historicism. The first maintains that the Bible is the Word of God and therefore an absolute and unconditioned authority. The virginal conception by Mary reinforces the divine character of the God-man Whose story the Bible tells, His authority undiminished by being born through natural biological processes. This way of conceiving absolute truth *beyond* the historical plane carries over into thinking about history or culture. What historians or anthropologists understand as historically produced and conditioned patterns of behavior and therefore malleable by human action may be understood by fundamentalists to be absolutely normative if thought to be similar to Biblical patterns. An approach to history that assumes no reality back of, above, or beyond the reality established by historical research is (theoretically) dispassionate about the fact of change in human relationships. The resulting corollary is a cultural, historical, and moral relativism anathema to fundamentalists.

When the sacred cosmos cannot be separated from cultural patterns, one can speak of "cultural" fundamentalism. The patterns of meaning and the institutions that sustain them—our culture—have in common with religious systems the ordering of values and the definition of meaning. Preference for traditional ways or mere resistance to change would not imply a "fundamentalist" response; but, if social or cultural change were to elicit an absolute commitment to cultural forms as sacred templates of reality, the response would be analogous to that of religious fundamentalists who, when challenged by the historicism of the nineteenth century, created an absolute, unambiguous, "inerrant" Bible. The sacredness attributed to cultural patterns can be inferred by the degree to which they are impervious to change. The most obvious example in the context of our discussion of ERA would be the feminist assault on gender and the amendment's effect on sex-neutral law. Both assault the unambiguous clarity of traditional form.

Cultural fundamentalism does not require a transcendent religious commitment, merely the belief in the absolute rightness of traditional (gender) roles as the key to social order. The belief would transform family, gender identity, and gender roles into a concrete model of what people ought to value and how they ought to behave against the defiling anomaly of equality. The concept of cultural fundamentalism is a way of underscoring the fact that the social creation of a sacred cosmos need not be limited to traditional religious institutions or ideas. Indeed, it can become so confused with patterns of culture not ordinarily thought of as religious as to create a new social-religious continuum from interaction between the subjective need for meaning and objective agents of change. Antiratificationists throughout the United States developed the theme that family and gender role were defiled by equality (anomaly). Against the threat of anomaly, they posed the purity of holiness, which requires, writes the anthropologist Mary Douglas in *Purity and Danger,* "that individuals shall conform to the class to which they belong. And holiness requires that different classes of things shall not be confused. . . . Holiness means keeping distinct the [sacred] categories of creation."[28] It means no women-who-want-to-be-men!

Cultural fundamentalism characterized opposition in Massachusetts as well as North Carolina. In the Bay State, for example, opponents also argued that ERA was a "cultural revolution" that would "restructure society" through assault on "the *immutable specialness* of the female role." The result would be "a unisex sameness." Using arguments from Sam Ervin and Phyllis Schlafly, anti-ERA Yankees said what North Carolinians said. Litanies of distorted marriages, dangerous integration, damaged privacy, and lost benefits all warned against changing the roles of mother and wife. The symbol of "forced busing" represented the mischief to be expected from an interventionist state imposing sexual equality. One never knew what to expect from government anymore—except the worst. "No one," wrote a past president of Massachusett's Essex County Women's Republican Club, "ever thought that the 14th Amendment's guarantee of 'equal protection of the laws' " would ever be interpreted to justify busing Boston public school children. "If the 14th amendment means busing, anything is in the realm of possibility for the 27th (ERA): . . . the possibilities are endless."[29] They would end in chaos.

Women throughout the nation responded to the Equal Rights Amendment as a dangerous attempt to to do the impossible. Whether encouraged by the Radical Right or manipulated by a variety of men for economic motives, women resisted the Equal Rights Amendment for reasons of their own. Their perception was not necessarily silly, precious, or *false.* Women inured by privilege to feminist analysis were joined in their opposition by working women who knew on the basis of personal experience that men could twist law to their own benefit. Women from both worlds had in common the belief that personal identity, social legitimacy, and moral order were rooted in conventional gender categories that would be subverted by "equality of rights." Wary of feminists who could not understand them, they feared the masculinity of aggressive women because it would get others into trouble. Personal experience had taught them that their primary protection was in acting like women—even in textile mills or cigarette factories. By doing this, they hoped to be treated "like women," prefecting the feminine side of the moral economy of gender relations. If men should fail to behave as they should, women still had something remaining of which men could not deprive them: their feminine understanding of who they and their daughters were, had been, and hoped to be. "A few women want to be men," a woman wrote her state senator, "don't force all of us to be one. Please!"[30]

7

Men Besieged "On Account of Sex": Bargaining in the General Assembly

ERA as an event began with women and ended with men—at least that was the way women ratificationists would remember it.

The amendment was killed in arenas where men contested with each other on a variety of issues of which the ERA was not a top priority. Ratification lost in the United States because the majorities of men favoring it were not distributed across enough states. Its loss in North Carolina can be inferred by stereotype. North Carolinians, like other Southerners, have historically defended elite male power by appeal to the myth of dependent women. Southerners are religious, and religion justifies the subjection of women. Southerners are traditionalists, and ERA challenged tradition. They are provincials opposed to cosmopolitan values and antistatists suspiciousof federal authority. Their state legislators reflect business values inimical to spending money on equality. Ratificationists understood the truth imbedded in these stereotypes but believed nonetheless that they could prevail. They failed to do so because other stereotypes—about women and gender—prevented them from doing so.

The "Honorables" of North Carolina

The North Carolina General Assembly that considered the Equal Rights Amendment in 1973 met with both the Democratic and Republican parties in disarray. For the latter, confusion resulted from its unexpectedly winning the governor's chair; Republicans—unlike Democrats—had not had seventy-two years of uninterrupted apprenticeship to help organize the executive branch. The victorious James E. Holshouser, who supported ERA, was not a forceful leader partially by temperament, partially by having no veto power, but mostly by having so few Republicans (50 in the 170-member General Assembly) to help him set a legislative agenda. The Democrats' disarray resulted from the nomination of George McGovern for president, which encouraged the party's nominees for governor and lieutenant governor to create their own organizations for the general election. Alienated from the national party, regulars throughout the state refused to help liberal Congressman Nicholas L. ("Nick") Galifianakis in his battle for the U.S. Senate against the right-wing television commentator and brand-new Republican Jesse Helms. Democratic voters as well as old-line activists apparently agreed with the slogan, "Jesse Helms—He's one of Us!" The carnage inflicted by Holshouser, Helms, and Richard Nixon left no seasoned political leader around whom to rally. The young lieutenant governor, Jim Hunt, was the highest ranking elected Democrat in the state, but to the Senate over which he presided he was an outsider. For their predicament, conservatives blamed new party rules and activists—women, blacks and young people.[1]

The ethos of the General Assembly was the one thing about which there was little confusion. It was oriented to business rather than the public trust, and that, said a political columnist for the Raleigh *News and Observer,* made it more conservative "than the progressive image of the state leads you to expect." It is, he thought, not concerned about poverty so much as whether "we live within our means." A "triple A bond rating" was more important than the health and education of children; business lobbies which reflected that standard shaped legislation because, the columnist pointed out, they controlled legislators' access to "relevant" information. The result could be impressive, such as defeat of a reform of collision insurance during the 1973 session when in one weekend banking lobbyists cornered thirty-nine of the Senate's fifty votes.[2]

A member of Holshouser's administration thought that legislators were "provincial." Only 38 percent of their constituents in 1970 lived in areas dominated by cities of more than 50,000 people; 44 percent lived in "places of less than 2,500." Critics from outside the state supported the sense of provinciality. In 1971 the Citizen's Conference on State Legislatures ranked the North Carolina General Assembly forty-seventh in the United States on professional competence, accountability, and representativeness. Chastened, legislators adopted annual sessions, electronic voting, standing committees, and research commissions, but problems of representativeness remained. In 1982 blacks and Republicans charged that North Carolina had violated the Voting Rights Act of 1965. With a population in 1982 that was 22.4 percent black, there were only 4 blacks in the 120-member House of Representatives. Arguments linking the Revolution of 1898 with contemporary politics convinced the U.S. Supreme Court that there were historical patterns of racial discrimination in North Carolina. Certain multiple-member districts "impaired," said the court, the ability of "black voters to participate equally in the political process and to elect candidates of their choice." The Court mandated selected single-member districts, but as of 1990 the legislature is still not proportionately representative of North Carolinians.[3]

The General Assembly was a part-time legislature. Members preferred to be described as "citizen-legislators," since legislating is a full-time job without any staff. The latter phrase also conveyed the concept of office as a "civic trust," but one defined by legislators' beliefs that the rules by which they have shaped their lives should be applied to legislative problems. Those can be inferred from the fact that in 1977, for example, 70 percent of them were self-employed as lawyers, farmers, realtors, and other businesspeople, while only 7 percent of the state's work force was then self-employed. Almost 100 legislators had household incomes greater than $30,000, when the average household income in the state was $9,500. Most legislators attended college (154 of 170), 115 graduated, and 76 did postgraduate work; Less than 15 percent of North Carolinians had a college degree. Most legislators were men, Democrats, and whites.[4] They governed by applying the rules of individual effort, personal independence, "sound" business practices, and tradition. The constituencies they took seriously were those representing business and financial interests and the influential people they knew in their hometowns.

The Legislative Culture of Bargain and Trade

Inhospitable as this situation seemed to be for ratification of the Equal Rights Amendment, there were other reasons to consider the General Assembly unfriendly territory. For one thing, proposed amendments to the U.S. Constitution are not the ordinary fare of state legislatures. They spend most of their time on the meat and potatoes of budgets and taxes, the bland fiber of government regulations, and the nutritious if distasteful substance of health and welfare. They always like to save something sweet for the "folks" back home. Even at the best of feasts, menus are often difficult to read, the courses not always consisting of items with which the diners are completely familiar. But with the help of legislative chefs who prepare the menu and wine-steward lobbyists, legislators get through sessions with a sense of accomplishment and pleasure—depending upon one's palate.

The meat of legislation was the budget. Public policy in the North Carolina General Assembly was shaped by fiscal conservatism and a line-item "ad hocism" that dictated low taxes and tight-fisted appropriations.[5] Between 1973 and 1982 a reluctant legislature presided over an ever-increasing and more intricate budget in the process becoming more independent of the executive branch—a shift that hurt passage of ERA. Before 1973 the governor had primary responsibility for the budget-making process. In 1971 the General Assembly created the Fiscal Research Division of the Legislative Services Office and in 1973 resolved to have annual rather than biennial sessions. Both actions gave the legislature greater control over the operation of state government as the Fiscal Research Division began to challenge the governor's figures. The appropriations process was further complicated when House Speaker Jimmy Green—a fiscal conservative—decided to withdraw representatives from joint meetings with the Senate Appropriations Committee. The committee had been appointed by Lieutenant Governor Hunt, who held a more generous attitude on public expenditures. The political differences between the two became more distinct throughout the session and were symbolized by the fact that the two houses developed budgets independently of each other.[6]

Those who wrote the budget were the most powerful men in the General Assembly. Since few legislators had the time and experience to become experts, this task was left to persons who had both because personal wealth allowed them the leisure to make lawmaking a full-time occupation. When Green became lieutenant governor and presiding officer of the Senate, he applied his economic conservatism and political savvy to appointing to the chief money committees chairs who shared his economic outlook. Because money is the root of power, and because Green's influence with the money committees was so secure, and because women proponents of ERA were so thoroughly isolated from the sources of influence in the General Assembly, they felt powerless to make a significant impact upon the most important decisions in the state. One of Governor Hunt's lobbyists for ERA pointed out that not having had the "big money" men with ratificationists in 1977 had crippled his efforts.

But something other than committee chairs was involved. Women, in studying the budget-making process and observing how legislators bargained for support, could only wonder what would have happened had pro-ERA partisans been in a position to bargain more effectively. Bargaining and trade were at the very heart of the legislative process. Moreover, the realities of government and the objectivity of the budget made the process by which it was created into a concrete example of what women did not concretely have. Bargaining "real" things like line-items provided access to power. Bargaining from positions of strength was what Trish Hunt wanted when she and her sister legislators talked about "hardball" in preparation for the lobbying campaign of 1979. She understood bargaining very well and remembered how she had stopped an attack on the Citizens' Advisory Council on the Status of Women by reminding a legislator of a favor done him, one he might not want his constituents to know about. Senator Wilma Woodard recalled a similar incident when she got funds for a study of discrimination against women in state government. But such incidents were few even after 1976, and while ERA was under consideration, women's position in the bargaining ethos was weak, symbolized by their distance from the budgetary process.

Except for discussion about how much passage of ERA would cost the state, the amendment did not fit into the mental framework of bargain and trade because it was an absolute restriction on the law. "It would have 'tied our hands,' " legislators said; freedom to maneuver was preferable to guaranteeing equality of rights. A person's commitment to "rights" would have to be taken on "faith" as he bargained from one legislative challenge to the next. One could not always apply principle—"doctrinaire equality"—to specific situations: for example, with regard to drafting women into the military. The metaphor of "tied hands" was consistent with the line-item "ad hocism" of the budgetary process and was not reserved for ERA. Environmental laws introduced in the early 1970s waxed and waned in strength as businesspeople and citizens' groups contended with each other. Insurance legislation was never decided once and for all; and education, health, and welfare programs remain continuing problems in the ebb and flow of economic prospects. If a lawmaker does not get all he or she wants at one time, there is always another opportunity or means to gain the desired end. Such caution dictates the postponement of hard issues from one session to the next while noncontroversial matters frequently receive swift resolution. Although legislators are sometimes willing to have their "hands tied," as were those who voted for ERA, the principle is usually reserved for abstractions rather than long-term public policy commitments.

How They Voted on ERA

Given the nature of the North Carolina General Assembly, it is reasonable to ask if the Equal Rights Amendment had a chance. From 1973 to 1977 the answer was yes. The momentum of ratification in other states dictated action. The cautious confidence of legislative sponsors suggested success. Activist women created a new constituency from which in 1973 Holshouser recruited women into government through his future secretary of cultural resources, Grace Rohrer. Based on the perception of female solidarity, "the numbers" seemed to stipulate ratification: 151 of 165 candidates for the General Assembly returning questionnaires said that they favored ERA. Of these, 21 were elected to the Senate, and to these were added the 5 men who signed on as sponsors in the earliest days of the session—giving proponents a majority. Senators had undoubtedly been impressed by the amendment's broad female support.[7]

Analysis of votes on ratification suggests a close enough balance between the two sides to have encouraged proponents' optimism.[8] In the Senates of 1973 and 1977 and the House of 1977, the two sides could be analyzed in terms of age, education, occupation, religion, religious activity, legislative experience, sex of children, membership in fraternal orders, and service in the military and General Assembly. In terms of age, legislators under age forty tended to vote for ratification, but above age forty there was no clear tendency. Formal education was equally insignificant, with the balance about even for high school graduates and people attending college. Those taking postgraduate work favored ERA 2:1. By occupation, small businesspeople opposed ERA, also 2:1, but business representatives divided equally. Lawyers favored ERA (3:2) as did teachers (4:1). By religion the two sides divided evenly in almost every category, one could not argue that one side was more religious than the other. Nor was legislative experience a significant indicator. When ERA was first introduced into the Senate in 1973, ten members in their fourth to tenth terms voted for ratification; twelve members in their fourth to ninth terms voted against. First and second termers divided, eleven for and ten against. The balance continued in later votes. In all other categories there seemed to be an equal balance on both sides. Of these differences, only formal education (for) and business experience (against) weighted a little one way or the other, but not to any significant degree. A realtor–insurance man like Livingston Stallings from the thinly populated east could vote for ERA as easily as George Miller, the Durham lawyer who sometimes represented insurance companies and always supported ratification. Lobbyists against ERA would still contact legisltors who studied beyond their bachelor's degree; lobbyists for ERA would still try to convert small town businessmen.

A clearer division between the two sides emerges from looking at legislators' constituencies. In the 1973 Senate seventeen of twenty-three proponents represented what the Department of Commerce calls a standard metropolitan area, and six represented small towns or rural districts. Twenty-two opponents came from small towns, although five came from cities; of the latter, four were conservative Republicans and one was a Democrat who had grown up in a small town. In the 1975 House, forty-four representatives from metropolitan areas voted for ERA; thirteen from small towns and rural districts joined them, and fifty-two representatives from small towns and rural districts opposed them; ten from urban areas joined them. Similar patterns held for the 1977 House and Senate, although ratificationists in the General Assembly had more help from small towns (twenty-seven) than their opponents did from cities (fifteen). In 1977, from fifteen districts of eastern North Carolina (Helms and George Wallace country) there were thirty-two representatives, of whom six voted for ERA.

Voting patterns cannot be interpreted, however, strictly as senators and representatives acting on behalf of their small town or city constituencies. Of the twenty-six senators who voted no in 1977, ten came from areas where members of the House voted yes. Moreover, of those twenty-six senators, four came from districts in which other senators voted yes. Earlier, in 1973, eight no senators came from districts where colleagues voted yes. In the 1975 House, twenty-one no votes came from districts where colleagues voted yes; the same held true for eighteen no votes in 1977. It should be clear, therefore, that with politicians reading their districts in diametrically opposite ways, something other than the popular will affected votes. This situation and the closeness of the division made proponents believe that they could win. The loss in the 1975 House was by five votes; a change in three votes would have meant victory—presuming other votes held. In 1977, with the House changing only a very few members, the vote in favor of ERA resulted from the proponents' gaining five votes and three opponents' arranging to be absent during the final vote.[9] There was reason to believe that ratification had a chance each time it was brought up in the sessions of 1973, 1975, and 1977. The closeness of the votes, the memory of a few conversions, and the votes of a few inner-directed individualists who confounded the logic of their background and environment kept ratificationists going.

Gaining Access and Credibility

Pro-ERA activists' major problem was how to gain access to the legislature. Access meant more than the textbook problem of finding a sponsor and writing a bill. It meant that even before introducing the bill, supporters should possess credibility. Greatest credibility lay with banking, real estate, textile, tobacco, and insurance industries. State government had credibility, as did the federal government when it applied economic pressure to desegregate higher education. The teachers' lobby had credibility when national focus on a crisis in education embarrassed miserly legislators. Credibility also came through persons who directed the budgetary process. Since credibility was affected by the source of proposed legislation, the Equal Rights Amendment was tainted almost from the very beginning because of its association with feminism. It would probably have passed if the women's movement had in fact included women whom legislators knew and with whom they were comfortable. The appeal was gone as soon as legislators discovered that women opposed ERA; they could then oppose or support ratification for the very same reason: to favor women.

Between the election of 1972 and the introduction of ratification in January 1973 assemblymen learned something else that diminished the credibility of ERA. A plausible argument could be made against it in the name of conservative values. That the argument was made by Sam Ervin probably helped settle the issue for some but there is no indication that Ervin radically changed anyone's mind. The only way Ervin could have done that was to have favored ratification; then his conservative credentials would have established the credibility required for passage. The conservative ambience even of certain legislators might have been exploited if proponents had not first expressed their goal as consistent with the values of the 1960s. Bill Whichard wondered in retrospect what would have happened if they had hired a lobbyist with impeccable conservative credentials, mentioning a redoubtable, clever, and respected former legislator. His taking on the task would have purged the proposed legislation of ideological "taint," giving ratification a better chance than that made possible by approaching the citadel of the General Assembly as part of a great crusade. But sponsors before 1979—in 1973, Whichard in the House and Charles Deane in the Senate; 1975, Herbert Hyde in the House and Whichard in the Senate; and in 1977, George Miller in the House and James B. Garrison in the Senate—were among the few authentic liberals of the General Assembly. In the 1979 Senate, ability, access, and credibility seemed to have fused when Craig Lawing became sponsor, but other factors confounded the issue.

In the 1975 Senate, ironically enough, Whichard, for all his youth and liberalism, would probably have been able to deliver. Lieutenant Governor Hunt, who favored ratification, was better able to direct legislatigve traffic after two years' experience in the president's chair, and Luther Britt, another proponent, would have chaired the Senate committee to which the ratification resolution would have been referred. According to two different reports from lobbyists Nancy Drum and Howard Twiggs there were, discounting lies, twenty solid votes for ERA in the Senate and two, possibly four (for total of twenty-seven) more who would join them.[10] But Whichard never had a chance to test his talents because ratifications did not control access in the

Sponsor of ratification bills in the N.C. House in 1973 and Senate in 1975, Willis Whichard typified male legislators whose support for equality was informed by a sense of history.

House. They went there first because they reasoned that victory in the House would be tantamount to ratification. Later they wondered if they should not have begun in more friendly territory in order to create momentum. It was a natural question, but in lawmaking, as in football, momentum is stalled by losing the "ball," and Hartwell Campbell would have stolen it.

As chair of the Committee on Constitutional Amendments, Campbell stretched out the legislative process while John Ed Davenport lobbied. Campbell knew where the votes were, and when he had what he needed—"I could count to sixty-one," he pointed out—he acted. Campbell's ability to control access lay in his position as committee chair, an appointment made by Speaker Jimmy Green. When Green was replaced in 1977 by Carl Stewart, who favored ERA, the latter's influence seemed to have helped get a slim proratification majority. The shift in the Senate from friendly to hostile territory seemed to result from the fact that Green was presiding officer in 1977—although that body had managed to defeat ERA in 1973 without him. Green always claimed to be evenhanded on the issue, and at least one anti-ERA legislator thought him far too neutral in 1975. Then, in a lapse of strategic planning, no proponents volunteered for the House Constitutional Amendments Committee, but Green balanced the committee among partisans from both sides. Later in 1977 he appointed a pro-ERA man to chair the Senate committee reviewing ratification. Green's attitude to ERA probably affected the failure of ratification, especially in 1975, but he was not the issue so much as what he symbolized—the proponents' failure to control access to the legislative process.

By 1978 ratificationists decided to gain access by entering electoral politics. This is not to say that all women who ran for the General Assembly did so as the result of working for ERA, but the women's movement had indeed been recruiting. "A Woman's Place," announced a bumper sticker, "is in the House." Nineteen seventy-eight, however, was too late because a conjunction of personal decisions, local political disputes, a national conservative trend, and the failure of Democrats to send a strong candidate against Senator Jesse Helms combined to return a more conservative General Assembly. Ironically, as women became a more effective presence in the General Assembly, ERA's support diminished. They built up seniority and earned the respect of men who had opposed them on ERA. Both Marshall Rauch and John Henley specifically singled out Senator Katherine Sebo as an outstanding example of someone who knew the rules, did her homework, pushed hard, and got most of her legislative agenda accepted. Others such as Louise Brennan, Ruth Easterling, and Carolyn Mathis of Charlotte, Helen Marvin of Gastonia, and Trish Hunt of Chapel Hill were acquiring the same reputation through experience and the amassing of "green stamps." They were learning how to play what Hunt called "hardball," but on the ERA, she and her associates lacked the numbers to "take the field" and pitch the first strikeout.

Lobbying for the People

The most direct access to the General Assembly from the outside is lobbying—and even here "insider" knowledge is important. The most effective lobbyists are those who become working colleagues of legislators whom they supply with information, technical advice, and explanations of how proposed legislation would benefit "the people of North Carolina."[11] Who "the people" are varies, of course; many presume to speak for "the people" in more sweeping terms than their special interests would suggest. *We the People of North Carolina,* for example, is the name of a magazine published by the North Carolina Citizens for Business and Industry—a group that worked unapologetically and effectively to kill a bill in the 1980s merely to study unequitable pay schedules in state government. The study would have been inconsistent with "free market forces."[12] The North Carolina Citizens' lobbyists have greater access to the General Assembly than, for example, volunteer lobbyists working on behalf of citizens who have greater claim to the title of "the people" but less capacity for maintaining a well-heeled presence in Raleigh. Even with solid financial support, however, lobbyists need to establish themselves as reliable.

Lobbyists for ERA faced the problem of reliability by claiming that they had a majority of the people with them. As we have seen, proponents based this claim on public opinion polls. There seemed to be concrete legitimization in the measurements reported by pollsters, but opponents claimed that "the people" were opposed to ERA even if the pollsters were not. Both sides tried to justify their position by constituent mail. By 1977 opponents were able to exploit direct mail appeals with the aid of conservatives' computers from outside the state, a tactic that ratificationists had neither the money nor the technology to use. The flood of mail seemed to sustain opponents' view of the popular will. Yet at the same time the way in which proponents relied on their interlocking networks of women's organizations to produce letters and cards also indicated popular support. What proponents thought was unassailable truth sustained by percentages and measurements was understood by legislators, however, as ambiguous enough to allow them some leeway in staking out their positions.

Legislators were rarely if ever forced into a position they did not want to take by "reading their mail." Charlie Winberry, Governor Hunt's liaison, who in 1977 did not believe that anyone would have lost a seat because of voting for ERA, thought that certain legislators believed that they would, and "there wasn't a lot you could do to dissuade them." The battle of constituents was not always about numbers; sometimes it rested on who one's constituents were. Someone thought that Bobby Lee Combs's most influential constituent was his mother; she was certainly more persuasive than President Carter. A politician pointed out that legislators who had little conviction one way or the other could be swayed by the women of their country club who taunted them for supporting "those crazy wimmin" in Raleigh. Some of the most effective lobbying may have been in the country clubs of towns in the eastern part of the state with the help of bourbon and the sweetly modulated tones of derisive feminine laughter. Proponents thought that the prominence of certain constituents should sway votes and recruited endorsements from them. The only prominent persons opponents could boast were Sam Ervin and former Chief Justice Susie Sharp; they probably did not need any others.

Lobbying also meant getting information to legislators. Partisans of both sides piled the desks of the General Assembly high with more material than a harassed public servant could look at, much less absorb. Ervin had already flooded the legislature with his speeches by Christmas of 1972, and he did not let up until after the legislative session of 1979. Much of the material coming into the Tar Heel State, such as statements from Representative Henry Hyde (R-Ill.) relied upon the country lawyer from Morganton who was probably the most quoted and plagiarized opponent of ratification. He was also most frequently criticized. There were more memorandums than one could keep track of explaining what was wrong with Ervin's arguments. The League of Women Voters, for example, had short point-for-point answers for every conceivable objection—serious or not; so did Common Cause; so did the Citizens' Advisory Council, the American Association of University Women, and Women's Global Ministries of the United Methodist Church. Opposition spokespersons who claimed that "there were too many unanswered questions" were either not paying attention to their mail and public testimony or trying to provide comic relief. Judging from the volume of material that erupted in Raleigh, combatants left no questions unasked or unanswered.

Information, mail, petitions, and pamphlets mean little if not backed up by personal contact. By 1977 proponents believed that they had become mistresses of the art—although they probably did not have quite the impact of those bourbon-inspired exchanges at the country club. One woman recalled how hard she had worked to establish personal relationships of trust and respect with members of the House. Senator Wilma Woodard later remembered lobbying as the beginning of her political career. She had learned a lot about the General Assembly and had gained invaluable experience in explaining a controversial issue to people who did not agree with her. She—like her activist colleagues—had been admonished "not to alienate" legislators. Since this should be a cardinal rule of any relationship, special attention to it beyond the canons of ordinary politeness is worthy of attention. That ratificationists cautioned each other to make a special effort revealed less about themselves than about assemblymen who were hypersensitive to interchange with pro-ERA women, possibly because of the ferocious scrutiny to which men had been put by feminism. Theirs was the kind of discomfort that Cornelia Jerman had found in the General Assembly while lobbying for woman suffrage.

If the major issue had been economic interest—ties with insurance companies and banks—such wariness on the part of women lobbyists would have made no sense. Conversation would have been limited to actuarial theory, economic gain, and the advantage of enlarging the labor pool of able people. But ratificationists knew male legislators better than that. They knew that the Equal Rights Amendment evoked scrutiny of relationships between men and women and sensed the discomfort of men when confronted by authoritative women other than their mothers—well, possibly also their mothers. They understood what was going on when emotions evoked by debate on ERA erupted from men who did not want women telling them how to run their lives, their business, and *their* legislature. One activist who had come from ERAmerica to help North Carolinians in 1979 throught that male legislators were responding to lobbyists as "uppity women." The observation was a feminist stereotype and ignores the fact that uppityness is the perception of an attitude not of gender: environmentalists and civil rights activists were uppity too—all of them persons stepping out of place.

The relationship betwen ratificationists and their legislative counterparts was sometimes imperfect. Far more than opponent women, those working on behalf of ERA wanted to be "in" on what was happening and called legislators to find out what was (and, more often, what was not) going on; One proponent legislator confessed sporadic resentment at Alexander Graham Bell. Ratificationists complained that legislators were not informed, not alert, not responsive, not in touch, or not committed. Some, like Wichard and Deane—sponsors in 1973—were criticized for being too young and too willing to cooperate with anti-ERA legislators. One woman accused Whichard of "having ambitions"—something not entirely unknown in politics, expressing a value judgment on and frustration with legislative culture. Her comment represented the tension between members of a political body accustomed to bargaining and participants in a movement with goals that cannot be compromised. Whichard understood this tension and responded to it with good humor and respect for the women who lobbied him, unlike the testiness of many opponent assemblymen. Gradually, proponents became sophisticated in their lobbying, learning to move comfortably in and out of legislative offices. A few, like Wilma Woodard and Ruth Cook, even won seats in the General Assembly, from which their lobbying for women's issues became even more effective.

Opponents' style differed from that of ratificationists, although Alice Wynn Gatsis had not intended it should do so. She was careful to explain to those who lobbied with her to treat assemblymen with respect—an attitude that Senator John Henley, for one, did not believe characterized proponents. Gatsis could not, however, control the actions of her allies, especially those whose religious orientation encouraged apocalyptic exaggeration. She as well as proponents could be uncomfortable with the tide of emotion unleashed in the Legislative Building—especially by those who brandished the Bible as if it were a battle flag. Yet it was the opposition—certainly *not* at Gatsis's behest—who held prayer meetings on Robert Davis's front lawn and who called him up in the middle of the night and said, "I knew you when you were little—how *could* you!?" It was the opposition who cornered legislators in their offices and berated them as homosexuals. "Lady," replied one angry legislator who later wished he had held his tongue, "I can give you the names of one hundred women in this town who can testify otherwise." It was the well-dressed and well-groomed opposition who surrounded Carl Totherow in the legislative cafeteria and loosed a torrent of emotion-charged accusations that ruined his appetite. It was not that he could not answer each objection or that the women shouted at him, but—surrounded by obviously hostile if smiling women—he was kept on the emotional defensive as television cameras recorded his discomfort for posterity. It was opponents who threatened the life of legislators, questioned their morality, and denounced their irreligion. Representative Margaret Keesee was unnerved by being surrounded in her cramped office by a group of praying women. It was opponents who claimed that the religious would oppose ERA; no proponent doubted the religiosity of the opposition, although a few commented on its modest commitment to sisterly love.[13]

Opponents invading the Legislative Building brought a jarring ferocity, lacerating ratificationists for believing marriage a "trap," children an "imposition," and sexual intercourse "rape." The women who spawned ERA were accused of raging at God for encasing them in the biology of their own bodies. The hostility behind these beliefs suffused the crowds who invaded legislators' offices and made them such a threatening presence that by 1979 citizen lobbyists seemed to be—at least to proponents and to those who shared the experiences of Keesee and Totherow—something like the plague. As opponents broke the rules of polite engagement in their crusade against deviants and infidels, legislators thought the more appropriate metaphor for the experience was a "bloodletting." Antiratification legislators who disapproved of such shenanigans warned sponsors of ERA in 1979 that if it went to the Senate floor without the votes to pass it, some of their colleagues would keep it there to allow "the rabble" to "bloddy" supporters. A few movement women wanted to go to the floor anyway in order to create a record to "use later," but sponsors thought it too cruel to submit their allies to an opposition skilled in the art of psychological intimidation. In the face of this kind of lobbying, it was difficult for pro-ERA women to forgive the accusation that ERA was done in by the "aggressive militancy of feminist leaders."

The most dramatic way of lobbying was to testify at public hearings. Unlike those held in relatively small rooms with a limited number of expert witnesses, hearings on ERA were held in the auditorium of the Legislative Building before committees of the General Assembly and perhaps seven hundred people. Most statements—with notable exceptions—were brief and resembled religious testimony rather than that of expert witnesses. Sam Ervin, however, got an entire hearing all to himself in 1975. Proponents counterbalanced Ervin with their own experts, but their arguments could not compensate for the senator's reputation. One of these, William Van Alstyne of Duke University, for example, in a later session discussed the difficult question of what the Supreme Court was likely to do about sending women into combat. His was, by the account of people themselves accustomed to public speaking, a virtuoso performance, leaving proponents stunned by the force of his clarity, reasoning, and eloquence. Legislatures were not impressed. A legislator later explained, "They weren't listening." Ervin carried more weight not because of what he said but because of who he was. Although legislators had undoubtedly heard Ervin's citation of cases before the Supreme Court demonstrating that ERA was "not needed," they could not have studied Ervin's statements carefully enough to know that he had contradicted himself.[14] They were not sufficiently interested in the issue to challenge authoritative citations by a man who was a public icon.

Thus it is difficult to determine what legislators actually learned in the hearings except that constituencies in North Carolina disagreed on what it was appropriate for women to do. Certainly it was obvious that opponents defended traditional gender roles as if the ERA would outlaw them. Except for Ervin, they did not talk about the law, and even Ervin was not so eloquent on the law as he was on gender. A professor from the University of North Carolina at Chapel Hill stepped out of his area of expertise (computer science) to say that ERA had to be rejected because men and women were psychologically different.[15] They thought differently—something few husbands in a fit of self-justifying anger would have denied. North Carolinians were different, too, the professor believed; they thought about gender differently from people living in Massachusetts and California. Although pro-ERA witnesses denied that traditional gender roles were the issue, they could not avoid the implication because they attacked the unfairness of continued sex stereotyping. As ministers, federal cabinet officers, bureaucrats, military officers, business executives, and state officials, they personified changes they approved. One anti-ERA legislator remembered proponents as highly articulate—he called them "vocal"—people apparently very much at home with expressing themselves. He was impressed, but he could not remember exactly what they had said.

In 1973 hearings undoubtedly conveyed information—about the rudeness of feminists; in 1975 they were part of the delaying tactics that defeated the amendment. By 1977 hearings had become a public ritual of symbolic expression for each side, and by 1979 they were larger-than-life events prefaced by hymns and patriotic songs. Legislators called them circuses. The display and hoopla, the sounds of people freed from ordinary constraints, the celebration of self before an approving audience—all combined to create an atmosphere of holiday in which anomaly and contradiction could sometimes provide comic relief. One eager woman wearing a STOP ERA pin almost bowled over an assemblyman as she hustled out of the men's rest room after a public hearing. "Hey," he said to himself, "I thought they were opposed to that sort of thing."

Jim Hunt (didn't do enough) and Jimmy Green (did too much)

The proponents' most important lobbyist in 1977, 1979, 1981, and 1982 was the governor. In 1973 and 1975 the governor of North Carolina, Republican Jim Holshouser, did little for ratification. Jim Hunt did much more. He had wanted to be governor for a long time, perhaps since the sixth grade.[16] Certainly he grew up to think of himself as a public servant, developing in his master's thesis at North Carolina State University what would one day become a major part of the tobacco acreage-poundage (subsidy) program of the U. S. Department of Agriculture. After earning his law degree at the University of North Carolina in Chapel Hill, he served as an agricultural adviser to the king of Nepal and in 1966 returned to his native county to practice law and politics. Elected president of the state's Young Democrats in 1968, he became assistant to the state party chair and head of a commission appointed to bring state party rules into line with reforms initiated by the national convention. He held hearings throughout the state, inviting new—young and black—people into the party. Hunt was no stranger to Democrats in 1972 when they made him lieutenant governor at the age of thirty-four. Four years later he was elected governor. Soon afterward at an academic ceremony in Chapel Hill, a distinguished historian of the South asked who that "nice young man" was who had helped him don his robes. "Oh," was the reply, "that's the governor."

Jim Hunt's power lay in his persuasiveness, patronage, and position. He had brought self-respect back to Democrats by bringing them back to the governor's chair, aided no doubt by renewed faith in a party that had had the good sense to nominate a Southerner for president. Since he brought many young activists into politics and seemed prepared to bring blacks and women into government, it is not surprising that older moderates and liberals should have thought him heir to their legacy. Democrats reluctant to identify with them nonetheless were impressed with an organization that could win. Inside state government, Hunt had the power to appoint political allies and particularly deserving people to important offices. Following four years of Republican government, as he did, he got legislation requiring state employees to have five years of service before achieving career status and made nine hundred jobs subject to patronage. He soon earned the reputation of using his appointment powers for "political purposes."

If supreme in the executive branch, the governor was exceptionally weak in his relations with the General Assembly. Unlike the other forty-nine governors in the United States, Hunt did not have the veto to facilitate achieving his goals. Rather than select one theme for his agenda, he tried to move on many fronts at once. In the first legislative session of his tenure as governor he won approval of plans to reorganize the executive branch. He got money for new reading and testing programs in the public schools, expanded facilities for mental patients, and camps for emotionally disturbed and delinquent youths. He also persuaded the General Assembly to provide representation for consumers before the state utilities commission, to establish a new Department of Crime Control and Public Safety, and to facilitate the administration of justice. His program moved easily through the legislature as befitted an assertive, young activist, although only one item was a major innovation. This was a new amendment to the state constitution to be submitted to the people to allow the lieutenant governor and governor to serve two successive four-year-terms rather than the traditional one. In retrospect it seemed that Jim Hunt could achieve anything—except ratification of the Equal Rights Amendment.

Not all Democrats applauded Hunt's activism. Jimmy Green, who was elected lieutenant governor when Hunt won the governorship, sat on his hands.[17] On the Equal Rights Amendment, as on so many other things, Green was the antithesis of Hunt—both men would probably consider this observation a compliment. No one would have mistaken Green for a "nice young man": certainly not the Japanese against whom he fought at Iwo Jima; certainly not his business competitors, one of whom tried to burn down one of his warehouses. Green learned all he needed to know about tobacco marketing to become a wealthy warehouseman with interests in several states. In 1963 he was elected to the House of Representatives, where he gained a reputation for being a "tough and feisty combatant." He also became popular enough to be elected Speaker in 1975. In contrast to the expressive and "open" Hunt, Green played politics like a "tough and calculating" poker player. Instead of trying to "play" the news media for his own advantage, as Hunt did, Green looked upon them as an enemy because of their persistent investigation into his affairs. When asked why he did not reply to his nemeses, he explained that he knew better than to get into a "contest with someone who buys his ink by the barrel."

Green's power lay in his ability to control a legislative body and his reputation for getting things done. When he organized the House in 1975, he placed supporters in strong positions while attempting to make each assignment fit the person's interests and abilities. He named Hartwell Campbell chair of the committee where he could do the most "good" on ERA, just as he assigned

Joy Johnson, a black representative from Robeson County, to chair the Human Resources Committee. He asked George Miller to chair the House Utilities Committee, not because he agreed with Miller's political philosophy but because he thought Miller was able to tackle a difficult and troublesome task. Green possessed an encyclopedic knowledge of state government, paid attention to detail, knew where every piece of legislation was, and possessed a superb sense of timing. In 1983, when he was under a great deal of personal pressure, he still knew the best time to move a bill for Senator Woodard on benefits for women in divorce proceedings. She claimed she could not have won without his help.

Whereas Hunt articulated public life in what one newsman called "glowing rhetoric" and "ambitious programs," Green growled in his "sandpaper voice" that he was not a "program" man. When he became presiding officer of the Senate it "was known" that he would be a loyal friend and ally in helping his colleagues provide all that the people of North Carolina needed. Green thought they did not need ERA. Legislators on both sides of the ratification fight credited him with defeating it in 1977, just as he had in 1975. Whereas ERA was the only issue for citizen lobbyists who loved to hate Green, it was obviously not Green's primary concern. If so, he would not have appointed Cecil Hill, a proponent, to chair the Senate Committee on Constitutional Amendments. A few ratificationists nonetheless thought that in 1977, over a three-day weekend, Green and a few lieutenants had hammered out strategy to defeat them. The story does not ring true in part because there were enough talented veteran legislators who opposed ERA to develop their own plan of action. A trio of veteran senators—Julian Allsbrook, Ollie Harris, and Harold Hardison—could hold their own very well indeed and Hardison's being chair of the Senate Appropriations Committee gave him considerable leverage.

Defeat of ratification was a defeat for the governor. Observers thought that Green used the issue to humble Hunt. Subsequently Hunt beat Green on gubernatorial succession and educational policy. His forces in the House also refused to go along with senators who voted to make the lieutenant governor chair of the State Board of Education and grant him other powers. In summing up what was considered by some to have been an "abrasive" session, observers noted that Green had "defeated" Hunt only once. Since Green had worked hard against succession, and since Hunt had won, this remarkable achievement, when contrasted with defeat of ratification, made a few women wonder if Hunt had really "done all he could have done." These doubts, also expressed after the disappointment of 1979, frustrated and angered the governor's advisers, who believed that he had done everything that he and they could think of. Hunt's successes after one notable failure evidently made him appear to be more powerful than he actually was, for he was highly motivated to win on ERA. It was the first major vote on his agenda, and he thought that he could win. He talked to senators on the telephone and at breakfast and explained the importance of equality. "It's for our daughters," he would say. "They should have as good a future as our sons." He talked to John Henley that way, and it did not work. When asked if Hunt was good one-on-one, Henley recalled that it seemed more like three-on-one. He remembered being extremely uncomfortable.

At the time Henley seemed to be caught between two equally persuasive forces. One of the governor's advisers remembered Henley as a man who thought that he would commit political suicide by voting for ratification. It was not Jimmy Green who "got to" Henley but constituents in Fayetteville. He had been "approached" by people who thought he ought to vote against ERA even though he had voted for it in 1973. He was "for equality," he said and had once thought that that was the issue. Since then, he claimed, legislative supporters of the amendment (referring to states rescinding ratification) and legal experts (of unknown reference) had changed their minds. At the time, Henley had explained that two judges had told him they opposed ERA and—respecting their "expertise"—he changed his mind. Ratification would have had too many "ramifications." "What ramifications?" "I wouldn't want to emphasize one over another." He had forgotten the intense lobbying by a strong STOP ERA cadre, based, said Sonia Johnson, a Mormon pro-ERA activist, on a local stake of the Church of Jesus Christ of Latter Day Saints. STOP ERA women were more respectful of legislators than were pro-ERA women, Henley recalled; in appreciation they had presented him with a dogwood tree. Proponents thought that the dogwood was also called a "Judas tree." Henley laughed.

Hunt's powers of persuasion may have been so great that, when he turned to the rest of his program, he found he had created a reservoir of uneasiness. Certainly his representatives played on this serious act of "disloyalty" to recruit support for the referendum on gubernatorial succession. Ratificationists were frequently frustrated, however, by Hunt's continued association with men who had broken with him on ERA. They thought that he should not be loyal to them after they had been disloyal. That he had to rely upon them for the smooth functioning of government and the success of his program seemed irrelevant; failing to cut off friends when they do not agree with you seemed peculiar. But the vote on the Equal Rights Amendment was only one vote to the governor, his associates, and members of the General Assembly; it was not an entire agenda. "Look," explained a seasoned lobbyist, "we need Henley in the Senate regardless of whether he is for ERA!" Hunt's priorities other than the Equal Rights Amendment meant, of course, a less than total commitment; but "total commitment" rarely gains political allies for more than one issue in legislative institutions if the "totality" of commitment demands excommunication. Political affability requires public concession of pure motives and reasonableness to people with whom one disagrees because in legislative culture there is always the possibility that tomorrow opponents may become allies. Legislators' letters to constituents reflected this ethos: "Even though we do not agree on this issue, I am certain we agree on many more . . ."

Hunt's ability to achieve ratification had other limitations as well. Even in his patronage power, he could not act as if ERA were the only issue that mattered. Charlie Winberry confessed with characteristic facetiousness that Hunt could not pass ERA because there just were not enough lawyers who wanted to be judges. Hunt was twice restricted on this point because he claimed to favor selection by merit. He contemplated making other appointments in exchange for votes in both 1977 and 1979 but found them inappropriate because he could not campaign on improving commissions only to appoint to them people in whom he did not have the greastest confidence. This connection may have been what one senator meant when he said that the "soft noes wanted too much." Other limitations on Hunt's power lay in the fact that for some legislators he was not as "authoritative" as former Chief Justice Susie Sharp. "Susie's a caller," observed a lobbyist. He could imagine a conversation with her senator: Are you going to pay attention to your constituents [Sharp] or that young whippersnapper over there who calls himself governor? In the end, the governor's liaison said, a major problem was that the General Assembly did not know that it was supposed to roll over and play dead now that a Democrat was governor.

But Hunt was always willing to do "what he could." When the gentleman's agreement of 1981 was signed, the governor canceled plans to travel to Washington, D.C., for a meeting of the Democratic National Committee. In late 1981, with the votes of the Senate still against them, leading proponents of ERA met with Hunt to plan one last time. The Democratic National Committee had made ERA a party issue, and, as a man whose vision and ambition embraced the nation as well as North Carolina, he had to act. If he could pull this one off he would have the reputation of being a "new face" and miracle worker. The Democratic party needed both. Hunt did everything he had done before—this time he did it in public. When the action moved from the governor's office to the Senate, it crashed into the reality of legislative elections. Voters had a part in defeating ERA—something ratificationists sometimes forgot. One woman politician, explaining that Governor Hunt failed to find votes in 1979 because his heart had not been in it, added as an afterthought, "We didn't do a good job in the [1978] elections."

Ratificationists did not do a good job in 1980, either. The most important race had been that for lieutenant governor between Green and House Speaker Carl Stewart of Gastonia in the Democratic primary. Stewart was for ERA—like Hunt; indeed, he was all too very much like Hunt to suit Green, who called him a "big spender" and a "liberal." Unlike the incumbent who knew all the reasons why government could not act on problems, Stewart was more than willing to seek solutions through public policy. He was also comfortable with a principled commitment to equality and had the aggressive support of ratificationist women throughout the state who gave time, money, brains, and legwork to defeating Jimmy Green. They did everything right, but Stewart lost. In the legislative elections, too, although proponents were more active than ever before, they did not win.

In the summer of 1977 a consultant for ERAmerica had gone through the entire list of North Carolina legislators with a well-informed political analyst. He targeted each possible change and even suggested a few excellent pro-

ERA people who would run well. But you have to run good people, he insisted; they have to be strong candidates. Little came of this discussion. After the 1978 election, nineteen no votes returned to the Senate; three new no votes won, and five no votes replaced yes votes. Fourteen yeses returned; five yeses replaced yeses; three yeses replaced noes, and one no switched—or would have if there had been a vote. The result would have been a 23–27 vote against ratification. The Senate returned in 1980 was similar to that sent in 1978—especially with regard to position on ratification—and killed ERA in 1982 by the same margin. The victors who sat in the state Senate of 1981–82 when an ultraconservative president was "booting" his program for the 1980s made passage of ERA in June 1982 impossible.

ERA: Principle, "Sex," and Symbolism

Reality lay not in lost elections alone. It lay also in the fact that ERA was not like other issues—almost everyone involved in the ratification process agreed on that. For one thing, it was a matter of principle. When ERA was first introduced into the General Assembly, it came as the principle of sexual equality. That was the way movement women understood it; that was the way Bill Whichard and Charles Deane understood it; that was what Deane and his colleagues in the Senate had said. Opponent women objected to the principle by appeal to their womanhood, a principle much less abstract than "equality." The General Assembly thus found itself with two lobbying groups that had symbolically invested the issue with their own collective identities. Since ERA symbolized "women" for each side, there could be no compromise, a concept foreign to legislative culture.

As a result, ERA had the capacity for creating conflicts that, by 1979, affected the smooth working of the General Assembly. It set at odds people who might otherwise have worked well together. Legislators who voted against ERA resented ratificationists' labeling their vote as being either against women or the ideal of sexual equality. The emotional intensity of the two sides seemed to be completely out of place, and lawmakers held proponents of ERA responsible for that feeling because it was they who "kept bringing the matter up." Many legislators were surprised and angry at the way in which partisans—some of them inside of the General Assembly but most of them outside—made a vote on the Equal Rights Amendment the sole standard of judgment on a total record of public service. They also were offended at the hostility. "They would not respect the other person's opinions," complained one man, referring to proponents.

The emotions evoked reminded a lobbyist of debates over the consumption of whiskey. "People are hypocrites on it," he said, and referred to some of the hard drinkers in the General Assembly who had opposed county option on serving liquor-by-the-drink. They "posed" on ERA because, like the drinking of whiskey, it was something about which voters seemed to be irrational, and they saw in the perceived irrationality an easily satisfied constituency. But

ERA was more complicated than whiskey because it was about sex and shrieked danger in response to scarcely shaped images of anomalous women undermining the verities of the past. Emotions instead of interest seemed to shape judgment. Senator Jack Rhyne's response to inquiries from ratificationists in 1972 was heated. "If some women want to be treated like men, then that's their bag! I won't vote for [ERA] regardless of the pressure that may be brought to bear. I suggest you concentrate your energies on someone with less sense and less compassion for Women!"[18] His anger at the challenge to social roles based on gender betrayed a defensiveness that shaped his self-justifying denial that ERA would benefit normal women. Another man was more to the point in his opposition in 1982 when he told a NOW interviewer, "Men are losing their manhood."

The word "sex" in the amendment elicited responses similar to those of antisuffrage legislators sixty years earlier. A representative from the eastern part of the state told a reporter, "I got a letter tellin' me to vote for the ERA and quit foolin' around. Evidently, the lady didn't know I was seventy-two years old." The comment captured the way in which many legislators approached even public matters relating to women if given the tantalizing cue, "on account of sex." The assemblyman belittled a constitutional issue as if it were nothing more substantial than the sexual nostalgia of aging men. That he should have thought of sexual intercourse when accused of not paying attention to business on a women's issue suggests something far more serious than bad taste. The woman's earnestness was "funny"; her choice of words was "naïve" enough to be the object of ridicule in rejecting her petition as a "joke." The legislator was like many of his male colleagues in the General Assembly. It was not his being Southern that made him say what he said, but his personal blending of experience, politics, humor, and what he took for masculinity. Attempts at humor reveal what people value and convey the sense of what is appropriate to an approving audience that shares the values of the "wit." It will understand implicit meanings as incongruous, ambiguous, naïve, or unexpected enough to punctuate a train of thought with laughter. ERA entered the General Assembly as incongruous, ambiguous, naïve, and unexpected. For men not accustomed to dealing with women as equals, ERA was something of a joke because "sex"—especially when it is anomalous or exaggerated—is something of a joke in male culture. The legislator thought that what he said was funny because the only audience that counted in talking about ERA was male—or perhaps women who thought "foolin' around" was funny.

Such remarks suggested that "sex"—as in "on account of sex"—had been transformed from a noun into a verb. It was a change that had consequences for the way in which public debate about ERA was conducted. In thinking about what he had been up against, George Miller ruminated about "sex." It was a "dirty word" to too many people, and he pulled letters out of his file to prove it. In them were denunciations of disorder, license, feminism, irresponsibility—and Miller's "manhood." He chuckled but immediately grew somber upon remembering a colleague who seemed to personify the problem. Whether discussing ERA or pornography, the taint of "sex" obsessed her; she believed pornography was evil because it was linked to sex rather than to exploitation or violence; the association made ERA pornographic, too. The connection was made all too clear during a demonstration three years later. Then, momentarily perplexed feminists confronted a group of anti-ERA teenage girls who—in happy and zealous insolence—shouted what they thought was an obscenity: "ERA sucks!"

For most legislators opposed to ERA, probably not sex, nor the amendment, nor feminists were pornographic. But Miller certainly believed that the word "sex" in the amendment was freighted with meanings which short-circuited political discourse. This result was evident among many male legislators who all too frequently did not know how to relate to women as colleagues. Reporters and women legislators observed that the social life of legislators in session suggested that females were part of the accoutrements of power. Interaction between male legislators and women who were not their colleagues was that of "superficial flirtation" or the casual categorization of women by calling out "hello, sweetheart." Representative Margaret Keesee-Forrester remembered one prankster who hummed, "I'm in the mood for Love" whenever she found herself near him at cocktail parties. One evening she turned around and in effect challenged him to "Put up or shut up." He changed his tune. Other women had had similar experiences. Jessie Rae Scott was angry with legislators who vowed to protect women from ERA but not the indignity of their own roving hands. Women legislators were no happier with George Marion's referring to them as "our girls."[19]

Such tasteless comments did not necessarily mean that the General Assembly was more "sexist" than other North Carolina institutions, although the experience of women legislators made them believe that it was. "I've never felt sex discrimination as I feel it as a member of this body," confessed Representative Louise Brennan.[20] And after many hours of lobbying and partying with male legislators, two homemakers from Winston-Salem observed that politics must be "smut." There was, they recalled, "an awful lot of talk about sex." The men seemed preoccupied with it, having apparently divided women into the categories of mother, sister, daughter, wife, and prey. In such an atmosphere, where political ego and male identity seemed so perfectly fused, it was natural to exclude women legislators from the informal interaction through which political bargaining was accomplished. The assemblymen would not have invited their wives, daughters, mothers, or prey either. That public issues were male issues meant that many legislators were unable to acknowledge the importance of public policy issues raised by women or on behalf of women.

Since women were identified with sex and since sex—rather than power—was thought by many male legislators to be the substance of rape, they naturally thought of that crime when ERA came up. That the amendment would mandate scrutinizing laws relating to rape somehow became evidence that it was dangerous, as Sam Ervin wad warned. That scrutiny would harm women was an automatic conclusion to people believing that laws designed to protect women actually did so. Believing that their understanding of "sex" was challenged by the amendment, opponents could single out "rape" as a test of appropriateness. If ERA required the rewriting of rape laws, they wanted nothing to do with it. This seemed to be the way in which Representative John Ed Davenport conceived of the problem. He wrote Ervin, "I have not done a great deal of research, but I cannot see how our present rape legislation . . . can be held constitutional, if the amendment passes."[21] Legal distinctions were justified by sexual distinctions. Women were "sex" first and "individuals" second. To think of women in any other way than personifying the social meaning of sex was to violate them.

Such thinking suggested that the stereotypes at which the amendment was aimed were shared by assemblymen called upon to pass it. This connection would affect the way in which legislators read "evidence." In 1975, for example, when Davenport worked so hard against ERA, he gave each member of the General Assembly copies of a report to the Virginia House of Delegates that he thought vindicated opposition. To a proponent, however, the report justified ratification. It found that ERA would impose strict standards of scrutiny to legal classification by sex; that sex-neutral language would replace that unable to pass new standards; that women's property rights would be secured; that courts would have to "accept a pattern of family living in which husband and wife are equal partners"; and that law would impose certain unspecified obligations on women.[22] Such changes would mean a "good deal of administrative work and organizational modification." To men whose values were measured in money and for whom government was a business rather than a guarantor of justice, the conclusion was negative.

With stereotypes so ingrained, it is not surprising that proponents should have found the image of women warriors so difficult to overcome. "Sex" meant male obligation as well as predation, and even men who voted for the amendment confessed dismay at the issue of draft and combat, even when they agreed with William Van Alstyne. Certain men in good conscience could not vote for ratification because they could not place women in harm's way in a world menaced by a "communist threat." They believed this even before the Carter administration suggested that women be registered for a possible draft in the wake of the Soviet invasion of Afghanistan. The argument that these two events affected the final vote on ERA was linked with the decision of the Supreme Court in *Rostker* v. *Goldberg* (1981) that the sexes were not similarly situated for purposes of military service. But none of these events affected the vote in North Carolina; no evidence links any of them with hardening positions in North Carolina. Louis Harris polls there in 1979 reported that 56 percent of those polled thought ERA meant women would "be drafted to serve in combat," whereas only 47 percent thought so in 1982. As for legislators, two no votes were ready to switch on this issue in the late spring of 1982.[23]

ERA was also unlike other issues in its symbolism. Hartwell Campbell, a graduate of Yale University's Divinity School and a former Baptist clergyman, captured that aspect of the controversy when he justified opposition by saying that women in North Carolina identified more with Anita Bryant than Gloria Steinem.[24] The Anita Bryant he had in mind was not the woman whose husband had victimized and exploited her for his own purposes until she divorced him. Campbell meant the other Anita Bryant, whose unhappiness was hidden behind the facade of domesticity, traditional values, and conservative politics. What ERA had to do with either Bryant and the New York feminist was as obscure as it was clear. It was obscure because equality of rights should have appealed to Bryant as much as to Steinem, to women who pushed orange juice as much as those who savored wine, to conservative Christians as much as to liberal Jews. It was obscure because the amendment was proposed to guarantee that both Anita Bryant and Gloria Steinem could appear before the law as individuals and not as representatives of a class.

At the same time, Campbell's statement was immediately clear. He meant to divide women along stereotypes (symbols) constructed by opponents of feminism. Steinem personified feminists in her dramatic fight for women's rights. She along with many other feminists seemed to have made all men the enemy of all women; conceding exceptions did not mitigate the style. Steinem had supported a long and complicated feminist agenda; identifying her with the most national, visible, ubiquitous, and politically coordinated feminist issue—ERA—indicated that antifeminists used ratification as an opportunity to hold a referendum on feminism. As the personification of ERA, Steinem represented a confrontational style and conveyed anger with men and therefore women who enjoyed their relationships with men. Identifying ERA with feminism, and ultimately *as* feminism, was easy. Anita Bryant's public campaign against rights for homosexuals in Miami symbolized "normality" and celebrated traditional relations between the sexes as a political statement. She was the perfect antifeminist—for a while.

The symbolic range of the Equal Rights Amendment was not limited to feminism because it was conceived as a human rights issue. Proponents mused about the meaning of ratification for securing the nation's commitment to equality. Indeed, activists whose public identities were caught up in ERA thought that the issue was important not because of any anticipated widespread gains for women—they believed that those would be hard to come by in any case. Women would still have to go to court; they would still have to battle against economic and cultural constraints. But if ERA passed, the principle of equality would have been made that much more secure for everyone as a reaffirmation of "our fundamental principles" of equality before the law against "senseless, arbitrary, capricious, unreasonable discrimination." The total impact of passing ERA included the law, to be sure—although many supporters thought that the impact would be as restricted as lobbyists had said in 1973—but it was, said one legislator, "almost entirely a symbolic issue and not substantive. But," he hastened to add, "symbols are important, we should not regard them lightly."

So inclusive was ERA in what it could be made to symbolize that one proponent called it a "garbage pail" issue. Some opponents included just about everything they were against: the Department of Health, Education, and Welfare, the Supreme Court, busing, civil rights, environmental protection, and the sexual revolution. "You know," observed George Miller, "we are still feeling the effect of integration; that has taken its toll in all this, too. It's not an outward thing, but . . . an inward thing. The role that it plays here is insidious." As one conservative legislator blurted out in free association, trying to explain his opposition to ERA, "I don't want another HEW hassle like the one with UNC—ordering us to do this 'n' that." He was referring to actions by the Department of Health, Education, and Welfare to force the University of North Carolina to integrate whites and blacks more effectively across the entire system. The orders seemed arbitrary and unreasonable. Life by the application of strict rules, he insisted, was unfair because it deprived people (like him) who really knew "the situation" of the right to use their discretion.

Playing Politics in the General Assembly

If ERA was a peculiar kind of issue, it was also enough like other issues that it could be exploited for political purposes. Indeed, proponent legislators believed that if the General Assembly had judged ratification on its merits hermetically sealed from "politics," most assemblymen (some estimated 60 percent) would have favored it. Since not even judges' chambers can be sealed in such a fashion, the observation seemed strange coming from politicians. But the point was that politicians make decisions in terms of the people they need to be associated with to attain their goals. Thus, when ERA came into the General Assembly in 1973, many people were undecided as to how it fit their own agendas. It attracted sponsors who would later vote against it because at first it seemed politically neutral enough to attract both conservatives and liberals. But sides began to harden when the opposition of women, the "acting out" of feminists, and the lobbying of a few opponents "delegitimated" it.

When ERA became controversial, it lost support from legislators who did not care about it one way or the other or who thought it was not important. Beth McAllister knew very well that many legislators simply could not absorb the issue into their political world view. It did not make sense to them because they did not see how, in a world where women had accomplished so much, they were in effect not already treated as well as men. Thus many votes on the issue were "soft," a fact that favored opponents because perceived change requires greater conviction than simply going along with things as they are.

This is not to say that men who voted against ERA had no convictions or were necessarily impervious to women's rights; Marshall Rauch was a case in point. He voted for ERA in the 1973 Senate and against it in 1977. Conversation with the Gastonia manufacturer revealed a man who was not a committed feminist, to be sure, but who was as proud of his daughters (one an account executive in New York, the other a state tennis champion) as well as his sons. As he talked about ERA, the conventional public wisdom of why he had changed his mind on ratification became an unspoken part of his own explanation. "Look," he insisted, "this is not about women . . . but political power." He did not mean powerful women; the feminist accusation that opponent men were afraid of powerful women was ludicrous when applied to Rauch. He meant power blocs within the Senate. On women, he argued that the passage of ERA would have made no difference in their status. The women's movement would continue to achieve its goals with or without ratification. We have already changed many laws discriminating against women, he insisted; "we" will pass more. On power, he talked about legislators' need to identify with colleagues who can help you "get things done." Once a commitment is made, you have to stick with your people, he pointed out.

Once Rauch had made his political bed, once convinced that ERA—not equality, not women, but ERA—was "not important," he could use the issue to jockey for position in the Senate. Perhaps, too, he could play a little legislative "tag" to the point of helping Craig Lawing believe the latter could win in 1979. Rauch emphasized that legislative politics was the extension of ego—it was a contest, perhaps a game. As in athletic contests, it was not the size or shape of the ball or what you did with it that counted, but how effectively you used it against the opposition. Rauch did not play on Jim Hunt's team. As lieutenant governor, Hunt had not rewarded the assertive and ambitious Gastonian with an important committee chair. The political scuttlebutt was that he had been offended needlessly, something, recalled Charlie Winberry four years later, that had been a mistake.

Rauch represented one way legislators could approach ERA once it became part of a pattern defining political allegiances. To be sure, not all Jimmy Green's friends voted against ERA any more than all of Hunt's friends voted for it. But it made little sense for Green to support an issue that gradually became identified with an activist role for government. And, as a presiding officer who had little patience with anything that interrupted the smooth functioning of the legislative process, he eventually became impatient with it. In 1979 he told one reporter early in the session, "All I want is a quick vote so we can stop all the hassle, get it over with and debate more important issues."[25] Even proponents had agreed by the end of the bitter fight in 1977. One admitted to voting against reconsideration then because, "I was tired of it—I think the whole Senate was tired of it. We haven't done anything except ERA since we've been here." The issue had become a nuisance.

Legislative opposition to ERA was linked also to a pervasive wariness of what ratification might bring from the federal government. The wariness could be laid to Southern white racism and left at that, but if such an explanation did not apply for understanding slaveholders' ideology, it scarcely suffices for people who want to know how—within the political culture of the late twentieth century—women's issues are likely to be understood by people to whom they are submitted for legislative action. There was a get-the-government-off-our-backs attitude that has as much saliency outside the South as within it. When opponent legislators told constituents that "ERA was not good for the state," what they meant is suggested by their connecting the amendment to the

Supreme Court and the federal bureaucracy. In 1975 they mentioned Congress's power to enforce the amendment in conjunction with then-current efforts by federal authorities to get Tar Heel educators to put a new school of veterinary medicine on a "predominantly" black campus. Sam Ervin had already observed that he did not want "Joe Califano telling the people of North Carolina how they're going to run their educational institutions." The linkage came immediately to mind like a motor reaction in the nervous system—as Hartwell Campbell indicated in the House during his speech against ratification. So did the notion of court-ordered busing, which represented the imposition by rules without accountability to the (white) people of the state. Even an anti-ERA black legislator, Peter W. Hairston, could emphasize "the danger of federal court interference in state and local affairs."[26]

Interference, intervention, intrusion—the words tumbled out naturally. When asked about the most significant changes in the General Assembly over the past generation, many legislators and observers mentioned one man–one vote. They were angry with federal restrictions on their right to rule within their own jurisdiction. In 1971 the Supreme Court in *Swann* v. *Charlotte-Mecklenburg School Board* broadened the powers of the federal courts to eliminate all vestiges of "state-imposed segregation." In 1973 the Supreme Court's decision on a woman's right to abortion negated the North Carolina legislature's own considerable wrestling with the issue. In 1975 HEW attempted to affect legislative appropriations to the University of North Carolina. By 1979 HEW could force the legislature to appropriate money ($40,000,000) for projects it had not initiated to remedy (to a limited extent) segregation in higher education.[27] To businessmen and like-minded lawyers who resented activist judges, affirmative action programs, the Equal Employment Opportunity Commission, federal environmental impact statements, and investigations by the Occupational Safety and Health Administration, a vote against the Equal Rights Amendment seemed a modest but necessary retaliation.

Was this racism? Undoubtedly—especially if racism is understood as institutionalized rather than personal. Legislators shared attitudes that characterized opposition throughout the state and nation. Sam Ervin's position on civil rights—while not shared by men like George Miller, Marshall Rauch, Bill Whichard, or Jim Hunt in North Carolina—was popular among conservative Democrats there who believed that governmental action on behalf of one group against another was contrary to the tenets of a "free" society. This attitude is as resonant with businessmen's values as it is racism and betokens a treasuring of one's own prerogatives against challenges from others. But the perception of challenge is often colored by political agendas. Many North Carolina legislators—undoubtedly most of those who voted for ERA—simply did not believe their "power" to be threatened by federal agencies or court orders. They did not always agree with decisions made by these bodies but trusted them nonetheless to act reasonably, most of the time, and for the best interests of most of the people.

Governing, to many opposition assemblymen, meant making only as many decisions as local elites thought necessary to allow individual enterprise a free hand in pursuing private goals. They wanted agencies from outside their localities to have as little say as possible in the regulation of their lives unless those agencies were accountable to them. They did not want to be accountable to agencies they could not control. Federal agencies and courts were beyond their control, and the constituencies of those bodies were outside legislators' jurisdiction with interests dissimilar, theoretically, to the interests of most of their constituents (who counted). They warmed to Sam Ervin's sarcastic comments on legislators who did not have enough confidence in their own ability to govern to enable them to resist encroachments from Washington. They were alarmed at the possibility that ERA would move responsibility for family law from their jurisdiction to the federal government—even though they were not particularly interested in family law. In justifying themselves, they claimed to represent the interests of "the people" against intrusion by outsiders not accountable to them. That the statements were self-serving and clothed in the mythological argot of elites resisting accountability to blacks, women, environmentalists, workers, consumers, and other outsiders does not detract from their credibility. The ideology of resistance to government, especially the federal government, was rooted deep in North Carolina soil, and Judson D. De Ramus, Jr., knew that. In a statement made in 1977, the proponent from Winston-Salem appealed to that ideology on behalf of ERA. The amendment would give to individuals "the free enterprise and exercise of their individual talents and abilities without interference from the General Assembly" or other state officials. It also restricts, he insisted, the federal government.[28] True enough, perhaps; he knew the catch words. But most opponents did not want anyone helping those individuals—unless it was the North Carolina legislature. That concession gave ratificationists an opening.

Bargaining for Women

Having said that there were things that the North Carolina General Assembly itself can do to meet specific "legitimate" grievances of women, opponents were challenged to be as good as their word. Senator Katherine Sebo was especially active in getting support for her bills to neutralize "sexist" language in the North Carolina code. Other legislators caught the implications of her concrete example and submitted their own emendations. In 1975 she was successful in sponsoring a resolution mandating the Legislative Research Commission of the General Assembly to establish a committee on sex discrimination to review the North Carolina code. The committee, which included legislators on both sides of the ERA debate, reported in 1977 and suggested legislation relieving legal discrimination against women in criminal, employment, tax, property, and family law. It could come to no conclusions on laws covering insurance; the matter was too complicated—and too much at odds with the interests of the insurance industry.[29] Not all of the committee's suggestions were enacted; but by 1979 thirty-three pieces of legislation dealing with various forms of sex discrimination had been discussed in the General Assembly, and nineteen laws had been enacted and eight amended to delete sexist language.

Sebo had proposed her committee for two reasons. Previously she and other proponents of the Equal Rights Amendment had opposed legislation on specific matters that did not include comprehensive revision of the law code. These specific cases had been seen as sops or cosmetic legislation; some were similar to Carl Hayden's reservations and Sam Ervin's exemptions. But now she concluded that the principle might not be detrimental to women's rights—given the political realities of the North Carolina General Assembly. Indeed, she thought that a large number of specific revisions would benefit women with respect to the areas affected. Moreover, she hoped that the long-term impact of a comprehensive revision of state law might eventually lead to ratification of ERA.

This approach was appealing to many men who had voted against ERA. They could demonstrate that their protestations of being for the goals of ERA without submitting North Carolinians to federal scrutiny were true. They could try to mollify the anger of disappointed ratificationists and avoid the penalties of being on the wrong side of a no-compromise issue. You see, they could—and some did—say, we are for the principle of sexual equality and are willing to apply it to specific issues consistent with the equal protection clause of the Fourteenth Amendment. They could improve the status of women without raising the ire of antifeminist women, and in doing so they would have adhered to their guiding principle of bargain and trade. I cannot be with you all of the time, they said; but I can be on certain specific and important matters. Hartwell Campbell, after he had left the General Assembly, insisted that the trial-and-error-method of piecemeal legislation and the case law being developed in the courts on such issues as sexual harassment in the workplace vindicated the defeat of ERA. The legal process was gradually improving women's position, he thought. Hearing of his argument, one ratificationist observed, "No thanks to *him*!"

The range of legislation, although relatively broad, was not an effusion of legislative goodwill: supporters worked hard to get the little they got. Their accomplishments included laws on sexual assault, displaced homemakers, funding of abortions for poor women, domestic violence, tenancy by the entirety, and the equitable distribution of marital property—with a bias in favor of equality. The law on domestic violence was important not only in itself but also for what it reveals about the way in which an issue important to women's rights can receive support among conservatives. Its passage in 1979 as one of several bills dealing with the family unit was a model of efficiency. Its intent was to promote a "confident rather than tentative" response to domestic violence by law enforcement officers. It provided "a civil remedy to permit victims of domestic violence to seek emergency relief on their own initiative and a criminal sanction" in response to family violence. After "impassioned floor debate," unmarried couples were included as well as married heterosexual couples living together.[30] The passion recalled George Miller's characterization of the way in which some Tar Heels conceived of normality: "Are you 'American,' or for this livin' together stuff?"

The bill was the child of Miriam Dorsey, chair of the North Carolina Council on the Status of Women, and Mary Susan Parnell, her assistant, who entered law school after her experience as a lobbyist. They began their campaign in 1978 with a mass mailing on domestic violence, emphasizing recent hearings on crime in which the issue was raised. It was a "crime" issue as well as a "woman's" issue, Parnell pointed out. Responses to the mass mailing convinced the governor to visit shelters for battered women; he was sobered by his experience, blurting out that something had to be done. Dorsey then contacted law enforcement officers throughout the state, recruited supporters, and had the attorney general, a former aide to Sam Ervin, write the bill. In the process of presenting it to the Judiciary [I] Committee of the Senate, supporters managed an important coup by recruiting its chair, one of the leading conservatives of the Senate, I. Beverly Lake, Jr., to be floor manager of their bill. (Lake himself would run against Hunt for governor in the 1980 election.) Asked about his commitment to the bill, he answered that his first response was to reject it because of the people who sponsored it. They were the same people who had sponsored ERA, and he was reluctant to join forces with them. "But," he observed, "there aren't very many battered men." He held hearings on the bill—but not in the auditorium. The attorney general came to testify on behalf of the bill; so did the chief of Greensboro's police and president of the Police Executive Association; so did representatives of other police departments across the state and the secretary of the Department of Crime Control and Public Safety, Herbert Hyde. There were questions about the mechanics of the bill, its wording, its implications, limitations, dangers, and advantages. It then received a favorable report and passed the General Assembly by votes of 37–5 in the Senate and 104–12 in the House.

The domestic violence bill passed for reasons that distinguished it from the Equal Rights Amendment. First of all, it was limited in scope and definite in focus, much more concrete and restricted than the abstract and broad range of ERA. Second, authorities whose expertise male legislators would respect—law enforcement officials—testified on technical matters that elicited questions and engaged the interest of committee members. Third, the hearings were low key and, although public by law, not attended by many people. Those who did attend did not oppose the bill; it was not controversial. Fourth, lobbyists created a constituency to support the bill based on authoritative action by a distinguished jurist and commitments made by the governor. Although feminists initiated the bill, it did not come into the General Assembly as a feminist measure and thus avoided the confrontational ambience attributed by many assemblymen to the women's movement. The bill also appealed to a male-authenticated value—the protection of women—rather than an abstract value—equality—of which many legislators were wary. Once out of committee, its floor manager in the Senate was a man of impeccable conservative credentials, thus depriving the bill of the liberal stigma associated with ERA. Something else may have favored passage: unlike the Equal

Rights Amendment, few legislators could imagine the law's being directed against themselves.

In evaluating the position of male legislators on ERA and women's issues, former Representative Ruth Cook agreed with former Senator Wilma Woodard and others who argued that the entire record of legislators has to be evaluated in judging their reliability on women's issues. "You do not understand the legislature if all you do is look at the bottom line," Cook observed. How someone gets to that bottom line may help you appeal to him on another issue in which he may side with you. The message was clear from other participants and observers as well. The implication was that identifying ERA with the entirety of feminism and condemning all men on the basis of one vote or one action was dramatic and helped to simplify keeping tabs on friends and enemies, but it was eventually self-defeating. Making ERA the symbol of all of women's rights played into the hands of doctrinaire antifeminists who opposed any special legislation for women. If politics is indeed about the past—defining positions from previous actions—it is also about the future: anticipating future victories fashioned not by the punishment of enemies but by the rewarding of (possible) friends. The most committed feminist legislator could make distinctions between obdurate, antiwoman colleagues and anti-ERA votes who could become allies under different circumstances. If the National Organization for Women had difficulty in making these necessary distinctions, the explanation lies not in its radicalism but rather in its distaste for the legislative culture of bargain and trade.

This is not to say that women benefited as much from piecemeal legislation as they would have from passage of the Equal Rights Amendment. At the present time it is difficult to judge just how much anticipating passage of the federal Equal Rights Amendment affected classification by sex or how much the actions of feminists have transformed American law simply by bringing their analysis and commitments to the task of writing and applying law. Certainly North Carolina legislators passed laws benefiting women as the result of feminist action. But they have also resisted doing so. And they can always roll back the clock. Indeed, by 1985 women seemed to be losing ground in North Carolina. In March of that year the House Judiciary [II] Committee killed authorization of a half-million-dollar appropriation to study state government salaries and develop a point system based on experience, skill, working conditions, and responsibility to avoid discrimination. Business lobbyists worked hard to defeat the appropriation because they thought salaries should be set by supply and demand.[31] Killing the appropriation suggested that equity was in short supply and only powerless people demanded it.

The Equal Rights Amendment in the North Carolina legislature failed for several reasons. It became a feminist issue instead of a woman's issue. As such it became controversial and generated intense conflict, an issue a majority of legislators came to view as dangerous to the political affability of the legislative culture of bargain and trade. It stalled as an issue after 1977 because it could not attract a winning coalition from among the increasingly conservative members of the North Carolina General Assembly. Once stalled, wariness of the corroding effects of controversy meant that ratification would be highly unlikely because proponents rather than opponents were assigned responsibility for the seeming irrationality of public discourse. The amendment became the symbol of federal intervention—a vague concept, to be sure, but nonetheless perceived as threat to business, family, locality, and the General Assembly's power to govern. Primarily, ERA failed because a large number of men did not like being challenged by women, and they did not believe that the things women were interested in were important. That women disagreed about the amendment seemed to diminish it even further. Worse, ERA was likely to be expensive. In the growing conservative climate of the late 1970s and early 1980s, when money was reserved for personal consumption or military weapons and not for achieving justice, it was not surprising that ERA should have been defeated. What was surprising was that, given its negative ambience, it came as close to passage as it actually did.

8

ERA and the Politics of Gender

Ratification of the Equal Rights Amendment was not like other political issues—all participants in the conflict could agree on that. The emotion, commitment, and apocalyptic confrontation it elicited endangered the political affability of the General Assembly. Gender differences gave the issue its texture and meaning. North Carolinians, like citizens throughout the country, had been at this point before. To be sure, the constitutional amendments for woman suffrage and equal rights differed. The former granted a specific right; the latter might have allowed broader implications by making classification by sex in law suspect. As historic events, however, conflict over these two amendments revealed how very much alike they actually were and suggests that, in the future, opposition to ERA may well be understood as similar to that against woman suffrage, with the same implication of regressive attachment to tradition.

In each case ratificationists argued that they wanted nothing more than what men already had, thus appealing to the symbol of equality. In making this argument, petitioners justified themselves by the contemporaneous unraveling of gender stereotypes that women had been taught to embroider as their own creation. Since both amendments were touted as bringing legal position into line with social trends, discourse was in many ways conservative, as befitted women who were predominantly middle class and who truthfully denied radical intent. In each case, too, legislators acted for reasons that had little to do with the merits of ratification; their personal attitudes toward women probably weighted their decisions. They evaluated their position in terms of their own agendas. In North Carolina that evaluation in both instances had racial implications and affected the way in which legislators perceived threats to their powers to govern. But legislators and women resisted ratification for another reason, one that became clearer as debate over political participation and legal equality slipped into a debate about gender relations. Both ratification campaigns became referenda on the changing social roles of women. Moreover, because both ratification movements represented suspect innovations and came at the end of a decade of change, they both became symbolic of other changes and therefore a referendum upon them as well. Opposition came in part from people who believed that in the midst of flux, fixed gender categories had to be maintained inasmuch as moral order and social cohesion were grounded in these categories. In the larger context of the twentieth century, both struggles—but especially that over ERA—illuminate the significance of fixed gender categories to persons discomfited by social change.

Movements identified with the two amendments mobilized women in unprecedented numbers. If one unanticipated result of this was a debate—probably more accurately it should be called a litany[1]—on gender relations, that litany politicized proponent women as the politics of the General Assembly became the politics of gender. Women of both generations gained greater insight into the meaning of male power and the political consequences of their own powerlessness. Since feminist discourse had prepared ratificationists of the 1970s to understand the nature of gender politics far better than their suffragist foremothers, the former acquired a feeling of intellectual—and perhaps "spiritual"—empowerment, even in defeat. The long-term implications of this dual politicization, while not clear by 1990 for pro-ERA activists, nonetheless seemed problematic. Although they have worked for specific laws benefiting women, the urgency identified with the ERA struggle is gone from electoral and legislative politics. Aggressive mobilization for candidates favorable to women's issues—begun in the election of 1978—has not continued in North Carolina. It is possible that an assault on both reproductive rights and the apolitical status of medical decisions may, however, provide the dramatic issues symbolic of a "woman's agenda" once again to call them to action.

This possibility suggests that a women's movement persists as an ideal of solidarity, promoting and justifying action along a broad front—in electoral politics, social work, public service, industry, the professions, and such innovative enterprises as the entrepreneurial organization of craftswomen. The numbers and influence of women in the North Carolina General Assembly seem to have reached a plateau; the legislature is still dominated by the business ideology of its male members. ERA remains a source of solidarity and its defeat a symbol of patriarchal control, but there appear to have been few attempts to reach out to anti-ERA women based on an understanding that sisterhood should transcend the divisions created by varied responses to gender. Perhaps such outreach could be based on the dispassionate, nonpartisan approach identified with coalition building and the League of Women Voters. But this tactic should not require the repudiation of the movement that sharpens issues by radical positioning and maintains the needed critique of patriarchy. Neither posture should require withdrawing from electoral politics for fear of co-optation but should encourage renewing efforts to act there on women's behalf. Perhaps the many different faces and tactics of the women's movement should be held in tension in order to avoid the political implications of cultural conflict that deprived the Equal Rights Amendment of its legitimacy.

If public conflict over the Equal Rights Amendment never lost the aura of gender conflict, neither did it lose the aura of opposition between *equality* and *difference*. That opposition was never so crudely put as by Sam Ervin, who defended a dual system of legal rights based on "physiological and functional differences" between men and women. His reservations to ERA that fell one by one in resounding defeat in the U.S. Senate represented traditional ways of making women different and restricting the application of equality to them. They were proof to anti-ERA partisans that ERA was intended to effect a gender revolution. Each reservation rested on traditional gender definitions that, by being identified with conservative issues, could link gender and politics in a new and potent way. Ervin—and state legislators who agreed with him—claimed that women would be penalized by a draconian equality imposed by ideologues who could not understand the meaning of physiological differences. Ervin was a living metaphor of male opposition to the ERA. Suspicious of equality and angered by the way in which the U.S. government had used the principle as a guide to public policy with regard to sex and race, he leaped at the chance to depict it as dangerous. He was protecting neither women (as he claimed) nor his own privileged position (as some feminists claimed) so much as a way of thinking in which the latter could be achieved by claiming to do the former. This way of thinking about responsibility defined by gender, when confronted by the words, "on account of sex," allowed him to impose upon the ERA debate a false dichotomy. He did not, of course, do this by himself; Phyllis Schlafly was able to use this polarity for her own political purposes because there were so many women who believed that resolving the dichotomy in favor of equality would harm them.

Suffragists had refused to make a political choice between equality and difference. They recognized almost intuitively and often explicitly that placing the two in an antithetical relationship limited their options to bad choices in the drive for emancipation (liberation). Equal suffrage associations were supposed to place women in the unassailably commonsense position of claiming equivalency with men for themselves as citizens. They did not propose to make men and women "the same" any more than equal suffrage had made all men the same in the past. That is, when property-holding qualifications were dropped, there were still poor and rich men. When racial qualifications were outlawed, the effect had not been magically to blanch dark-skinned men or to darken the fair. Thus equal suffrage when applied to sex could not have been used figuratively to make women into men, driving them to the same duties and responsibilities. The issue was not sameness but equivalency. Equality, as two scholars have recently pointed out, assumes a social agreement to consider obviously different people as equivalent, not identical, when their citizenship is taken into account.[2]

The issue of equality versus difference was not so deftly managed by proponents of the Equal Rights Amendment. They seemed at times to accept the polarity and to chose equality over difference because the latter, they believed, had been used to deprive women of their rights. Nowhere was this more clear than in the issue of women warriors, in which proponents insisted on placing equality above difference through strategic use of sex-neutral language. Jane Mansbridge believes that difference with reference to women warriors would probably have been sustained by the legislative history of ERA and the popular dispositions of the American people. Assignment to combat would have been up to military commanders.[3] North Carolina ratificationists embraced both positions not from confusion but from an intuitive if rarely stated belief that equivalency rather than a hard choice between equality and difference was at issue. They had rejected women in combat as a significant issue because they did not want to back down before the posturing of male opponents. They had asked William Van Alstyne to reassure the General Assembly because they did not want the symbolic issue to pose difference against equality. If the first position begged the issue of equivalency, it nonetheless challenged opponents' contention that ratificationists were attempting to impose sameness as the meaning of equality. That Van Alstyne's arguments made little impression upon opponents suggests that the symbol of women warriors was serviceable because of what it said about gender conventions instead of military policy. Even when they understood that difference and equality were not necessarily in opposition to one another, proponents could not prevent anti-ERA activists from making them polar opposites and benefiting rhetorically from the contrast.

The polarity was not entirely the invention of opponents. ERA entered North Carolina, as it did other states, as a feminist issue. It came as one of the "new creative forces"[4] that would subvert the sex-gender system, with interpreters emphasizing that difference—identified with stereotype—would be redefined with no reference to sex. There was a heady feeling of inexorable progress that encouraged women who thrilled at Helen Reddy's singing "I am Woman, I am Invincible!" to hiss at law professors who lectured them on being grateful to Tar Heel men. The language of rights had captured their imagination, and they could not conceal their satisfaction at the prospect of lawyers' formulating sex-neutral language to place women before the law as individuals. Purging the law of classifiction by sex meant they could at last claim equality. Since this repudiation of stereotype was so pervasive among proponents of ERA, those who thought stereotype to be the received wisdom of the culture were understandably perturbed. That this anxiety was not limited to North Carolina was suggested in the U.S. Congress in 1983 when representatives debated resubmitting the ERA for ratification. Then, writes the historian Mary Frances Berry, "it quickly became evident that the issues aired during the ratification debates had poisoned the congressional well. Not the principle of equality of rights but the argument for traditional male and female roles and family values would be the focus of debate." When the vote was taken on November 14, the majority was six votes short of what it had to be to send ERA on to the Senate. Berry goes on to point out with references to both successful and failed bills in Congress that bills reflecting women's traditional roles were more likely to succeed than those not doing so. Moreover, issues such as equitable pay, which were easily identified with feminism or were likely to be interpreted as not reflecting traditional roles, were almost certain to be defeated. Opinion polls suggested a popular, social basis for this resistance.[5]

Even as ERA entered North Carolina a feminist issue, however, it was not thought of as the whole of feminism. Indeed, the extent to which ERA was a feminist issue was possibly better understood in retrospect than in 1973. Working for ERA was the social experience through which women acquired a feminist consciousness necessary to understand the nature of power. At first partisans thought of the Equal Rights Amendment as a legal issue best explained in terms of legal change by experts in command of relevant facts. Since the amendment had been identified with historic expectations of women for equality, the radical implications with which it came to be freighted were not altogether clear even to people who read the long article in the *Yale Law Journal.* Or if those implications were understood, they were thought to be so consistent with the way in which history was moving anyway, and so important to women as a whole, that ERA should be embraced by all women.

Unlike suffragists, ratificationists had not been alerted to the need for an ERA much before it passed the U.S. Congress. They did not have networks in place to lobby for ratification. Those upon which they came to rely were built from a coalition of existing organizations, most of them created for non-feminist purposes. The grass-roots mobilization they envisioned seemed to be just beyond their grasp, although the National Organization for Women seemed to be extraordinarily adept at canvassing in 1982. In retrospect, despite their sense of personal and group failure—which probably reflected a healthy-minded attitude toward the responsibility of citizens to work for the public good—North Carolina women managed to create one of the best pro-ERA public networks in the United States. The ratification process engaged women in public life, made them activists in ways that were frequently surprising to them, and taught them a great deal about the politics of gender as well as of the General Assembly.

The people who went out of their way to emphasize ERA as a feminist instrument of gender revolution were anti-ERA women. The connection was not the result of skillful manipulation on the part of Phyllis Schlafly and the minions of the Radical Right. It was actually made before Schlafly had really shifted her STOP ERA machine into "drive" by women who wrote Sam Ervin objecting to their impending masculinization at the hands of feminism. This anxiety was the basis upon which Schlafly and her North Carolina counterparts, Dorothy Malone Slade, Alice Wynn Gatsis, and Bobbie Matthews, and others, were able to build a constituency for anti-ERA legislators. Their alarmist accusations seemed so inappropriate to what ratificationists had actually expected of the Equal Rights Amendment that the latter were understandably confused. But for a great many women the accusations were meaningful and elicited a positive response. This constituency was never so thoroughly involved in the day-to-day action of the General Assembly as proponents who seemed to think of pro-ERA legislators as their surrogates and the bills of ratification as their own bills. Although anti-ERA women did become involved in enough electoral politics to win seats in the General Assembly, they were a tiny minority. Most opponents preferred to remain in a supporting role. Through style, image, and words they represented themselves as being unfairly threatened by feminists and the federal bureaucracy—a dragon the St. Georges of the General Assembly could easily identify.

Both proponents and opponents thought that they represented "the people." The former were part of networks with which scholars and politicans are quite familiar: formal organizations of activist women ranging from Church Women United and the League of Women Voters to the Women's Equity Action League and the National Organization for Women. They eagerly discussed their experiences as activists, sharing with others what they had learned, how they had changed, where they thought they were going, and how they thought women would benefit from collective action on their own behalf. They persistently referred to public opinion polls suggesting that a majority of both men and women favored ERA in North Carolina as well as in the nation at large. This legitimating "fact," when filtered through a commitment to helping women on a broad range of issues never part of their opponents' political universe and given impetus by the essential justice of ERA, made them believe that "the people" were with them.

Their female opponents were often those who demurred from the activism of even such nonpartisan groups as the League of Women Voters. They were frequently less comfortable than proponents in revealing themselves to people they did not know—although there were impressive exceptions. They were often private women enticed into action by activists refreshingly unlike the feminists they had seen on the television screen. They were frequently people whose denominational spokespeople supported ERA and worked hard on its behalf, but who themselves were suspicious of such activities. The Methodist bishop Robert Blackburn of Raleigh, for example, might indeed support ERA, and a lay activist, Tibbie Roberts of Morehead City, could stump the state among her co-religionists; but anti-ERA petitions would still come from small country Methodist churches or disaffected members from the city. Reports of antiratificationists' activities sometimes seemed to them to be accusatory even though none had anything to hide. Some did structure their networks on previous contacts within the John Birch Society. And members of the Church of Jesus Christ of Latter Day Saints were indeed active in opposition. Phyllis Schlafly did help organize STOP ERA groups, and the Eagle Forum did provide networks of anti-ERA activists. Birchers, Mormons, and right-wing women may have been new to the public forum, but they had as much right to it as their counterparts from a different political culture. Their sacrifice in the "cause" was just as admirable as the ratificationists' sacrifice. The work of Dorothy Malone Slade revealed a person of remarkable energy who gave her every moment to defeating the despised "libbers." Notes on her reports to the State Board of Elections reflect a fervent belief that hers was the cause of the people; she felt it in her bones after long hours of exhausting work.

The different styles of the contending parties were seen in their mobilization and public discourse. Proponents came from networks of organizational women comfortable with a range of voluntary, professional, business, and political associations. They were comfortable, too, with the sciences of society, polling, and psychology and were fascinated if often dismayed at the way in which masculine politics worked. The General Assembly, one of them could remember in the fall of 1988, had been "painfully male and mostly mediocre." Their experiences in the ratification struggle—defeat notwithstanding—seemed to have been positive. Indeed, for many women active in the broad range of women's politics in the 1980s, the ERA movement had been the beginning of public service. Their opponents came from domestic networks such as that structured on Slade's Christmas card list or from her contacts in the Birch Society; they were connected, too, by the petition campaigns of North Carolinians against ERA. They were distinctly suspicious of the social sciences and placed great emphasis on "feminine" values. The derisive laughter of feminists in 1973 was counterbalanced by their own shrieks at being "desexegrated" in 1977. They were situated in country clubs as well as country churches and included women who brandished Bibles as regimental colors as well as those who hoped that they would be spared such embarrassing displays in the future.

The major difference in style, of course, was that of public discourse. The insight of a former proponent suggests the impact of that style: "We thought we were engaged in a political issue," she remembered, "and then all of a sudden it was total war." The reaction of opposition women to the Equal Rights Amendment was one of the earliest indications that an apocalyptic style had once again entered mainstream politics.[6] This style sharpens political debate as if it were conflict between good and evil; it makes moral statements by transforming metaphor and simile into empirical fact. Thus the feeling that ERA would be like raping women becomes the alarmist accusation that it would decriminalize rape.[7] The style of popular anticommunism in the 1950s was especially successful in transforming statements of meaning into statements of fact. Liberals who were "like" communists because theoretically what they wanted was like communism were transformed into communists. Southern white supremacy campaigns such as that of the North Carolina Revolution of 1898 screamed "Negro Domination" because the success of dissidents who relied on Afro-American votes was like Afro-American rule, at least for Democrats. The racism of the culture sustained the easy transformation of simile into fact. Traditional ways of thinking about gender sustained the same kind of rhetorical vivisection among women who opposed ratification.

The apocalyptic style of the opposition lay not in an irrational shriek of anguish or in the predictable cadence of fundamentalist preaching that identified feminists with the Whore of Babylon. It lay rather in the perception of Tommy's mother and of all the women who did not write but felt as she did when contemplating the meaning of equality and its imposition upon her world of difference. That perception was that the U.S. Constitution as the supreme law of the land was about to be changed by women who had transcended gender definitions and were about to impose their own ideology upon girls and women whose sense of personal and collective identity were bound up in those definitions. That they attributed so much power to a constitutional amendment may have resulted from an assumption that the Constitution is in effect the documentary embodiment of American society with considerable moral as well as legal effect. Certainly they felt that there was a coercive element in ERA which neither they nor their children could successfully withstand. To be sure, such feelings were encouraged by ultraconservative publicists for their own purposes, but the widespread confessions ranging from dismay to grief over a possible gender revolution suggest that not a few women thought something important was being lost to the feminist revolution—the loss was apocalyptic.

The apocalyptic had been explained as not wanting to be men. The plaint seemed hardly worthy of notice—except that it was so pervasive and so heatedly expressed. Being forced to become men has to be understood as an apocalyptic event—the "actual" reality behind the surface phenomenon of appealing to what one scholar has pointed out is "an emancipatory vision of natural rights."[8] The appeal to individual rights and individual self-fulfillment was especially troubling to anti-ERA women because of their understanding of women's capacity to function competitively in a male-dominated society and that understanding did not make rights emancipatory.

Anti-ERA women were not foolish in expressing their doubts. Rights consciousness has been linked historically with a tradition of individual autonomy that, argued the definitive *Yale Law Journal* article, connected the Equal Rights Amendment with American values.[9] Within that tradition, humans are envisioned as rational individuals freely pursuing their self-interest consistent with the rights of others engaged in the same enterprise. Equality of opportunity, however, has always involved competition among persons who are socially and economically unequal because of differences in ability and the distribution of resources. The result of competition celebrated under the aegis of equality has ironically been and continues to be *in*equality. As conservatives, antiratificationist women were unlikely challengers of that reality because they believed it to be the result of natural differences heedless of class. As women, however, they understood that inequalities are compounded by male advantage in this society and that women as a group lack the facilities to compete with the advantaged group on its own terms. Male terms were thought to be inherent in the gender-neutral ethos associated with ERA. Sensitive to their fragile position and the ways in which the idea of woman's work guaranteed them a job, they feared that equality would force them them to "compete" as "men" and thus deprive them of whatever claim they had to economic independence. The appeal to individual rights that elicited from ratificationists a liberating vision evoked from their opponents different images—women *forced* to be men and thus bereft of the female relationships that had defined and sustained them.[10]

Antiratificationists' discomfort with individual rights involved more than a concern about competition in the workplace. In assuming the role of family-oriented women, they seemed to be affirming the affiliative characteristics that the psychologist Carol Gilligan believes have helped girls and women develop an ethic less comfortable with abstractions than with human connections.[11] The point was made by an opponent who charged that those fighting for "rights" seemed to be "an ego trip." The inflation of ego seemed to be beside the point and the association inappropriate until the phrase was identified with revulsion at the self-dramatization of the 1960s. It had to be understood as a double entendre interrupting a train of thought about "rights." Rights and ego were overlaid with reference to public demonstrations and men who thought that rights were more important than relationships. This linkage of rights and individual autonomy with (male) ego and masculine attributes suggests that antiratificationists grasped intuitively the same deficiency that feminist scholars argue characterizes liberal ideology: the absence of affiliative virtues associated with the private sphere of home and family.[12]

Anti-ERA women did not need feminist political theorists to tell them that eighteenth-century liberal ideology is grounded in patriarchal structures and that the language of equality and individual rights is "masculine."[13] They could hear it for themselves; and what they heard was not the traditional female language of suffragists seeking to legitimate *women's* action in the polity but feminist demands for equality *with men,* using men as the measure of female value and action.[14] The more ERA partisans resorted to the language of liberalism in situating the amendment to defuse charges of cultural change, the more persistent and intense those charges became. As women still identified with the private sphere, antiratificationists were simply positioned to hear the masculine language of equality and rights with greater clarity than were their pro-ERA counterparts. The difference was one of life experience and commitment as well as the lack of an alternative language with which to make claims for parity between the sexes.[15] Many ratificationists, socialized as "men" in the public sphere, had become committed to equality as liberals and as a result had become "men." The visceral response of opponent women therefore made sense.

This resistance to "masculinization" is illuminated by the historian Linda K. Kerber in a recent discussion of separate spheres. Opponent women seemed to have an acute awareness of the extent to which even in the modern world of overlapping male and female spheres, private and public space is still defined by gender. They seemed to be pointing out that the public space of the workplace is male space—even when there are women attached to or placed within it. The ease with which rape was identified with integrated public rest rooms—placing women in more public space—suggests the threat and danger to women. Battlefields are, of course, the ultimate symbol of male, public, and apocalyptic space. Opponent women understood—perhaps were awed by—the way in which economic and social contexts within which they lived were masculine. The "gendering" of space and the power relationships inherent in that activity suggest that opponent women had a firm grasp of reality, one born of a consciousness that, if not shared by most ratificationists and liberal feminists, nonetheless bore remarkable similarities to that of feminists in other times and places.[16]

The impact of the apocalyptic style on campaigns for ratification was primarily to prevent either side from understanding what its counterpart was actually saying. Debates between women representing the two sides were discontinued by proponents because they believed the opposition simply took too many liberties with the truth, distorting the issue with half-truths and alarmist accusations. At first, proponents could not "process" what opposition women were saying, and they tended to dismiss or discount arguments presented in an apocalyptic style—it seemed too much like hysteria. Later, when confronted with plausible explanations as to why women would oppose the ERA, they were no better able to counter with convincing arguments. ERA had become a symbol of irrational politics for too many people. The emotional intimidation of pro-ERA legislators by people unaccustomed to conceding good motives and honest differences to people with whom they disagreed meant that the psychological pressure was intimidating. Apocalyptic denunciation by anti-ERA activists "bloodied" legislators and made them wary of continuing the fight without guarantee of winning. In what proponents thought was a double standard, they found themselves held accountable for the emotionalism because they kept bringing the matter up. They and ERA became a nuisance.

Understanding the story of ERA in North Carolina complements studies of ratification elsewhere. Lobbying strategies varied in North Carolina as in Illinois, but Tar Heel ERA women seemed to have gotten higher marks for effectiveness from the General Assembly than those elsewhere. Both Illinois and North Carolina proponents eventually thought that direct participation in electoral politics earlier would have improved their chances. As it was, the ERA coalition in North Carolina established, exploited, and maintained ties among a broad range of women. It helped to make women's issues important among League of Women Voters members as well as United Methodist Women and Business and Professional Women. "There is a lot of quiet feminism in my church," one woman said; "we will not let it die." It is a feminism with different styles and emphases—as Mansbridge found in Illinois, and links women executives, ministers, civil servants, academics, social workers, and volunteers in a common enterprise.[17] As for opposition women, there was no one in North Carolina quite like Phyllis Schlafly, but Alice Wynn Gatsis and Dorothy Slade were important surrogates. Their presence in the ratification struggle deprived the ERA of the legitimacy of being supported by all women. Now that ERA is defeated for the foreseeable future and proponents can evaluate the ratification process from a perspective that transcends blame and accusation, perhaps they can understand what opponent women were concerned about and learn from the latters' experiences as well as their own.

Nothing said about the Equal Rights Amendment in North Carolina differed from what was said elsewhere. The issue of women warriors was, as Mansbridge and Gilbert Steiner point out, one of the most damaging arguments against ratification. Proponent Tar Heel legislators thought it did them in. There is no reason to believe, however, that the Afghan conflict made Southern legislators more susceptible to the argument than non-Southerners—military tradition or not. In fact, some North Carolina legislators were prepared to change their vote to favor ERA even after President Jimmy Carter suggested registering young women. The issue of women in combat would have been raised even if every feminist in North Carolina had been able to convince unsympathetic state legislators that William Van Alstyne was correct and spoke for her. As it was, anti-ERA state legislators paid no heed to Van Alstyne. Perhaps Sam Ervin's reservation preventing the use of women in combat would have reassured a few if it had passed the Senate. And "a few" would have changed the vote. But once women in combat had been introduced as a plausible issue—even if employing what Mansbridge calls the "deferential" argument,[18] as ratificationists did sporadically in North Carolina—and once legislators were reminded that there were already women at risk in the armed services, the possibility—even if remote—of extending that risk to include women they knew was not likely. The debate over the Equal Rights Amendment would not have been affected by tinkering with the issue of women warriors as long as there were women already serving in the armed services.

There are three reasons for believing this. They involve the Supreme Court, the governing powers of state legislatures, and gender. First the Supreme Court: That institution was thought by many to be unreliable when it came to allowing people to live the way they wanted to live in their hometowns. And when ratificationists said that they wanted to give the Court a mandate to do something that the Court by itself had not thought of yet, many legislators thought of court decisions on behalf of one man–one vote and busing to achieve racial balance; they also thought of decisions opposed to prayer and Bible reading in public schools. The first decision was mentioned by opponent legislators as indicative of the present Court's thinking. The reference to one man–one vote was made often enough to force attention to the way in which opponent legislators thought ERA to be the next step taken by the federal government in diminishing the powers of the states. The concept was not states' rights but state government. North Carolina legislators were not alone in believing that passage of the Equal Rights Amendment would detract from the power exercised by state government and the sense of local responsibility.

The most critical issue in the campaign for the Equal Rights Amendment, however, was "gender" as embedded in the words "on account of sex." "Gender" is the way in which humans have fabricated the social meaning of sex in spinning their culture into webs of meaning. Although the Equal Rights Amendment could not have effected a cultural revolution, it nonetheless came to represent widespread subversion of gender roles. This understanding developed not because feminists, or even ratificationists, thought of the amendment as a strategic means of achieving gender change. Nor was it because the amendment would actually have had a broad cultural impact. Nor did the representation occur because the opposition hit upon a devilishly clever way of diverting attention from "real" to imaginary or symbolic issues.[19] It happened because the immediate, "knee-jerk" response of large numbers of women was to perceive threat from the specific act as a creation of people thought to have declared war upon them. It happened, too, because a large minority of men were also unsettled by changes identified with gender roles, changes that could in turn be identified with other changes, all of which could be perceived as threatening.

The two sides of the issue were not symmetrical, that is, for every argument supporting ratification, there was not a counterargument addressed to that particular way of understanding the issue. Both sides talked past each other. Arguments representing strands within the webs of meaning were held in place by whether or not the people making them thought that traditional ways of thinking about gender were adequate for the equitable treatment of women. In some ways, women on each side shared a skeptical attitude toward men and masculinity, an attitude that may be the means of future cooperation among women. But especially for female opponents of ERA, the issue of ratification represented an entire value system, a sense of self-hood, a way of thinking about the patterns that make up the culture—all matters that are not subject to the ordinary politics of bargain and trade. ERA came to represent a "way of life," ultimate principles, indeed the very (gender-defined) nature of the cosmos. Irrationality attributed by proponent women to their counterparts was also attribued in the reverse direction. Equality that represented positively perceived universal principles to one side represented a negative masculine bias to the other, one that was more comfortable with asserting feminine solidarity and identity. Proponents were comfortable with malleable gender roles; opponents felt such malleability to be threatening. Cultural patterns, the latter believed, were part of the "givens" of life; the belief was the essential core of feminine resistance to the ERA.

This belief in fixed, universal patterns of behavior and type was a cultural fundamentalism that resisted the implications of a modern understanding of life. The malleability of human culture was rejected in much the same way an immanent God has been rejected by religious fundamentalists. The malleability of gender role was, in theory at least, rejected in much the same way that religious fundamentalists had rejected ethical relativity. The apocalyptic references and assumptions sensed in warnings of (sexual) chaos were very much like the warning of final judgment that has characterized religious fundamentalists for more than a century.[20] The apocalyptic warning that denying difference in favor of equality would be devastating to women was not merely "expressive" politics in the service of the patriarchy—although it was surely that in part. It represented, in addition, an existential cry of pain because of the way in which gender has been presented as a given from generation to generation. Its grounding in culture made it so easy to believe. Understanding that fact is surely one achievement of feminist scholarship, and responding to it is surely one responsibility of feminist theory.

The politics of gender was stained with racism. The ambience was clear from the beginning of the ratification process in non sequitur, humor, analogy, and parody, in suspicion, anger, and defensiveness. The overlay of desegregation and "desexegration" was too easily effected, the connection of "forced women" and "forced busing" too naturally made, the absurdity of equality too carefully contrived to allow anyone to ignore the way in which resentment at actions mandating change for blacks suffused resistance to the Equal Rights Amendment. Men who had denied that Afro-Americans had grievances appropriately addressed by law now rushed to deny the same for women. Resistance was based in each case upon the presumed innocence of those holding power.[21] Sam Ervin "knew" that whites had done nothing to harm blacks and that men as a class had not harmed women. He did not understand that by being positioned to withhold or inflict harm, whites and men exercised government over blacks and women at least as tyrannical as a federal government heedless of a Bill of Rights. Claiming innocence for whites and men in the face of just objection from blacks and women—even if all of the latter did not object—was a form of racism because racism assumes the innocence of an entire class when its elites are challenged by representatives of another.

This is not to say that all opponents of the ERA adopted their position because they were prejudiced in some way against blacks, or women, as a class, although there were individual instances of this attitude. Rather, racism lay in the holding and witholding of power. The Revolution of 1898 in North Carolina had manipulated race prejudice to achieve power. Heirs to that tradition—Senator Ervin among them—repudiated the ERA because of its association in their own minds with *Brown* v. *Board of Education* (1954), one man [*sic*]–one-vote, *Swann* v. *Charlotte-Mecklenburg School Board* (1971), and directives from the Department of Health, Education, and Welfare—all of them challenges to power and most of them associated with race. Although one black representative did vote against ratification, it is a fair assumption that had the General Assembly been more representative of the black population of North Carolina the amendment would have passed because of the resonance between Afro-American aspirations and the principle of equality. The presence of more blacks would have changed the rules of bargain and trade, challenging the values of an institution in which decisions were based on what was "good" for business rather than for people regardless of sex, race, and investment. (This would have been less true had women been more numerous because of the appeal of antifeminism.) Linking "equality of rights . . . on account of sex" with federal attempts to make society more equitable regardless of race was natural for conservatives of the General Assembly; making equality dangerous for those to whom it was to be applied—women—made it dangerous as a principle, thus vindicating resistance to it in all instances. In doing this, anti-ERA legislators implied the moral limits of their claim to represent "the people." Business values based on competitive market forces divide people into winners and losers, whereas in a true republic everyone should "win," that is, receive justice and equity. Powerful people less comfortable with equality than penurious government in a society stained by injustice assume responsibility for that stain.

Ratification of ERA failed for several reasons. It failed in the nation because the majority of men favoring it was not large enough and because the majority of states ratifying was not large enough. It failed because it became a conflict based on gender differences, and many state legislators did not like to take a stand on issues that radiated such heat. It failed because that conflict could be symbolized by the specific as well as symbolic issue of women in combat, which came to represent a threatening gender revolution. It failed because it lost its legitimacy as an issue supported by (all) women. It failed because of political opposition to applying the principle of equality in public policy and the racism this resistance implied.

To ratificationists the loss was bitter, especially since a majority appeared to be with them. As symbolic feminism, ERA became popular feminism as had no other issue since suffrage, and as such it mobilized American women in the millions, especially in those states where the ratification drama played the decade. Politicized at multiple levels, women became feminists and in some cases politicians, creating in the process key linkages with other women and with the government. That they were predominantly middle class and white, and that the politics they practiced were mobilizational and electoral in the traditional sense, should not lead feminists to discount the significance of their recruitment. For ratificationists and their staunchest male allies, as perhaps for no other group in America, sex, gender, and the Constitution were not subjects of academic discourse but, taken together, a matter that engaged minds, galvanized action, and transformed lives. One woman spoke for an entire generation: "I participated in small group discussions, and talked about women's liberation, but in fighting for ERA I became involved with other women working *for* women—it was the women's movement for me!" Was she a feminist? "I am *now*!" Was fighting for "equality of rights" a mistake? The question made absolutely no sense to her.

"Equality," she said, "is never a mistake."

Acronyms

AAUW	American Association of University Women
ACLU	American Civil Liberties Union
AFL-CIO	American Federation of Labor–Congress of Industrial Organizations
BPW	Business and Professional Women
DU	Duke University
EEOC	Equal Employment Opportunity Commission
ERA	Equal Rights Amendment
HEW	Department of Health, Education, and Welfare
HOW	Happiness of Womanhood
LCY	Library of Congress
NAACP	National Association for the Advancement of Colored People
NAWSA	National American Woman Suffrage Association
NCAE	North Carolina Association of Educators
NCAERA	North Carolinians against ERA
NCC	North Carolina Collection, University of North Carolina at Chapel Hill
NCDAH	North Carolina Division of Archives and History
NCUERA	North Carolinians United for ERA
NCWPC	North Carolina Women's Political Caucus
NEA	National Education Association
NOW	National Organization for Women
NWPC	National Women's Political Caucus

PAC	political action committee
SDS	Students for a Democratic Society
SHC	Southern Historical Collection, University of North Carolina at Chapel Hill
SNCC	Student Nonviolent Coordinating Committee
WEAL	Women's Equity Action League
YWCA	Young Women's Christian Association

Notes

Preface

1. Molly Yard to Friend, June 1988, authors' files.

2. Janet K. Boles, *The Politics of the Equal Rights Amendment: Conflict and the Decision-Making Process* (New York: Longman, 1979); Gilbert Y. Steiner, *Constitutional Inequality: The Political Fortunes of the Equal Rights Amendment* (Washington, D.C.: Brookings Institution, 1985), p. 49.

3. Mary Frances Berry, *Why ERA Failed: Politics, Women's Rights, and the Amending Process of the Constitution* (Bloomington, Ind.: Indiana University Press, 1986); Jane J. Mansbridge, *Why We Lost the ERA* (Chicago: University of Chicago Press, 1986).

4. Theodore S. Arrington and Patricia A. Kyle, "Equal Rights Activists in North Carolina," *Signs* 3 (Spring 1978): 666–80; David W. Brady and Kent L. Tedin, "Ladies in Pink: Religion and Political Ideology in the Anti-ERA Movement," *Social Science Quarterly* 36 (March 1976): 564–75; Carol Mueller and Thomas Dimieri, "The Structure of Belief Systems among Contending ERA Activists," *Social Forces* 60 (March 1982): 657–73; Kent L. Tedin, "Religious Preference and Pro/Anti Activism on the Equal Rights Amendment Issue," *Pacific Sociological Review* 21 (January 1978): 55–67; Ira Deutschman and Sandra Prince-Embury, "Political Ideology of Pro- and Anti-ERA Women," *Women and Politics* 2 (Spring–Summer 1982): 39–55. There are many more articles on the subject—probably more than one hundred—but these are the ones that helped us in the early stages of writing.

5. Bruce Jennings, "Interpretive Social Science and Policy Analysis," in *The Social Sciences and Policy Analysis,* ed. Bruce Jennings and Daniel Callahan (New York: Plenum Press, 1983), pp. 83–111.

6. See, for example, Carroll Smith-Rosenberg, *Disorderly Conduct: Visions of Gender in Victorian America* (New York: Oxford University Press, 1985); Natalie Zemon Davis, *Society and Culture in Early Modern France* (Stanford, Calif.: Stanford University Press, 1975).

7. To allow for a degree of anonymity, quotations from interviews, like this one,

are not cited. For a list of interviews, see app. 4. In addition to those listed, interviews were conducted with persons who wished to remain anonymous.

Chapter 1. Parable: Woman Suffrage and "Manhood" in North Carolina

1. Studies of woman suffrage that have helped give substance to the following discussion include Ellen Carol DuBois, *Feminism and Suffrage: The Emergence of an Independent Women's Movement in America, 1848–1869* (Ithaca, N.Y.: Cornell University Press, 1978); Ida Husted Harper, ed., *History of Woman Suffrage* (New York: National American Woman Suffrage Association, 1922), 6:490–500; Alan Grimes, *Puritan Ethic and Woman Suffrage* (New York: Oxford University Press, 1967); Eileen Kraditor, *The Ideas of the Woman Suffrage Movement,* (New York: Columbia University Press, 1965); Elizabeth Cady Stanton, Susan B. Anthony, and Matilda Joslyn Gage, ed., *History of Woman Suffrage* (Rochester, N.Y.: Susan B. Anthony, 1886), Vol. 3; Susan B. Anthony and Ida Husted Harper, eds., *The History of Woman Suffrage* (Rochester, N.Y.: Susan B. Anthony, 1902); Carrie Chapman Catt and Nettie Rogers Shuler, *Woman Suffrage and Politics: The Inner Story of the Suffrage Movement* (New York: Charles Scribner's Sons, 1923); Mari Jo Buhle and Paul Buhle, eds., *The Concise History of Woman Suffrage* (Urbana, Ill.: University of Illinois Press, 1978). A. Elizabeth Taylor's two articles on "The Woman Suffrage Movement in North Carolina," which appeared in the *North Carolina Historical Review* 38 (January, April 1961): 45–62, 173–89, were helpful overviews. Taylor also wrote articles on woman suffrage in Georgia (*Georgia Historical Quarterly* 28 [June 1944]: 63–79, and 42 [December 1958]: 339–54) and Arkansas (*Arkansas Historical Quarterly* 15 [Spring 1956]: 17–52).

2. This discussion is based on Albert Coates, *By Her Own Bootstraps: A Saga of Women in North Carolina* (Chapel Hill, N.C.: Privately printed, 1975), pp. 3–36; Emily Clare Newby Correll, "Women's Work for Woman: The Methodist and Baptist Woman's Missionary Societies in North Carolina, 1878–1930" (Master's thesis, University of North Carolina at Chapel Hill, 1977); Sallie Southall Cotten, Scrapbook 1, Sallie Southall Cotten Papers, Southern Historical Collection, University of North Carolina at Chapel Hill (hereafter SHC); Josephine K. Henry, "The New Woman of the New South," *Arena* 11 (June 1895): 355; Nell Battle Lewis, in Raleigh *News and Observer,* May 3, 1925; Raleigh *News and Observer,* May 21, 1896; *North Carolina Christian Advocate,* April 15, 1896; Stanton, Anthony, and Gage, eds., *History of Woman Suffrage,* 3:825–28.

3. Rev. 3:8, cited in Correll, "Women's Work for Woman," p. 33.

4. Cotten, Scrapbook 1, esp. pp. 130–48, Scrapbook 2, passim, Cotten Papers; SHC; Anastatia Sims, "Sallie Southall Cotten and the North Carolina Federation of Women's Clubs" (Master's thesis, University of North Carolina at Chapel Hill, 1976); Bruce Cotten, *As We Were: A Personal Sketch of a Family Life* (Baltimore: Privately printed, 1935), North Carolina Collection, University of North Carolina at Chapel Hill (hereafter NCC).

5. Outlines of Addresses, in Charles D. McIver to Elvira E. Moffitt, June 20, 1902; enclosure of letter from Edith Royster Judd to Elvira E. Moffitt, July 26, 1916, both Elvira E. Moffitt Papers, SHC; James L. Leloudis II, "School Reform in the New South: The Woman's Association for the Betterment of Public School Houses in North Carolina, 1902–1919," *Journal of American History* 69 (March 1983): 886–909; "The Woman's Association for the Improvement of School Houses," *North Carolina Journal of Education* 1 (September 1906): 16.

6. Raleigh *News and Observer,* November 24, 1904.

7. Edith Royster, "President's Address," *Proceedings and Addresses of the Twenty-ninth Annual Session of the North Carolina Teachers Assembly . . . November 27–30, 1912* (Raleigh: State Superintendent of Public Instruction, 1913), p. 76. See also pp. 75, 10–11, for the Resolution of the Teachers Assembly asking for changes in the legal status of women teachers: "The cause of education and the children in North Carolina have sustained and are sustaining a great loss through the legal limitations to women's services in this state." See also Royster's address the following year in the *Proceedings* for 1913 (Raleigh: Edwards and Broughton, 1914), pp. 100–2.

8. Sallie Southall Cotten, A Line a Day Book, May 3, 5, June 6, 7, August 3, 1912, Sallie Southall Cotten Papers, NCC.

9. This discussion is based on Robert Howard Wooley, "Race and Politics: The Evolution of the White Supremacy Campaign of 1898 in North Carolina" (Ph.D. diss., University of North Carolina at Chapel Hill, 1977); George Howard to Henry G. Connor, November 14, 1898, H. G. Connor Papers, SHC; Richard L. Watson, "Furnifold Simmons: 'Jehovah of the Tar Heels'?" *North Carolina Historical Review* 44 (Spring 1967): 166–87.

10. Daniel Jay Whitener, *Prohibition in North Carolina, 1715–1945,* James Sprunt Studies in History and Political Science, 27 (Chapel Hill: University of North Carolina Press, 1945); Heriot Clarkson, "The Drink Evil," July 13, 1904, Heriot Clarkson Papers, SHC; *Raleigh News and Observer,* February 22, 1903, January 20, 1905, May 24, 1908.

11. Walter Clark, Jr., and J. Melville Broughton, Jr., *The Legal Status of Women in North Carolina* (Raleigh: North Carolina Federation of Women's Clubs, 1914); Raleigh *News and Observer,* January 18, June 5, 6, 8, 26, 1913, June 5, 1914.

12. Raleigh *News and Observer,* April 19, 26, May 6, November 10, 12, 1914; Barbara Henderson to Walter Clark, February 23, 1915, Walter Clark Papers, North Carolina Department of Cultural Resources, Division of Archives and History, Raleigh (hereafter NCDAH); *Proceedings of the Second Annual Convention of the Equal Suffrage Association of North Carolina* (Henderson: Jones-Stone, 1915), pp. 7–10.

13. Raleigh *News and Observer,* February 3, 5, 10, 1915; *Proceedings of the Second Annual Convention,* pp. 7–10.

14. Raleigh *News and Observer,* February 13, 19, 1915, May 10, 1925.

15. Raleigh *News and Observer,* January 18, 26, 31, February 4, 7, 10, 1917.

16. Raleigh *News and Observer,* February 18, 22, 23, 27, 28, March 2, 1917. Ellis Gardner of Yancey County submitted a bill to allow women to vote for or against suffrage in the next Democratic primary; it was tabled. See also *Proceedings of the Third Annual Convention of the North Carolina Equal Suffrage League,* November 6, 1917 (Goldsboro: n.p., 1917).

17. Mrs. R. H. Latham to Mrs. John S. [Otelia Carrington] Cuningham, October 12, 1917; Mrs. L. D. Bulluck to Cuningham, February 3, 1918, both Mrs. John S. [Otelia Carrington] Cuningham Papers, in Otelia Connor Papers, SHC; Gertrude Weil, form letter, December 8, 1917, Gertrude Weil Papers, NCDAH; Weil to Laura Jones, July 27, 1917, Southgate Jones Papers, Duke University Manuscript Department, William L. Perkins Library, Duke University, Durham (hereafter DU); Julia McGehee Alexander to Walter Clark, June 17, 1919, Walter Clark Papers. See also Laura Holmes Reilly to Temporary Chairmen (county committees form letter), June 18, 1917; Reilly to D. H. Hill, June 10, 1917, both Military Collection, World War I Papers, North Carolina Council for Defense, box 39, Woman's Committee, NCDAH; Katherine S. Reynolds to Cuningham, October 15, 1917; Reynolds to County Chair-

men, October 22, 1917; Helen Guthrie Miller to Cuningham, March 16, 1917, all Otelia Connor Papers. See also William J. Breen, "Southern Women in the War: The North Carolina Women's Committee, 1917–1919," *North Carolina Historical Review* 55 (July 1978): 251–83, esp. pp. 260–62, 272.

18. Eugenia Clark to Laura Jones, February 5, 1918, Southgate Jones Papers; Eugenia Clark to Cuningham, February 3, 1918, Otelia Connor Papers. See also Gertrude Weil, form letters to the Equal Suffrage Association of North Carolina, January 19, February 5, 1918, Weil Papers; Raleigh *News and Observer,* January 11, 1918.

19. Carrie Chapman Catt to Cuningham, January 31, 1918: Eugenia Clark to Cuningham, February 3, 1918, both Otelia Connor Papers; *Charlotte Observer,* February 14, 1918; Weil to Laura Jones, February 26, 1918; Walter Clark to Laura Jones, February 10, 12, 1918, both Southgate Jones Papers. See also letters from Laura Jones to Walter Clark, February 14, 1918, copy; Weil, February 15, 1918, copy; Anna Forbes Liddell, February 15, 1918, copy; William P. Few, February 14, 1918, copy; Helen H. Gardner to Laura Jones, March 11, 1918, all Southgate Jones Papers.

20. *Greensboro Daily News,* April 10, 1918, January 11, 12, 1919; Harriot W. Elliott, ed., *The North Carolina Federation of Women's Clubs, 1918–1919* (Greensboro: n.p., 1918), pp. 37, 44, 71.

21. Raleigh *News and Observer,* March 6, 7, 1919; *Greensboro Daily News,* March 6, 7, 1919; Walter Clark to Laura Jones, March 17, 1919, Southgate Jones Papers. See also William T. Bost, "Heroes and Serpents," clipping from *Woman Citizen,* April 19, 1919, Mary Cowper Papers, DU; Alexander to Walter Clark, June 17, 1919, Walter Clark Papers.

22. Certificate, February, 16, 1920, O. Max Gardner Papers, SHC.

23. Raleigh *News and Observer,* March 28, 1920; *Greensboro Daily News,* March 31, April 4, 1920; J. J. Mackay to Josiah William Bailey, April 1, 1920, Josiah William Bailey Papers, DU; Weil to Elsie G. Riddick, March 28, 1920, Elsie G. Riddick Papers, SHC.

24. Carrie Chapman Catt to Weil, March 29, 1920; Weil to Catt, April 3, 1920, both Weil Papers; *Greensboro Daily News,* April 3–11, 1920; *Greensboro Patriot,* April 8, 12, 1920; Raleigh *News and Observer,* April 4–10, 1920; Josephus Daniels to Catt, telegram, April 9, 1920, copy, Josephus Daniels Papers, Library of Congress, Washington, D.C. (hereafter LC). Professor Charles Eagles of the University of Mississippi graciously facilitated research into Daniels's activities with suggestions that paid rich dividends.

25. The following discussion is based on an analysis of private and public statements in letters, circulars, broadsides, articles, periodicals, speeches, and public testimony reported in *Raleigh News and Observer, Greensboro Daily News,* and *Charlotte Observer.* The newspapers are on microfilm. The other material resides in the Equal Suffrage Association of North Carolina files, in Equal Suffrage Amendment Collection, NCDAH; Weil Papers; Cowper Papers; Southgate Jones Papers; Bailey Papers; and Otelia Connor Papers.

26. Kate Gordon to Walter Clark, August 26, 1915, Walter Clark Papers. Insights into the connections between racism and suffrage can be gleaned from pamphlets and broadsides in the Equal Suffrage Amendment Collection, including "The Three Immediate Women Friends of the Anthony Family. See Biography of Susan B. Anthony, Page 1435, by Mrs. Ida Husted Harper" (Montgomery, Ala.: Brown Printing Co., n.d.); James Callaway, "Some Strange History" (n.p., n.d.), an extract from the *Macon Telegraph,* May 26, 1918; "The Nineteenth Amendment: A Socialist Measure!"

(Boston, Mass.: Massachusetts Public Interests' League, May 1920); "How Woman Suffrage Was Imposed on New York State" (n.p., n.d.), associated with materials published by the National Association Opposed to Woman Suffrage; "The Feminist and Our Bible" (Raleigh: Capital Printing Co., 1920); "The Woman's Bible" (n.p., n.d.); "Opposing Woman Suffrage" (n.p., n.d.); Mrs. Mary Green, West Plains, Mo., to Representative Harry P. Grier, North Carolina legislator, July 27, 1920, a form letter written in 1918 and sent to legislators around the country; James Callaway, "Susan B. Anthony Eulogizes Ben Butler" (n.p., n.d.); "We Plead in the Name of Virginia Dare, That North Carolina Remain White" (Montgomery, Ala.: Brown Printing Co., n.d.); "Suffrage Democracy Knows No Bias of Race, Color, Creed, or Sex.—Carrie Chapman Catt" (n.p., n.d.); "The Suffrage Amendment Is Not Inevitable" (n.p., n.d.). Nancy F. Cott, "Feminist Politics in the 1920s: The National Woman's Party," *Journal of American History* 71 (June 1984): 53, reports that North Carolina members of the National Woman's party refused to register in the 1921 convention because of the presence of black women. This action obviously reveals a deep-seated racist etiquette. White North Carolina women suffragists included racists in their number; there is little if any evidence to suggest, however, that in public they emphasized blatant racist appeals to win the vote in the period between 1914 and 1920. In letters among themselves they seem simply to have forgotten black people, which in the aftermath of the Revolution of 1898 was reprehensible enough, to be sure. Some of them could applaud (only privately) prosuffrage statements by black men, however, and all seem to have avoided the rhetoric of disfranchisement because at least some of them thought that it, too, was reprehensible. It must be conceded, however, that North Carolina white suffragists separated themselves from black women and equal suffrage for blacks. They also identified white supremacists as favoring woman suffrage—Woodrow Wilson, Josephus Daniels, and Furnifold Simmons (the reference to Simmons was not altogether true, as he had not asked the legislature to ratify). See pamphlets, broadsides, and posters in Equal Suffrage Amendment Collection.

27. *Raleigh News and Observer,* November 15, 1914.

28. William T. Bost, "Senator Overman's Vacation," clipping, August 10, 1918; Marjorie Shuler, "Strenuous Recess of North Carolina Senators," August 10, 1918, both Cowper Papers.

29. Alexander to Walter Clark, June 17, 1919, Walter Clark Papers. Taylor, "Woman Suffrage Movement in North Carolina," pp. 173–89, argues that North Carolina women opposed the militancy of the Congressional Union. Some of them did; many did not. Help from the union's lecturers was gratefully accepted by at least a few suffrage leaders. See Mrs. Josephus [Addie Worth Bagley] Daniels to Catt, July 12, 1920, in which the wife of the secretary of the navy observed that the National Woman's party was sending what seemed to her to be a large amount of printed materials to North Carolina. She offered to go to Raleigh to help in the ratification drive—but not to fight the party. Copy, Josephus Daniels Papers.

30. Mrs. Josephus Daniels, "The Justice, Expediency and Inevitableness of Ratification," *Everywoman's Magazine,* special session ed. (July–August 1920): 3; Josephus Daniels to My Dear Blessing, April 25, 1920, Josephus Daniels Papers.

31. Walter Clark, Jr., *Address by Chief Justice Walter Clark before the Federation of Women's Clubs, New Bern, N.C. 8 May, 1913* (N.p., n.d.), p. 17, NCC.

32. Mary Hilliard Hinton wrote one legislator pleading with him for "protection from political responsibility." Hinton to Colonel Benehan Cameron, June 25, 1920, Cameron Family Papers, SHC. Antisuffrage women elsewhere thought "political responsibility" included the range of citizens' responsibilities: jury duty, campaigning for

public office, and military service. Many women opposed to suffrage thought their influence would be diminished if they were no longer perceived as disinterested.

33. Broadside placed on legislators' desks in 1920 special session, Equal Suffrage Amendment Collection.

34. See, for example, Raleigh *News and Observer,* July 18, 1920.

35. Clipping from *Woman Patriot,* May 1–15, 1920, Equal Suffrage Amendment Collection; Brian Howard Harrison, *Separate Spheres: The Opposition to Woman Suffrage in Britain* (London: Croom Helm, 1978). Southern and English antisuffragists were not unique in their belief that woman suffrage threatened elite control necessary for "good government." See Grace Duffield Goodwin, *Anti-Suffrage: Ten Good Reasons* (New York: Duffield and Company, 1913).

36. *Greensboro Daily News,* February 11, 1919; "Suffrage Democracy Knows No Bias of Race, Color, Creed, or Sex"; "Thomas Nelson Page" (N.p.: Southern Rejection League, n.d.), both Equal Suffrage Amendment Collection; L. J. H. Mewborn to Weil, August 26, 1920, Weil Papers.

37. Walter Clark, *Ballots for Both* (Raleigh: Commercial Printing Company, 1916), p. 10.

38. Weil to Laura Jones, July 5, 1920, Southgate Jones Papers. For the rhetorical excesses of the campaign, see Gardner, Scrapbook, 1920, Gardner Papers.

39. Catt to Weil, July 1, 1920; Weil to Dear Friend, June 21, 1920, in which she also told workers that the key to success is "PUBLICITY." "Get material into newspapers, hound the editors until they do it—AT ONCE and WITHOUT FAIL," both Weil Papers. See also Weil to Catt, July 9, 1920, Weil Papers.

40. Cornelia Jerman to Laura Jones, July 31, 1920, Southgate Jones Papers; Bailey, ms, July 21, 1920; Riddick to Mrs. Josiah Bailey, August 9, 1920, both Bailey Papers.

41. North Carolina Legislators to the Tennessee Legislature, Telegram, August 11, 1920, printed copy, Equal Suffrage Amendment Collection; *Greensboro Daily News,* August 8, 1920; Raleigh *News and Observer,* August 8, 1920. See also Raleigh *News and Observer,* August 12, 1920.

42. Raleigh *News and Observer,* August 13, 1920.

43. Mary Hilliard Hinton to ——— Houk, July 30, 1920, National American Woman Suffrage Association Papers, LC: Raleigh *News and Observer,* August 13, 15, 1920. See also Hinton to Henry G. Connor, July 6 [4?], 1920, H. G. Connor Papers, SHC; Hinton to Romulus A. Nunn, July 12, 1920, Romulus A. Nunn Papers, NCDAH.

44. William O. Saunders, in Elizabeth City *Independent,* August 20, 1920, pp. 1, 4; Joint Resolution of North Carolina Legislature in re: Newspaper Article in Pasquotank County, N.C. *The Independent* Edited by a Member of the General Assembly, Equal Suffrage Amendment Collection.

45. The following discussion is based on Raleigh *News and Observer,* August 14, 1920; *Greensboro Daily News,* August 14, 1920; Sanford Martin, comp., *Letters and Papers of Governor T. W. Bickett, 1917–1921* (Raleigh: Edwards and Broughton, 1923), pp. 60–63.

46. The following discussion is based on Raleigh *News and Observer,* August 18, 1920; *Greensboro Daily News,* August 18, 1920.

47. There is little concrete evidence to substantiate the legend that Lindsay Warren locked a prosuffrage senator in the men's rest room to prevent his voting. A reference appears in Joseph L. Morrison, *Governor O. Max Gardner: A Power in North Carolina and New Deal Washington* (Chapel Hill: University of North Carolina Press, 1971), p. 34, no. 14, but the citation to the *Charlotte Observer* for March 26, 1961, is in error.

48. Catt to Weil, August 26, 1920, Weil Papers; *Greensboro Daily News,* August 27, 1920.

49. Cuningham to My Dear Mr. Editor, August 21, 1920, Otelia Connor Papers.

Chapter 2. "Physiological and Functional Differences": Sam Ervin on Classification by Sex

1. For a fuller discussion of postsuffrage feminism and ERA, see Susan D. Becker, *The Origins of the Equal Rights Amendment: American Feminism between the Wars* (Westport, Conn.: Greenwood Press, 1981); Berry, *Why ERA Failed;* Loretta J. Blahna, "The Rhetoric of the Equal Rights Amendment" (Ph.D. diss., University of Kansas, 1973); Boles, *Politics of the Equal Rights Amendment;* Nancy F. Cott, *The Grounding of Modern Feminism* (New Haven, Conn.: Yale University Press, 1988), chaps. 2–3; Joan Hoff-Wilson, "The Unfinished Revolution: Changing Legal Status of U.S. Women," *Signs* 13 (Autumn 1987): 7–36; Stanley J. Lemons, *The Woman Citizen: Social Feminism in the 1920s* (Urbana, Ill.: University of Illinois Press, 1975); Mansbridge, *Why We Lost the ERA;* David George Ondercin, "The Compleat Woman: The Equal Rights Amendment and Perceptions of Womanhood, 1920–1972" (Ph.D. diss., University of Minnesota, 1973). For activities of the North Carolina League of Women Voters, see Marion W. Roydhouse, "The 'Universal Sisterhood of Women': Women and Labor Reform in North Carolina, 1900–1932" (Ph.D. diss., Duke University, 1980).

2. Cott, *Grounding of Modern Feminism,* chap. 4.

3. "Look Twice at the Equal Rights Amendment" (Washington, D.C.: National League of Women Voters, March 1943); editorial, *Washington Post,* February 10, 1938, copy; "Equal Rights or Human Rights" (n.p., n.d.); Paul Freund, "The Following Statement on Legal Implications of Proposed Federal Equal Rights Amendment" (Washington, D.C.: National Committee on the Status of Women in the United States, n.d.), all League of Women Voters Papers, LC.

4. Freund, "Legal Implications of Proposed Federal Equal Rights Amendment"; *Muller* v. *Oregon,* 208 U.S. 412 (1908).

5. U.S. Congress, Senate, 79th Cong., 2d Sess., July 18, 1946, *Congressional Record,* 42:9401. See also U.S. Congress, Senate, 81st Cong., 2d sess., January 23, 24, 25, 1950, *Congressional Record,* 96:738–44, 758–62, 809–13, 861, 873.

6. On the origins of the feminist movement and especially the importance of demographic change that preceded it, see William Chafe, *The American Woman: Her Changing Social, Economic, and Political Roles, 1920–1970* (New York: Oxford University Press, 1972); Ethel Klein, *Gender Politics: From Consciousness to Mass Politics* (Cambridge, Mass.: Harvard University Press, 1984); Valerie Kincade Oppenheimer, *The Female Labor Force in the United States: Demographic and Economic Factors Governing Its Growth and Changing Composition* (Berkeley, Calif.: University of California Press, 1970).

7. For an account of women who maintained their feminist commitment in the postsuffrage decades, see Leila Rupp and Verta Taylor, *Survival in the Doldrums: The Women's Rights Movement, 1945 to the 1960s* (New York: Oxford University Press, 1987). An excellent account of the President's Commission on the Status of Women and the feminists in government who played such a pivotal role in the founding of NOW is provided by Cynthia Harrison, *On Account of Sex: The Politics of Women's Issues, 1945–1968* (Berkeley, Calif.: University of California Press, 1988), pp. 109–91.

See also U.S. President's Commission on the Status of Women, *American Women* (Washington, D.C.: Government Printing Office, 1963); North Carolina Governor's Commission on the Status of Women, *The Many Lives of North Carolina Women* (n.p., 1964).

8. Jo Freeman, *The Politics of Women's Liberation: A Case Study of an Emerging Social Movement and Its Relations to the Social Policy Process* (New York: McKay, 1975), pp. 53–55; Harrison, *On Account of Sex,* pp. 187–97.

9. Freeman, *Politics of Women's Liberation,* pp. 35–37; Mississippi legislator, quoted in Carolyn Hadley, "Feminist Women in the Southeast," *Bulletin of the Center for the Study of Southern Culture and Religion* 3 (1979): 10.

10. Quoted in Sara Evans, "Tomorrow's Yesterday: Feminist Consciousness and the Future of Women," in *Women of American: A History,* Carol Ruth Berkin and Mary Beth Norton (Boston: Houghton Mifflin, 1979), p. 396. This paragraph and the following one rely heavily on Sara Evans, *Personal Politics: The Roots of Women's Liberation in the Civil Rights Movement and the New Left* (New York: Alfred Knopf, 1979).

11. Evans, "Tomorrow's Yesterday," p. 400.

12. For a discussion of women's liberationists' activities and ideas, see Marian Lockwood Carden, *The New Feminist Movement* (New York: Russell Sage Foundation, 1974), chaps. 8–10; Gayle Graham Yates, *What Women Want: The Ideas of the Movement* (Cambridge, Mass.: Harvard University Press, 1975), chap. 2.

13. Mary O. Eastwood and Pauli Murray, "Jane Crow and the Law: Sex Discrimination and Title VII," *George Washington Law Review* 34 (December 1965): 232–65. At the time of the article's publication, Murray was simultaneously pursuing another strategy by helping to draft a brief through the American Civil Liberties Union for a 1965 Alabama case that she, Eastwood, and others hoped would generate a Supreme Court decision extending the equal protection clause of the Fourteenth Amendment to women. Following a ruling by the federal district court, the case was not appealed to the Supreme Court, so the decision never materialized. Harrison, *On Account of Sex,* pp. 183–84.

14. Harrison, *On Account of Sex,* pp. 183–84.

15. *Hoyt* v. *Florida,* 368 U.S. 57 (1961): *Mc Donald* v. *Board of Elections,* 394 U.S. 802, 809 (1969). For a more detailed discussion of the equal protection clause and its application to women, see Leslie Friedman Goldstein, *The Constitutional Rights of Women: Cases in Law and Social Change* (New York: Longman, 1979), pp. 66–163.

16. U.S. Congress, Senate, *The "Equal Rights" Amendment,* Hearings before the Subcommittee on Constitutional Amendments of the Committee on the Judiciary; U.S. Congress, Senate, 91st Cong., 2d sess., on S. J. Res. 61 . . . May 5, 6, 7 1970 (Washington, D.C.: Government Printing Office, 1970); U.S. Congress, Senate, *Equal Rights for Men and Women,* Senate Report 92-689, 92d Cong., 2d sess., 1972.

17. Samuel J. Ervin, Jr., to Harry Vander Linden, January 22, 1957, Samuel J. Ervin, Jr., Papers, SHC.

18. Remarks Prepared by Senator Sam J. Ervin, Jr. (D-N.C.) for Delivery to the Senate on August 21, 1970, MS, Ervin Papers, ERA General Subject Files, box 98, folder 255, p. 7.

19. U.S. Congress, Senate, *Equal Rights, 1970,* Hearings before the Committee on the Judiciary, U.S. Congress, Senate, 91st Cong., 2d sess., September 9, 10, 11, 15, 1970 (Washington, D.C.: Government Printing Office, 1970) (hereafter *Equal Rights, 1970*).

20. The responses, all in Ervin Papers, box 398, folder 269, include correspon-

dence from Clarence A. H. Meyer, Nebraska, August 31, 1970; Jack P. F. Gremillion, Louisiana, September 1, 1970; James M. Jeffords, Vermont, September 2, 1970; Arthur K. Bolton, Georgia, September 3, 1970; Slade Gorton, Washington, September 4, 1970; Ray Shollenbarger, New Mexico, September 4, 1970; B. J. Jones, Missouri, September 9, 1970; Daniel R. McLeod, South Carolina, September 10, 1970; Paul Sand, North Dakota, September 10, 1970; Joseph W. McKnight, Texas Family Code Project, September 11, 1970; Robert B. Morgan, North Carolina, September 14, 1970; Robert Woodahl, Montana, September 16, 1970; Louis J. Lefkowitz, New York, September 16, 1970; James S. Erwin, Maine, September 17, 1970; Gary K. Nelson, Arizona October 8, 1970; Thomas N. Biddison, Jr., Maryland, October 16, 1970; Andrew P. Miller, Virginia, October 16, 1970; Earl Faircloth, Florida, October 16, 1970; Warren Felton, Idaho, October 19, 1970; G. Kent Edwards and M. Gregory Papas, Alaska, November 9, 1970. Other letters, addressed to Emanuel Celler, also Ervin Papers, are from Paul Sand, North Dakota, July 31, 1970; Leslie Hall, Alabama, August 3, 1970; W. J. Monroe, Oklahoma, August 3, 1970; Donald L. Painter, Wyoming, August 4, 1970; Walter W. Andre, South Dakota, August 5, 1970; Paul W. Brown, Ohio, August 5, 1970; John B. Breckenridge, Kentucky, August 6, 1970; George H. Hawes, Nevada, September 4, 1970; Patricia J. Gifford, Indiana, September 30, 1970.

21. C. C. Small, Jr., to Ervin, September 15, 1970, Ervin Papers.

22. The following paragraphs are based on U.S. Congress, Senate, 91st Cong., 2d sess., October 8–November 16, 1970, *Congressional Record,* 26:35621–28:37280, esp. 27:36265–78, 36299–315, 36372–3, 36448–51, 36478–505, 36862–66, 28:37268–9. See also *Equal Rights, 1970,* pp. 4–6.

23. The following discussion is based on Paul R. Clancy, *Just a Country Lawyer: A Biography of Senator Sam Ervin* (Bloomington, Ind.: Indiana University Press, 1974), pp. 171–96; Dick Dabney, *A Good Man: The Life of Sam J. Ervin* (Boston: Houghton Mifflin, 1976), pp. 179–81, 213–22. See also Ervin to Mrs. A. V. Anderson. August 2, 1971; Ervin to W. L. Owens, December 22, 1971, in which Ervin boasted of his persistent opposition to civil rights acts, the Warren court, the Department of Health, Education, and Welfare, school desegregation, and busing, both Ervin Papers.

24. See Ervin's remarks in U.S. Congress, Senate, 92d Cong., 2d sess., February 15, 1972, *Congressional Record,* 4:3863–66, citing Title VII of the Civil Rights Act of 1964, which Ervin opposed, and the powers of the federal government to deal with job discrimination under EEOC, which Ervin also opposed. Also see form letters, October 18, November 9, 1971; Ervin's suggestion to Mrs. J. L. McCabe, Jr., November 10, 1971, that she file complaints about sex discrimination not with his office but EEOC, all Ervin Papers.

25. Sam J. Ervin, "Freedom in Peril," speech to the Tennessee Bar Association, January 24, 1964, Johnson City, Tenn., Henry Brandis to Ervin, March 3, 1964, both Ervin Papers.

26. Ervin to Hugh T. Lefler, February 20, 1964, Ervin Papers. See also letters from Ervin to William E. Lovell, February 10, 1964; Grace E. Paw, April 12, 1964; William M. Alston, Jr., May 19, 1964; and others in Civil Rights folders, Ervin Papers. Ervin believed that women who favored the Equal Rights Amendment were also "snivelers." When Elizabeth Caviness wrote him in 1979 that women did not always have the funds to go to court to redress grievances and asked if he would be willing to represent them free of any charge, he replied, "I have some obligations of my own and cannot devote my time to acting as attorney for all women who may be dissatisfied with their life." February 21, 1979, Ervin Papers.

27. Ervin to William E. Lovell, Duke University Divinity School, February 10, 1964, Ervin Papers. See also Ervin to Rev. David E. Miller, December 13, 1963, Ervin Papers.

28. Neil Spain to Ervin, August 8, 1971, Ervin Papers. There are literally hundreds of letters in the Ervin Papers on the problems Cary faced. See, esp., Ervin to A. V. Anderson, August 2, 1971; letters to Ervin from Thomas S. Hughes, August 25, 1970; E. L. Geiersbach, August 3, 1971; Robert Havens, August 3, 1971; Leslie F. and Marie E. Hardy, August 5, 1971, all Ervin Papers.

29. The following discussion is based on Clancy, *Just a Country Lawyer,* pp. 200–21.

30. Ervin to Rev. Claude Logan Asbury, May 28, 1971; Ervin to Mrs. H. F. Kimbrough, November 26, 1974, copy, both Ervin Papers.

31. Barbara A. Brown, Thomas I. Emerson, Gail Falk, and Ann E. Freedman, "The Equal Rights Amendment: A Constitutional Basis for Equal Rights for Women," *Yale Law Journal* 80 (April 1971); 871–985, reprinted in U.S. Congress, Senate, 92d Cong., 1st sess., October 5, 1971, *Congressional Record,* 27:35012–41 (hereafter referred to in text as Yale article).

32. For a brilliant analysis of the analogical possibilities of comparing the place of blacks and women in American culture, see William H. Chafe, *Women and Equality* (New York: Oxford University Press, 1976).

33. Brown, Emerson, Falk, and Freedman, "Equal Rights Amendment," pp. 873, 876, 882, 890, 893, esp. pp. 884, 885.

34. Ibid., pp. 872, 884, 885, 890–92, 905–7, 939.

35. See chap. 5.

36. Sam J. Ervin, "Excerpts from *Yale Law Journal* Report Projecting Effects of Equal Rights Amendment" (n.p., March 1972); "The Equal Rights Amendment: A Potentially Destructive and Self-defeating Blunderbuss," reprint from the *Congressional Record,* March 20, 1972; "The Equal Rights Amendment: Excerpts from the Minority Views of Senator Sam J. Ervin, Jr. (North Carolina)" reprint of "Excerpts from *Yale Law Journal* Report," published as a pamphlet by the National Defense Committee of the Daughters of the American Revolution, Ervin Papers. See also "The Legislative History of ERA," *Phyllis Schlafly Report* 10 (November 1976): 1–3.

37. Ervin to Mrs. J. L. McCabe, Jr., November 10, 1971, Ervin Papers.

38. California women workers and unions in the early 1970s soundly supported ERA only after they pushed for and got authorization for extension of some protective legislation to men by the California Industrial Welfare Commission. Although the authorization was passed by the state legislature, it has been effectively eviscerated by business interests. See Barbara Balser, *Sisterhood and Solidarity: Feminism and the Modern Labor Movement* (Boston: South End Press, 1987), pp. 103–8. It is probable under the impress of eight years of the Reagan administration that such resistance to protection will continue. Ervin and Freund were not out of date and out of touch on this particular point.

39. Ervin's exceptions are in box 398, folder 262, Ervin Papers. His amendments became the basis for a STOP ERA handout. See Joan M. Grubbins, "The Equal Rights Amendment and the Intent of Congress" (Reidsville: North Carolina STOP ERA, n.d.), Ervin Papers.

40. Dorothy M. Slade to Ervin, April 28, 1972, Ervin Papers. For contacts with anti-ERA activists from other states, see anti-ERA files, including letters to Ervin from Zilla Hinton, STOP ERA coordinator for South Carolina, April 24, 1973; Mary D. Turnipseed, Atlanta, Ga., February 21, 1973; Mrs Pat Ormsby, Happiness of

Womanhood of Kingman, Ariz., September 28, 1972, all Ervin Papers. See also a HOW membership card in the men's auxiliary with Ervin's name on it, Ervin Papers.

41. Phyllis Schlafly to Ervin, February 19, July 3, 1972, April 20, 1973, September 16, 1974, all Ervin Papers. See also Phyllis Schlafly, *A Choice Not an Echo* (Alton, Ill.: Pere Marquette Press, 1964).

42. Schlafly to Ervin, July 3, 1972, Ervin Papers.

43. See Phyllis Schalfly to Gay Holliday, January 19, 22, 25, 26, 29, 1973, April 20, 30, 1973; Holliday to Schlafly, February 8, 1973, copy; Holliday, memorandum, February 7, 1973, all Ervin Papers. See also Holliday to Carol Grayson, March 21, 1973, identifying the national movement against ratification with Ervin and claiming that by now the press has begun to believe that opposition to ERA is "defendable," Ervin Papers.

44. Lola D. Bruington to Holliday, September 25, 1972, Ervin Papers.

45. Ervin to Mrs. Frederick J. Maas, September 19, 1972, copy; Ervin to Lyle E. Cox of Indiana, September 25, 1972, both Ervin Papers.

Chapter 3. "We Just Didn't Realize It Was Going To Be That Difficult": Political Socialization the Hard Way

1. Information is this and the following paragraph is based on participant observation beginning in 1969, discussions with Sara Evans and Paula Goldsmidt, formerly of Chapel Hill, and NOW Papers in the possession of Miriam Slifkin. See Slifkin to Becky Patterson, May 11, 1973; Slifkin to Jackie Frost, December 27, 1973, both Slifkin Papers, Chapel Hill. See also NOW newletters, also Slifkin Papers, for accounts of NOW's North Carolina activities.

2. Participant observation. For a list of the questions put to gubernatorial candidates, see the NCWPC Questions for Gubernatorial Candidates, Slifkin Papers.

3. See chap. 2; Freeman, *Politics of Women's Liberation,* pp. 209–29; Emily George, *Martha W. Griffiths* (New York: University Press of America, 1982), pp. 169–82.

4. Martha McKay, "NCWPC Chairwoman to the Membership," in NCWPC newsletter, summer 1972, NCWPC Papers, in Florence Glasser Papers, Chapel Hill.

5. The above points are developed in greater detail by Berry, *Why ERA Failed,* chaps. 4, 6; Janet K. Boles, "Systemic Factors Underlying Legislative Responses to Woman Suffrage and the Equal Rights Amendment," *Women and Politics* 2 (Spring–Summer 1982): 5–22.

6. See, for example, *NOWSLETTER,* the newsletter of the Raleigh NOW chapter, n.d. [ca. October 1972], Slifkin Papers; *Action Line,* January 1973, Slifkin Papers and Tennala Gross Papers, Greenville, temporarily in authors' files. Common Cause also contributed a pamphlet entitled "Equality of Rights Shall Not Be Arbridged on Account of Sex" as well as an "Action Program," which included information on ERA, its legislative history, and detailed instructions on media, public relations, lobbying, coalition building, including sample press release, speeches, testimony, quotations from the opposition, all Gross Papers. Extensive materials were also solicited from the Citizens' Advisory Council on the Status of Women, including memorandums on the probable impact of ERA on a wide variety of issues ranging from alimony and child support to sex-based discrimination in prison sentencing. See esp. "The ERA—What It Will and Won't Do: A Memo on the Proposed Interpretation of the ERA in Accordance with Legislative History," Glasser Papers. Typical of the many speeches given by

Elisabeth S. Petersen on the impact of ERA is that reported in *Durham Morning Herald,* November 19, 1972.

7. See Mrs. Robert A. ("Charlie") Griffiths to Petersen, November 13, 28, 1972, January 4, 1973, Elisabeth S. Petersen Papers, Chapel Hill. Griffiths, a North Carolinian, was a member of the Citizens' Advisory Council on the Status of Women.

8. For sample questionnaires, letters from various members of the General Assembly, and the final tally, see Petersen Papers. BPW also polled candidates.

9. Senator Charles B. Deane, Jr., to Petersen and Patricia Locke, September 22, 1972; Sidney S. Eagles, Jr., to Representatives Marilyn R. Bissell, Jo Foster, and Lura Tally, December 5, 1972; Petersen to Deane, January 1, 1973; Petersen to Tally, December 14, 1972; Tally to Willis P. Whichard, December 20, 1973; Petersen to Deane, November 27, 1972, all Petersen Papers.

10. Petersen to Elizabeth ("Beth") Geimer, n.d.; Jane Patterson to Petersen, n.d., both Petersen Papers. Jane Patterson and Petersen also discussed the need to mobilize support form the Democratic Women's Clubs, prominent Democratic women, black women's professional groups, and such organizations as the League of Women Votes and the North Carolina Federation of Women's Clubs. See Petersen to Jane Patterson, n.d., Petersen Papers. For efforts to enlist Republican support, see Petersen to Grace Rohrer, n.d., Petersen Papers.

11. Other members included the North Carolina Civil Liberties Union, the Professional Women's Caucus, the North Carolina Chapter of the Women's International League for Peace and Freedom, and the North Carolina Federation of Women's Clubs. Federally Employed Women in North Carolina also supported ERA.

12. Betty Wiser to Marcia Herman-Giddens and Petersen, November 12, 1972, Petersen Papers. See also Petersen to Deborah Henderson, December 14, 1972; Petersen to Geimer, n.d.; Geimer to Petersen, n.d.; Petersen to Margaret Harper, n.d.; Petersen to Stephanie Carstarphen, n.d.; Nancy Brock to Petersen, December 12, 1973, all Petersen Papers.

13. *Charlotte Observer,* January 30, 1973; *Greensboro Daily News,* January 1, 1973. As president of the Winston-Salem Chapter of NOW, Nancy Drum helped develop a coalition that included the League of Women Votes, Church Women United, North Carolina Women's Political Caucus, the YWCA, and the American Civil Liberties Union. The Winston-Salem coalitions met with candidates and held lobbying workshops for ratificationists as well as workshops on the impact of the amendment. See Drum to Locke, January 3, 1973; Meyressa H. Schoonmaker to Eloise D. George, January 15, 1973, both Petersen Papers.

14. Petersen to Brock, May 31, 1974, Petersen Papers.

15. Slade to Ervin, April 28, 1972, Ervin Papers. Other North Carolinians had been in touch with Ervin for material with which to acquaint Extension Homemakers' Clubs about ERA. See Ruth Futrell to Ervin, August 29, 1972, Ervin Papers.

16. *Asheville Citizen,* January 23, 1973. For the full texts of HR 2 and S 8 and list of legislative sponsors, see Bills and Resolutions, N.C. General Assembly, House (1973); Bills and Resolutions, N.C. General Assembly, House (1973); Bills and Resolutions, N.C. General Assembly, Senate (1973), both NCC. Governor James H. Holshouser's endorsement came in his State of the State Address but was not followed up by substantial lobbying. *Greensboro Daily News,* January 1, 1973. For editorial endorsements, see esp. *Charlotte Observer,* January 14, 1973; *Asheville Citizen,* January 22, 23, 1973. Thorough coverage was provided by the Raleigh *News and Observer* and the *Greensboro Daily News;* the *Greensboro Daily Record* also provided coverage. An

occasional article appeared in the *Wilmington Morning Star,* usually in its "women's section."

17. Ervin to Marilyn Dawson, February 7, 1973, Ervin Papers. For accounts of antiratificationist lobbying, see *Charlotte Observer,* January 24, 1973; *Greensboro Daily Record,* January 26, 1973. Television stations throughout the central piedmont also gave extensive coverage to anti-ERA women and their ministers, especially Charlotte's CBS affiliate WBT and Greensboro's WFMY. Prior warnings of antiratificationist efforts had come from McKay and also from Ann C. Scott, NOW's national vice-president for legislation who addressed the state convention of NOW in Chapel Hill. Scott spoke extensively about the tactics of groups such as Happiness of Womanhood while referring also to the opposition of Schlafly and Ervin. The opposition of the latter she astutely noted arose not on constitutional but on personal grounds. See an extensive account of Scott's remarks in *Chapel Hill Newspaper,* January 17, 1973.

18. Raleigh *News and Observer,* January 28, 1973; *Greensboro Daily News,* January 29, 1973.

19. *Charlotte Observer,* February 2, 1973; *Greensboro Daily Record,* February 2, 1973; Raleigh *News and Observer,* February 2, 1973. The reaction of ERA partisans, estimated at more than five hundred, was as riveting as the earlier appearance of Bible-carrying anti-ERA women. Their response, especially to Lee's exasperated observation that they were angry at men because they had to bear children, was duly broadcast by television reporters covering the hearings, thereby reinforcing the image of booing, hissing pro-ERA "militants." Recognizing that the merits of the issue were in danger of being obscured, the state's leading newspapers tried not only to cover the hearings but also to inform the public on the impact of the amendment, usually in a format that presented arguments of both sides based on interpretations put forth by Ervin and by the Yale lawyers. See *Greensboro Daily News,* February 4, 1973; Raleigh *News and Observer,* February 4, 1973; *Asheville Citizen,* February 4, 1973.

20. Slade to Ervin, May 11, 1973, Ervin Papers. For fuller reports of the hearings, see the February 9, 1973, issues of Raleigh *News and Observer, Greensboro Daily News, Charlotte Observer, Wilmington Morning Star;* Institute of Government, University of North Carolina at Chapel Hill, *Weekly Legislative Summary,* February 9, 1973.

21. *Charlotte Observer,* February 17, 1973; Raleigh *News and Observer,* February 19, 1973.

22. *Greenboro Daily Record,* February 24, 1973. Interview with Martha McKay by Belinda Rigsbee, quoted in notes on the 1973 struggle for ratifiction, April 22, 1973, authors' files.

23. Raleigh *News and Observer,* March 1, 1973; *Asheville Citizen,* March 1, 2, 1973; *Greensboro Daily News,* March 2, 1973; *Charlotte Observer,* March 4, 1973.

24. Slade to Ervin, March 6, 1973; Schlafly to Holliday [and Ervin], April 30, 1973, both Ervin Papers.

25. In the weeks after the defeat of ERA, opponent senators proposed a state ERA with the qualifying phrase that "the rights, benefits, and exemptions now conferred by law upon persons of the female sex shall in no way be impaired." Rejecting the qualification, McKay and NCWPC countered with a proposal for a state equal employment act that would extend coverage to employees not covered by the federal act and also establish a fair employment practices commission with investigatory and enforcement powers. See Statement by Martha McKay, Legislative Affairs Chairperson, NCWPC, n.d., Petersen Papers.

26. Quoted in Lisa Cronin Wohl, "White Gloves and Combat Boots: The Fight for

ERA," *Civil Liberties Review* 1 (Fall 1974): 78–80. Connections to the Radical Right were indeed present, and subsequent commentators have written about the women of the Right who opposed ERA with considerable insight. See, for example, Barbara Ehrenreich, "Defeating the ERA: A Right-Wing Mobilization of Women," *Journal of Sociology and Social Welfare* 9 (September 1982): 391–98; Andrea Dworkin, *Right-Wing Women* (New York: G. P. Putnam's Sons, 1983); Rebecca Klatch, *Women of the New Right* (Philadelphia: Temple University Press, 1987).

The temptation for pro-ERA partisans to emphasize the far-right coloration of opponent women was consistent with their own understanding of themselves as representing the women's movement—that is, all women—and as being "grass-roots people," a perception that required them to depict the opposition and especially opposition women as an unaware (and misled) minority. Students of rhetoric as well as of collective behavior have frequently commented on the dynamics at work in political conflict that has become rancorous and symbolic and requires contending parties to differentiate, creating and manipulating images of self and of opponents that serve to legitimate the former and "de-legitimate" the latter. The problem with such rhetorical strategies, however, is that while they are understandable in terms of movement imperatives, they often obscure a realistic assessment of the opposition, a theme that will be developed further subsequently. For analysis that speaks to this issue, see, for example, Susan E. Marshall, "Ladies against Women: Mobilization Dilemmas of Antifeminist Movements," *Social Problems* 32 (April 1985): 348–62; Sonja K. Foss, "Equal Rights Amendment: Two Worlds in Conflict," *Quarterly Journal of Speech* 65 (October 1979): 275–88. The interpretation of conflict over ERA as symbolic conflict was first elaborated by the authors in "ERA in North Carolina: A Case Study in Opposition to the Equal Rights Amendment," paper presented to the Center for the American Woman and Politics, Eagleton Institute for Politics, Rutgers—the State University, New Brunswick, N.J., September 29–30, 1977. Other scholars developing this argument include Carol Mueller, "Rancorous Conflict and Opposition to the E.R.A.," *New England Sociologist* 3 (1981): 17–29. Wilbur J. Scott also sees conflict over ERA as symbolic (and cultural) conflict. "The Equal Rights Amendment as Status Politics," *Social Forces* 64 (December 1985): 499–506.

27. For a case study of the problems of ERA coalition building from the vantage point of an AAUW vice-president and a member of the board of ERAmerica, see Nancy Douglas Joyner, "Coalition Politics: A Case Study of an Organization's Approach to a Single Issue," *Women and Politics* 2 (Spring–Summer 1982): 57–70. For a very perceptive early analysis of problems affecting coalition building in North Carolina—the jockeying among NOW, the league, and NCWPC, the overrepresentation of small groups such as NOW and the underinvolvement of larger, more traditional organizations such as the North Carolina Federation of Women's Clubs and especially BPW, the difficulty of finding leaders able to work effectively with diverse groups with varying levels of commitment to ratification, and others—see Petersen to Brock, May 31, 1974, Petersen Papers; Brock to Petersen, June 9, 1974, Nancy Dawson-Sauser (formerly Drum) Papers, Winston-Salem (hereafter Drum Papers).

28. Petersen to Brock, [January 1974], Petersen Papers. Not until April did ratificationists meet to form a new coalition. See Brock, Chair NCWPC ERA Committee to Organizations and Individuals Interested in Ratification of the Equal Rights Amendment to the US Constitution, n.d., Gross Papers.

29. See correspondence referred to in nn. 27, 28; Petersen to Brock, January 27, 1974; Brock to Jane Patterson, May 31, 1974, both Petersen Papers. Note, too, Minutes of Meeting Called to Form a North Carolina Equal Rights Amendment Ratifica-

tion Coalition in Greensboro, April 21, 1974; Proposed PR Plan for the ERA Coalition, April 21, 1974, Wilma Davidson Papers, Charlotte, temporarily in authors' files. See also Minutes of the Field Service Task Force, May 2, 1974, Petersen Papers.

30. For an excellent discussion of Howard Twiggs's perception of the issue, the strategy he would employ, and the image he believed ratificationists should cultivate, see Linda Grimsley to Brock and Jane Patterson, June 14, 1974, Drum Papers. On the question of clerical help, Nancy Brock, who had taken the initiative in 1974 until moving to Virginia in the late summer, calculated that she could have "easily trebled" the amount of work had she had clerical help, reaching many more individuals and groups in much less time. Paid help, she believed, was essential to ensure "dependability." She also argued that the coordinator's job should be paid because as the key organizer and administrator the individual so designated had an " 'on call' job that can easily require *much* more than 40 hours a week, *and* can involve a good deal of travel *and* quite irregular hours and much upset to family life and normal living routines." Brock to Drum, August 13, 1974, Drum Papers.

31. Drum to Wilton Hartzler, [summer 1974], Drum Papers.

32. See the four-page undated memorandum of fund raising issued by ERA United, Gross Papers; Minutes of the ERA United Advisory Board Meeting, July 13, 1974, Davidson Papers; Nancy Klein to Drum, October 14, 1974; Drum to Douglas Bailey, December 6, 1974, both Drum Papers.

33. Gladys Tillett was, as younger colleagues noted, a "great person" with "great contacts and good advice" but not as "chipper" as she had been in 1973 and not to be trusted with administrative matters because she was so forgetful. Klein to Drum, July 21, 28, 1974, Drum Papers. As for field services, Drum's concerns were shared by other coalition leaders. Although contacts had been found for the major congressional districts and almost every county, county organization had proceeded so slowly that the recommended hiring of a full-time paid organizer was not acted upon because of financial constraints. See Minutes of the Field Service Task Force, May 2, 1974, Petersen Papers; Minutes of the Meeting of the Board of Directors, ERA United, July 28, October 13, 1974, Davidson Papers. See also Maria Bliss Papers, Asheboro, temporarily in authors' files. Enlargement of the coalition had been Drum's special concern, and she and others made efforts to recruit diverse groups.

34. The preceding paragraphs on opponent activity are based primarily on interviews. For press accounts, see Raleigh *News and Observer,* January 7, February 25, 1975; *Charlotte Observer,* January 7, 1975; *Durham Morning Herald,* January 13, 1975. For a sample of opponent letters denouncing ERA, see *Charlotte Observer,* January 19, 1975; *Durham Morning Herald,* January 5, 7, 1975; *Asheville Citizen-Times,* January 12, 13, 17, 18, 19, 22, 28, 1975, Raleigh *News and Observer,* January 15, 23, 25, 26, 1975. Such letters focused on adverse economic effects (wives would be deprived of legislation requiring spousal support and widows denied social security benefits), the deleterious effect on family life (women would be no longer have the option of remaining at home and caring for their children). The abolition of protective labor legislation, loss of draft exemption, and loss of privacy, as well as the increased power of the judiciary and increased federal intervention, were also frequently cited as consequence of passage. Other letters asserted that the rights of women who do not want to compete with men should be protected, that male superiority was biblically ordained, that ratification would mean the transformation of the United States into a "unisex" society, and that ratificationists were "a small core of women's libbers who hate men and are out to destroy marriage, morality, and the family," "libbers" who are "determined to crush the male." See, respectively, Raleigh *News and Observer,* January 26, 1975;

Asheville Citizen-Times, January 13, 1975. In reply supporters objected to arguments "too absurd to require comment"; one ratificationist felt obliged to explain that supporters were not "leftists or lesbians." Raleigh *News and Observer,* January 26, 1975; *Asheville Citizen-Times,* February 4, 1975. The use of letters to the editor to publicize opponent charges against ERA was part of antiratificationist strategy and not a reflection of anti-ERA sentiment on the part of the state's major newspapers. Whether letter opinion is congruent with public opinion when one side self-conciously uses letter columns as a forum is a question public opinion experts now argue. See Emmett Buell, Jr., "Eccentrics or Gladiators? People Who Write about Politics in Letters to the Editors," *Social Science Quarterly* 56 (December 1975): 440–49; David B. Hill "Letter Opinion on ERA: A Test of Newspaper Bias Hypothesis," *Public Opinion Quarterly* 45 (Fall 1981): 384–92. Hill argues for congruence.

For more on the emerging opposition of the Mormon Church and its impact, see John Unger Zussman and Shauna M. Adix, "Content and Conjecture in the Equal Rights Amentment Controversy in Utah," *Women's Studies International Forum* 5, no. 5 (1982): 475–86, esp. pp. 476–81.

35. For initial press accounts of the campaign in the General Assembly, see the January 17, 1975, issues of *Asheville Citizen-Times,* Raleigh *News and Observer, Charlotte Observer, Greensboro Daily News, Greensboro Daily Record,* and *Durham Morning Herald.*

36. On the referendum ploy, see Raleigh *News and Observer,* February 8, 1975; *Greensboro Daily News,* February 8, 12, 1975. On A. Hartwell Campbell's delaying tactics, see the February 26, 1975, issues of *Greensboro Daily News, Greensboro Daily Record,* and Raleigh *News and Observer.* See also *Asheville Citizen-Times,* March 9, 1975; Raleigh *News and Observer,* March 8, 1975.

37. *Charlotte Observer,* March 5, 1975. For accounts of the hearings, see the March 5, 1975, issues of Raleigh *News and Observer, Charlotte Observer, Greensboro Daily Record, Asheville Citizen-Times,* and *Durham Morning Herald.* Primary witnesses were Howard Twiggs and Gladys Tillett. Also giving testimony were Betty Barber, director of the State Commission on the Education and Employment of Women, and Grace Rohrer, secretary of the Department of Cultural Resources and the only woman in Holshouser's cabinet. Joining them were the Reverend William Finlator of the U.S. Civil Rights Commission, Jane Patterson, chair of Guilford County Democrats, President Grimsley Hobbs of Guilford College, and Alfreda Webb, vice-chair of the Democratic State Committee.

38. Raleigh *News and Observer,* March 7, 12, 1975; *Greensboro Daily News,* March 8, 16, 1975. Institute of Government, University of North Carolina at Chapel Hill, *Weekly Legislative Summary,* March 14, 1975. See also *Durham Morning Herald,* March 8, 11, 1975; *Greensboro Daily Record,* March 8, 1975; *Charlotte Observer,* March 12, 1975.

39. *Charlotte Observer,* March 20, 1975; Raleigh *News and Observer,* March 14, 16, 20, 1975. Proponents also stepped up efforts to respond to the barrage of anti-ERA letters to the editor appearing in the press. For examples of the former, see *Charlotte Observer,* March 25, 1975; *Durham Morning Herald,* March 25, 1975; Raleigh *News and Observer,* March 16, 1975. For examples of the latter, see *Asheville Citizen-Times,* March 23, 27, 29, 1975; Raleigh *News and Observer,* March 16, 1975; *Durham Morning Herald,* March 13, 23, 1975; *Greensboro Daily News,* March 19, 23, 1975. Proponents sought especially to dispel fears with respect to loss of privileges and privacy and to assure antiratificationists that ERA would not change male-female relationships.

40. Raleigh *News and Observer,* April 10, 1975; *Asheville Citizen-Times,* April 10, 1975.

41. *Greensboro Daily Record,* April 9, 1975; Raleigh *News and Observer,* April 10, 1975; *Charlotte Observer,* April 10, 14, 1975; *Asheville-Citizen Times,* April 14, 1975.

42. Raleigh *News and Observer,* April 13, 16, 1975; *Greensboro Daily Record,* April 15, 1975; *Asheville Citizen-Times,* April 16, 1975; *Charlotte Observer,* April 16, 1975.

43. For the full text of the address on which this and subsequent paragraphs are based, see Patricia S. Hunt, speech to North Caolina House of Representives, April 16, 1975, Patricia S. Hunt Papers, Chapel Hill, temporarily in authors' files.

44. Raleigh *News and Observer,* April 17, 1975; *Greensboro Daily Record,* April 16, 17, 1975; *Durham Morning Herald,* April 17, 20, 1975; Winston-Salem *News,* April 16, 17, 1975; *Charlotte Observer,* April 17, 1975.

45. *Greensboro Daily Record,* April 17, 1975.

46. Wohl, "White Gloves and Combat Boots," pp. 77–86. Support for ERA and for feminism reached a new high point in 1974. See Mark R. Daniels, Robert Darcy and Joseph W. Westphal, "The ERA Won—At Least in the Opinion Polls," *PS* 15 (Fall 1982): 578–84; Keith T. Poole and L. Harmon Ziegler, "The Diffusion of Feminist Ideology," *Political Behavior* 3, no. 3 (1981): 229–56. Poole and Ziegler attribute the decline in support after 1974 to the effectiveness of Schlafly's charges. With respect to the one hundred groups supporting ERA as members of coalition, Boles rightly notes that less than a dozen were actively involved in the ratification effort, and, among most of those, ERA was only one of several priorities. Boles, "Systemic Factors," p. 17. For the Scott analysis, see Wohl, "White Gloves and Combat Boots," pp. 83, 86. For an excellent account of the Florida ratification struggle, see Joan S. Carver, "The Equal Rights Amendment and the Florida Legislature," *Florida Historical Quarterly* 60 (April 1982): 455–91. In Florida proponents in 1974 had also conducted a low-key campaign and lost in the Senate because of defections.

47. Report of the Structure and Strategy Committee, n.d.; Report of the Strategy Committee, February 28, 1976; Minutes of the Board of Directors, ERA United, February, 28, 1976, all Bliss Papers; North Carolinians United for ERA: An Overview, n.d., Davidson Papers. The eight standing committees covered budget, fund raising, public relations, resources, speakers, legislative matters, field service, and research and information.

48. The official name change did not occur until the following June. See Minutes of the Board of Directors, NCUERA, June 19, 1976, Bliss Papers.

49. For a discussion of the ratification coalition at the national level and the evolution of strategy, see Janet K. Boles, "Building Support for the ERA: A Case of 'Too Much, Too Late,' " *PS* 15 (Fall 1982): 572–77. The defeat of state ERAs in New York and New Jersey demonstrated the effectiveness of the opposition in states that had initially ratified the federal amendment. Of the three highly urban, liberal northeastern states in which ERA was submitted to voters in the mid-1970s, only Massachusetts voted for a state ERA. Opposition arguments, note Carol Mueller and Thomas Dimieri, did not differ from those used against the federal amendment, and the level of resources mobilized was also similiar. With respect to the Massachusetts struggle, Mueller and Dimieri observed that although the campaign involved a state rather than a federal amendment, the issues raised and the resources mobilized did not differ. A close reading of the Massachusetts and North Carolina campaigns indicates that the opposition used identical arguments, the primary difference being that opposition

leadership drew from conservatives in the Catholic church rather than fundamentalist Protestants, as was the case in North Carolina. See Mueller and Dimieri, "Structure of Belief Systems," p. 673 n. 2.

50. For an account of the New York campaign, see Lisa Cronin Wohl, "The ERA: What the Hell Happened in New York?", *Ms,* March 1976, 64–66, 92–96; Martha Weinman Lear, " 'You'll Probably Think I'm Stupid'," *New York Times Magazine,* April 11, 1976, pp. 30–31, 108–21. See also NCUERA Public Relations Committee report, Setting the Tone for the Campaign, May 1, 1976, Davidson Papers.

51. Setting the Tone for the Campaign, May 1, 1976; Minutes of the Meeting of Board of Directors, North Carolinians United for ERA (ERA United), June 19, 1976, both Davidson Papers. See also Media Campaign Options, n.d.; NCUERA Overview, n.d., all Davidson Papers.

52. For an early account of the Legislative Committee's objectives, see Report of the Strategy Committee, February 29, 1976, p. 4, Bliss Papers. For a much more detailed report that also outlined legislative strategy and conveyed important information on the House and Senate, see North Carolinians United for ERA: Legislative Report, [September 1976], Davidson Papers.

53. For the initial statement of field service objectives, see Report of the Strategy Committee, February 29, 1976. See also Liz Wolfe to Field Service Committee Members, March 18, 1976; Wolfe and Hope Williams to Field Service Committee Members, April 1, 1976, both Davidson Papers. The packets dispensed to district conveners accompanied by a cover letter from Bliss dated July 1976 are in Bliss Papers. For a detailed and candid assessment of field services efforts in 1976–77 on which this and previous paragraphs rely, see Lora Lavin to Jane De Hart, July 13, 1977, authors' files. Districts where conveners were still needed as late as October are listed in Lavin to the NCUERA Board and Constituent Organizations, October 9, 1976, Davidson Papers.

54. Finances were a constant problem. A tentative budget of $50,000 was approved in June 1976 with the stipulation that $15,000 be raised by August 1 and that the budget be reviewed on that date. Revised downward to $41,000 in August, it was revised upward again to $50,000 in October to permit hiring Thomas Barringer as a lobbyist. Minutes of the Meeting of Board of Directors, June 19, 1976; Minutes of the Meeting of the Board of Directors, October 9, 1976; Minutes of the Meeting of the Steering Committee, October 9, 1976, all Davidson Papers. Ultimately the League of Women Voters contributed $10,500—the largest single amount poured into any state by the league. Constituent organizations at the state level put in small sums, the largest being the North Carolina League of Women Voters, $1,495, and BPW $595; most organizations contributed $100–200. See Treasurer's Reports, Davidson Papers; Raleigh *News and Observer,* January 14, 1977.

55. NCUERA Overview, n.d., Davidson Papers. Tibbie Roberts had been active in the 1975 campaign as Nancy Drum also made an effort to enlist liberal religious activists. See *Chapel Hill Newspaper,* September 19, 1976, and *Charlotte News,* January 5, 1976, for reprints of typical addresses by William B. Aycock. The *Chapel Hill Newspaper* issued a special reprint of a December 2d speech, which ratificationists circulated widely.

56. Ratificationists stated that their first priority after evaluating the primary and runoff primary elections in August and September was the opening of headquarters in Raleigh to "centralize and better effect campaign coordination." "The campaign," however, seems to have focused as much on "mobilizational" politics as on electoral politics. See North Carolinians United for ERA: Campaign Projections, [October 1976], Bliss Papers.

57. NCUERA's newsletter carried a front-page article in its September 1976 issue on the election and especially the lieutenant governor's race. Reporting on the different positions taken on ERA by Green and Lee, ERA supporters were urged to work for the candidate of their choice—with the caveat that "NCUERA does not endorse candidates"—despite the headline "Sept. 14th Run-Off Crucial to ERA." The title of NCUERA's newsletter varies; copies are in authors' files. Green had been urged by his campaign manager to claim a position of neutrality on ERA, toning down his opposition. Raleigh *News and Observer,* November 22, 1976.

After the elections NCUERA leaders reported that proponents had picked up one more House vote than expected and that they were "very optimistic." Campaign Update, n.d., Davidson Papers. Nationally, both sides expressed confidence based on election returns, Schlafly claiming that momentum was on the side of the antiratificationists, Sheila Greenwald of ERAmerica pointing to referenda victories in Massachusetts and Colorado. Winston-Salem *Sentinel,* December 31, 1976.

58. Schlafly to Ervin, January 20, 1977, Ervin Papers.

59. Slade to North Carolina STOP ERA Members, n.d., Ervin Papers.

60. For the characterization of Hunt, see NUCERA Legislative Report, n.d., Davidson Papers. A typical speech by Hunt in support of women's issues, including ERA, is his address to the North Carolina BPW, Winston-Salem, June 11, 1976, summarized in a news release, Davidson Papers.

61. Newspapers surveying legislators were the Raleigh *News and Observer, Charlotte Observer, Asheville Citizen-Times,* Winston-Salem *Journal, Greensboro Daily News, Durham Morning Herald,* and *Fayetteville Observer.*

62. NCUERA's call for letters would bring in a minimum of fifteen to twenty thousand letters in the month of January alone if county and district organizers followed through as instructed. Lillian Woo to District Conveners and Support Organization Chairmen, December 16, 1976, Davidson Papers; Lavin to NCUERA District Conveners and Support Organization Leaders, December 10, 1976, Bliss Papers. See the Raleigh *News and Observer*'s account of the advisory board and its endorsement, January 2, 1977. Schlafly countered with the observation that the tide had shifted in the opponents' favor. The comments about elitism were those of North Carolina antiratificationists.

63. For the wording of the bill and a list of sponsors, see Bills and Resolutions, N.C. General Assembly, 1977, House, NCC. See also Raleigh *News and Observer,* January 12, 22, 26, 27, 1977.

64. See the January 27, 1977, issues of *Charlotte Observer,* Raleigh *News and Observer,* and *Greensboro Daily News.*

65. The account of proponent testimony in this and the following paragraphs is based on newspaper accounts appearing on January 28, 1977, in *Charlotte Observer,* Raleigh *News and Observer,* and *Greensboro Daily News.* For the entire testimony, see ERA Proponent House Testimonials, George W. Miller Papers, Durham.

66. Following the hearings, opponents who had been inundating legislators with mail and personal visits often led by fundamentalist ministers once again resorted to their letter-writing strategy with respect to the press. Typical of the comments that rehashed arguments about the amendment's presumably disastrous effects were the remarks of an irate correspondent who quipped, "Let HEW run the nation and be done with it!" and another who charged that lawmakers had been taken in by "militant females." Raleigh *News and Observer,* January 30, 1977; *Greensboro Daily News,* February 3, 1977. For ratificationists' responses, see, for example, Raleigh *News and Observer,* January 30, 1977; *Charlotte Observer,* February 2, 1977; *Asheville Citizen-*

Times, February 1, 1977. Although the Raleigh *News and Observer* devoted a full page to the ERA controversy, commenting on, among other things, the correlation between the areas of the state where opposition was intense and where support for Ronald Reagan was greatest, the number of letters to the editor and accounts of local debates on ERA carried in the press in 1977 was definitely lower than in 1975, as was press coverage generally. For John Gamble's response to antiratificationist efforts to delay the bill in the Constitutional Amendments Committee, see *Asheville Citizen-Times,* February 1, 1977; *Charlotte Observer,* February 2, 1977; *Greensboro Daily News,* February 3, 1977; Raleigh *News and Observer,* February 3, 1977. For other developments in the House, see *Asheville Citizen-Times,* February 7, 8, 9, 1977. For additional coverage of the House debate and vote, see the February 9, 1977, issues of Raleigh *News and Observer, Charlotte Observer, Greensboro Daily Record,* and *Greensboro Daily News.*

67. See February 10, 1977, issues of *Greensboro Daily News, Charlotte Observer,* and Raleigh *News and Observer.* See also the Institute of Government, University of North Carolina at Chapel Hill, *Weekly Legislation Summary, Assembly* February 11, 1977. The summary noted that the volume of letters from constituents was such that many House members had stopped trying to answer ERA mail. Also noted was the introduction of identical versions of eleven bills designed to eliminate sex discrimination in North Carolina statutes. The bills would extend to both sexes benefits applicable only to one or remove provisions applying to one sex only in criminal laws as well as laws relating to property rights, income tax, divorce, child support, and employment. An outgrowth of the Legislative Study Committee established in the wake of ERA's defeat in 1977, the bills were part of the proposed legislation contained in Legislative Research Commission, *Report to the 1977 General Assembly of North Carolina: Sex Discrimination* (Raleigh: North Carolina General Assembly, 1977), NCC.

68. *Greensboro Daily Record,* February 11, 1977; *Asheville Citizen-Times,* February 13, 1977. See n. 61 for the list of newspapers surveying legislators. On the Senate Constitutional Amendments Committee's actions, see *Charlotte Observer,* February 18, 1977; Raleigh *News and Observer,* February 18, 1977. Opponents tried to get additional public hearings in order to gain time to pressure wavering senators, Raleigh *News and Observer,* February 18, 1977. For accounts of the Senate hearings, see *Greensboro Daily News,* February 20, 22, 1977; *Charlotte Observer,* February 22, 1977; Raleigh *News and Observer,* February 23, 1977.

69. Schlafly to Ervin, February 12, 1977, Ervin Papers.

70. Schlafly to Ervin, January 29, 1977, Ervin Papers. For the mailing by Richard Viguerie, see the special appeal from Schlafly with postcards attached, Ervin Papers.

71. For accounts of other opponent activity, see *Charlotte Observer,* February 22, 23, 1977; *Asheville Citizen-Times,* February 17, 1977; *Greensboro Daily News,* February 23, 1977. For a graphic account of the rally and especially the Ervin-Schafly interchange, which was taped by Albert Coates, professor emeritus of law at the University of North Carolina, see Coates's Senate testimony, January 27, 1977, pp. 11–14, Miller Papers. Schlafly was particularly effective in her use of Van Alstyne's comment that to submit ERA to the voters in a referendum would be to refer the decision to the "lowest common denominator." Playing on the elitist image of ratificationists and their "experts," she proclaimed to her fervent followers: "and you know who the lowest common denominator *is!*" The roar of the crowd left no doubt that they did. Schlafly, speech at Dorton Arena, Raleigh, February 13, 1977, tape, authors' files.

72. *Charlotte Observer,* February 25, 26, 1977; *Greensboro Daily News,* February 24, 25, 26, 1977; *Greensboro Daily Record,* February 24, 1977; Raleigh *News and Observer,* February 25, 26, 1977.

73. According to the *Greensboro Daily News,* March 9, 1977, Marshall Rauch, who received phone calls from President Jimmy Carter and from White House counsel Robert Linshutz, especially resented the use of a Jewish aide to lobby a Jewish senator. John T. Henley, who received a call from Rosalynn Carter, conceded that "she was the perfect lady, of course." Clearly, however, the First Lady was no match for Susie Sharp. Legislators, according to columnists Rowland Evans and Robert Novak, were puzzled by Carter's intervention in what they considered a purely state matter. The Carters also talked to other senators, only one of whom, Joe B. Raynor of Fayetteville, voted for ERA. *Charlotte Observer,* March 1, 1977.

74. See the March 2, 1977, issues of *Greensboro Daily News, Charlotte Observer,* and Raleigh *News and Observer.*

Chapter 4. "Idealistic Sisterhood Has Gotten Us Nowhere": Dressing for Political Combat

1. Ann Scott, quoted in Wohl, "White Gloves and Combat Boots," pp. 77–86; D.C. activist, quoted in Patricia Beyea, "ERA's Last Mile," *Civil Liberties Review* 4 (July–August 1977): 45–49; statement by Attorney General's Office, in "Controversy over the Equal Rights Amendment," *Congressional Digest* 56 (June–July 1977): 168; ERA spokeswoman, quoted in Nadine Cohadas, "The Elusive Three States: With Only Two Years to Go, ERA Still FAces an Uphill Fight and Murky Legal Questions," *Congressional Quarterly Weekly Report* 38 (June 28, 1980): 19813–15.

2. Louis Bolce, Gerald De Maio, and Douglas Muzzio, "ERA and the Abortion Controversy: A Case of Dissonance Reduction," *Social Science Quarterly* 67 (June 1986): 304. For the role of New Right groups in establishing those linkages, see Pamela J. Conover and Virginia Gray, *Feminism and the New Right* (New York: Praeger, 1983), pp. 140–43, 167. For the significance of the abortion linkage, see Steiner, *Constitutional Inequality,* pp. 58–66.

3. Alice S. Rossi, *Feminists in Politics: A Panel Analysis of the First National Women's Conference* (New York: Academic Press, 1982), pp. 224–31.

4. For a fuller discussion of changes in the Supreme Court's handling of sex-based discrimination, see Mansbridge, *Why We Lost the ERA,* pp. 45–59.

5. Issues raised by the boycott are addressed by Alan J. Greiman and Harold A. Katz, "Is the ERA Ethical?" *Illinois Issues* 4 (December 1978): 18–22, 9. For further discussion of the merits of extension, see Steiner, *Constitutional Inequality,* pp. 75–95. The best assessment of shifts in ratificationist strategy at the national level is Boles, "Building Support for ERA," pp. 572–77.

6., Schlafly to Ervin, March 7, 1977; Ervin to Schlafly, March 16, 1977, both Ervin Papers.

7. Slade to STOP ERA Workers, July 29, 1977, Ervin Papers.

8. The impact of defeat in 1977 on the North Carolina partisans far surpassed that felt in the wake of prior and subsequent defeats. Many were literally unable to talk about it for months.

9. For ratificationists coming out of the radical wing of the feminist movement, reluctance to engage in electoral politics derived not from historic nonpartisanship but from disdain for electoral politics itself. For an excellent discussion of feminists' indif-

ference to or skepticism about the power of the vote, see Mary Fainsod Katzenstein, "Feminism and the Meaning of the Vote," *Signs* 10 (Autumn 1984): 4–26.

10. For a short history of NCWPC, see Bobette Eckland to Iris Mitgang, July 1, 1979, Bobette Eckland Papers, Chapel Hill; Mary Hopper, 1977 Annual NCWPC Report; The Plan, submitted by Eckland to NCWPC Policy Council, September 18, 1977, all Ekland Papers. This document was followed by an October 22 Call to Action from Eckland. As president of NCWPC in 1978, Eckland has in her possession papers that are especially illuminating for this period. For the criteria of endorsement, see NCWPC Endorsement of Candidates, Carolyn Highsmith Papers, Winston-Salem; Endorsement Questionnaire, Eckland Papers.

11. Meyressa Schoonmaker's candidacy illuminates the dilemma confronting the NCWPC for she was not only a feminist with impressive credentials but an evocative candidate who activated women previously uninvolved in politics. Heightening the dilemma was the fact that Schoonmaker's appeal was not limited to feminists. Her record of civic involvement and her position on a variety of issues such as utility rate reform, statewide land-use planning, and tax reform won her the endorsement of the Winston-Salem *Sentinel* (April 27, 1978) which declined to renew its backing of ERA-supporter Carl Totherow, who seemed to be too cosy with banking and insurance interests to suit the paper.

12. Information on caucus efforts in this and previous paragraphs is based on interviews and on the 1978 campaign files of NCWPC, then in the possession of Betty Ann Knudsen, Raleigh. We are indebted to Knudsen and her associates for allowing us access to confidential district-by-district assessments during the campaign.

13. Leading the campaign to defeat James McDuffie were Marsha Newton-Graham, president of the Charlotte caucus; Jackie Frost, a businesswoman and president of the Charlotte chapter of NOW; caucus member Fran Wells, a homemaker; Marilyn Williams, a city employee; and Ann Wood, also a businesswoman. Because the McDuffie challenge was so clearly a cooperative effort involving caucus activists as well as NOW leader Frost, it was illuminating to find in the NOW Papers at the Schlesinger Library, Radcliffe College, Cambridge, Mass., that McDuffie's defeat was claimed as a NOW victory. N.C. Caravan, 1978, box 38. This claim suggests that the NOW Papers, as valuable as they are voluminious, should nonetheless be used with care. Caucus leadership and funds were acknowledged in a contemporary account appearing in *Ms* magazine. William Arthur, Jr., "North Carolina: Got Their Man," *Ms,* September 1978, p. 91.

14. For thorough coverage of the Winston-Salem races, see Winston-Salem *Journal,* April 22, 28, May 31, November 8, 1978; Winston-Salem *Sentinel,* April 19, May 3, 26, 31, October 6, November 2, 8, 1978. The characterization of Helms is that of Vermont Royster, former editor of the *Wall Street Journal,* who, upon his retirement to Chapel Hill, became an avid observer of North Carolina politics. The quotation is cited by William D. Snider in *Helms and Hunt: The North Carolina Senate Race, 1984* (Chapel Hill: University of North Carolina Press, 1985), p. 39.

15. For Mildred ("Millie") Jeffreys's and Schlafly's assessments of election results on ratification nationally, sees "Equal Rights Fight," *Editorial Research Reports* 2 (December 15, 1978): 928. For results in North Carolina, see Raleigh *News and Observer,* November 9, 1978. For the NCWPC's assessment, see NCWPC newsletter, December 1978, Highsmith Papers. ERA advocates had mobilized support for twenty-five senate candidates in the general election, seventeen of them incumbents who were reelected; among the vacated seats, however, they won only two, gaining only one

anti-ERA seat. Ratificationists were able to maintain their majority in the House. The combined total spent amounted to $85,465, with more than $60,000 of that sum contributed by NWPC, which targeted the state, and by NCWPC. NOW contributed $10,000 and the North Carolina Association of Educators, a political action committee, approximately $12,000. Contribution of in-kind services to pro-ERA candidates was estimated to have amounted to another $60,000. Less partisan groups, such as AAUW, the League of Women Voters, and the North Carolina Council of Churches, conducted numerous workshops aimed at increasing constituent involvement in campaigns. For ratificationists' contributions, see Jessie Rae Scott and Wanda Canada to ERA Legislative Delegation Chairs, January 1, 1979, Miller Papers.

STOP ERA and other antiratificationist groups were also active and donated generously to candidates. PAC financial records in the State Board of Election's Office of Campaign Reporting indicate numerous contributions from out-of-state residents, many of them from California, to various STOP ERA PAC. Since such contributions represent only the tip of the iceberg and since figures comparable to those cited above in connection with ratificationist campaign efforts were not made available by antiratificationists, it is virtually impossible to assess expenditures. Money to anti-ERA candidates could also be channeled through the coffers of Jesse Helms's National Congressional Club, which raised enough money in connection with the senator's reelection to finance the most expensive campaign as of that date in the state's history. In some instances, funds were contributed from the club directly to Slade and STOP ERA. See, for example, Carter Wrenn to Alex Brock, August 24, 1979, Office of Campaign Reporting, State Board of Elections, Raleigh, which reported a contribution of $2,600. Begun as a fund-raising group in 1972 to pay off Helms's campaign debt, the club, aided by direct mail mogul Richard Viguerie, had become by 1981 the nation's largest political action committee. Snider, *Helms and Hunt,* p. 52. Although there is widely believed to have been corporate financing, especially from the insurance industry, in anti-ERA coffers, many of the PAC contributions were from individuals who raised money from church donations, bake sales, and similar activities, confirming antiratificationists' populist image. See for example, Vickie Harvey to Slade, June 3, 1977; Report of STOP ERA PAC 1, January 1978, both Office of Campaign Reporting, State Board of Elections. Mrs. Harvey, a Missouri resident, had raised $1,000 at an Easter bake sale "to help out our friends in critical states." Other contributions to anti-ERA candidates came through groups such as the Home and Family Committee of Mecklenburg County. See files in Office of Campaign Reporting, State Board of Elections.

16. Scott's belief that pressure from the White House on individual legislators was counterproductive was shared by many politically knowledgable ratificationists.

17. Scott's regard for Whichard is evident from a letter thanking him for his efforts in 1979 on behalf of ERA, to which the senator replied: "If my grandchildren should ever read my ERA files, they will know that I received many unkind letters regarding my position on this question. I hope it will make them feel as good as it did me to read one as kind as yours." Whichard to Jessie Rae Scott, February 22, 1979, Willis P. Whichard Papers, SHC.

18. The so-called confidential letter detailed the campaign support provided by ratificationist women. For an overview of strategy, see North Carolina ERA Ratification Plan, attached to Scott and Canada to ERA Legislative Delegation Chairs, January 1, 1979, Miller Papers. For press assessments, see *Greensboro Daily News,* January 10, 1979. Raleigh *News and Observer,* January 10, 1979; Winston-Salem *Journal,* Janu-

ary 10, 1979. The press also carried articles on the new "get-tough" policy of ratificationist women in the legislature. See, for example, *Greensboro Daily News,* January 11, 1979.

19. Despite the party circuit atmosphere associated with the arrival of various celebrities, the visits of Judy Carter, Liz Carpenter, Erma Bombeck, and Alan Alda were important for energizing and rebonding ratificationists, given the demoralized state of the coalition. Although Bliss had been determined to hold the coalition together following the 1977 defeat, she had great difficulty accepting the fact that, while she herself had enjoyed considerable credibility among national leaders in ERAmerica, her two closest associates did not. More important, Washington ratificationists, themselves disillusioned by losses, were unwilling to entrust leadership of future campaigns to state-based coalitions. Miriam Dorsey, who as representive of NWPC to ERAmerica was fully aware of Washington sentiments, tried with a few other North Carolina leaders to convince Bliss to form a separate strategy committee that would also be closely associated with ratificationist women in the legislature, who had felt underutilized and underinformed in 1977. Although Dorsey's efforts were persistent, they were unsuccessful. Bliss's later resignation, followed by the brief, ill-fated presidency of a former legislator, Mary Odum, meant that prior to the 1979 campaign NCUERA had had little of the attention so essential to generating an all-out effort once the legislature convened.

The Judy Carter trip to Raleigh thus provided a much-needed occasion for the faithful to rally and for veteran legislators and activists to dispense information on campaigning for pro-ERA candidates, along with the standard instructions on organizing and lobbying at the local level once the General Assembly convened. Alda's Raleigh rally, on the other hand, was a different kind of rally—and no less important. A cross between an ecumenical religious service and a political convention, it encompassed a religious invocation, jazz performances, banners, appeals to human rights and simple justice, quotations from the theologian Paul Tillich, exhortations from Representative George Miller and other political notables, and, of course, a stirring speech by Alda himself. The diversity of the women present—evidenced by age and dress—and their enthusiasm and commitment suggested the extent to which ratification had itself become both a movement and an expression of popular feminism for thousands of women who might not otherwise have enlisted in feminist ranks. For related activities sponsored by various of NCUERA's constituent organizations, see NCUERA mailings such as Beth McAllister, ERA Up-Date to N.C. ERA Supporters, [December 1978], authors' files.

20. *Raleigh Times,* January 11, 1979; Raleigh *News and Observer,* January 12, 1979; *Wilmington Morning Star,* January 12, 1979; *Greensboro Daily News,* January 13, 1979; *Asheville Citizen-Times,* January 13, 1979.

21. *Durham Sun,* January 25, 1979; *Wilmington Star,* February 1, 1979; Raleigh *News and Observer,* February 4, 1979; *Durham Morning Herald,* February 4, 1979; Winston-Salem *Journal,* February, 6, 1979.

22. *Greensboro Daily Record,* January 26, 1979; *Greensboro Daily News,* February 3, 1979; *Durham Morning Herald,* February 10, 1979; Raleigh *News and Observer,* February 10, 1979; *Charlotte Observer,* February 10, 1979. Walter C. Cockerham did have a certified public accountant attempt to verify ballots. After discarding 1,461 ballots of the 6,271 cast that could not be verified, the breakdown for the remaining 4,810 was 2,479 against ratification and 2,331 for. However, the accountant concluded his report to the senator with the statement: "Because an unusually large percentage of the ballots contained errors and reflected discrepancies, 23% per Exhibit A, I am unable to certify and attest to the results reflected above." The so-called poll thus

proved as unreliable as critics charged, further contributing to the cynicism about the genuineness of Cockerham's uncommitted stance. Ulbert L. Frost to Senator Walter C. Cockerham, February, 14, 1979, copy, authors' files. See also Mansbridge, *Why We Lost the ERA,* pp. 8–28.

23. Raleigh *News and Observer,* February 10, 1979; *Washington Post,* February 12, 1979.

24. *Greensboro Daily Record,* February 12, 13, 1979; *Charlotte Observer,* February 13, 1979; Raleigh *News and Observer,* February 14, 1979; *Greensboro Daily News,* February 14, 1979; Winston-Salem *Journal,* February 14, 1979; participant observation. Estimates of the size of the crowd ranged from Winston-Salem *Journal*'s high of five thousand to the *Greensboro Daily News*'s count of twenty-five hundred to the Raleigh *News and Observer*'s two thousand.

25. *Charlotte Observer,* February 14, 1979; Raleigh *News and Observer,* February 14, 1979; *Asheville Citizen-Times,* February 19, 1979. Opponents staged hearings on ERA and on state funding of abortions for poor women on consecutive days, North Carolina being one of the few states that provided such funds in the wake of the Hyde Amendment denying them federal funding. The linkage was reinforced by opponent signs that read: "STOP ERA STOP ABORTION" AND "STOP ERA ABORTION KILLS." The most elaborate was a handmade sign: "MOMMY, PLEASE DON'T HAVE AN ABORTION AND KILL ME. NEXT YEAR I WANT TO BE YOUR VALENTINE." Opponents contrasted the elitism of ratificationists with their own grass-roots base. Participant observation.

26. Raleigh *News and Observer,* February 15, 1979; Winston-Salem *Journal,* February 15, 1979; *Greensboro Daily Record,* February 15, 1979; *Charlotte Observer,* February 19, 1979.

27. *Greensboro Daily News,* February 16, 1979; Raleigh *News and Observer,* February 16, 1979; Winston-Salem *Journal,* February 19, 1979.

28. *Raleigh Times,* February 16, 1979; Raleigh *News and Observer,* February 16, 1979.

29. *Greensboro Daily News,* February 16, 1979; *Wilmington Morning Star,* March 11, 1979.

30. The account of the vigil in this and the following paragraphs is based on participant observation and a copy of the sermon supplied at the authors' request entitled "Know Your Enemy: A Sermon Preached by Steven Shoemaker at the Service of Prayer for People of Faith for ERA," February 26, 1979, Raleigh. Shoemaker, a Presbyterian minister with a Ph.D. in religion from Duke University, headed the Raleigh Ministerial Council and the Presbyterian University Ministry at North Carolina State University in Raleigh.

31. For an account of the Washington conference, see NCUERA newsletter, September 1979.

32. Boles, "Building Support for ERA," p., 57. For full documentation of NOW initiatives, see NOW-Papers. Although not yet cataloged by the Schlesinger Library, the papers are sufficiently well cataloged by NOW so that appropriate files can be examined on these tactics as well as earlier tactics, such as extension and the boycott.

33. *New York Times,* October 14, 1979.

34. Included were John Wilson, a national board member of the politically active National Education Association, and Collins Kilburn, a longtime lobbyist for the North Carolina Council of Churches. They recruited top-level Hunt allies, influential male legislators, and Democratic party activists. Apart from interviews, the best writ-

ten statement on ERA PAC activities is the report entitled NC ERA Update, March 15, 1980, copy, N.C. Financial and PAC files, box 39, NOW Papers. NOW activities in the campaign were also impressive, indicating extensive information gathering on incumbent and potential candidates, effective targeting, and extensive work on behalf of pro-ERA candidates. See also N.C. 1980 House and Senate Interviews, box 38, NOW Papers.

35. Raleigh *News and Observer,* April 25, 29, 30, 1980.

36. Raleigh *News and Observer,* April 4, 14, 27, 1980.

37. For the Birch Society's actions, see Raleigh *News and Observer,* October 24, 1980.

38. Raleigh *News and Observer,* October 23, 26, 1980. The conservative celebrities included Howard Philips, Chair of the Conservative Caucus; Paul Weyrich, executive director of the Committee for the Survival of a Free Congress; and Connaught Marshner, coordinator of the profamily movement.

39. Raleigh *News and Observer,* February 24, March 16, April 6, 13, 24, 27, May 4, 7, 8, October 19, 1980.

40. For the North Carolina races, see Raleigh *News and Observer,* November 5, 6, 1980. For expenditures of the National Congressional Club, which led in the amount of money spent to elect conservative candidates—$43 million—and other conservative PAC, see Raleigh *News and Observer,* April 22, 1981; *New York Times,* May 31, 1981. Political scientists point out that the Radical Right could not "buy" all it had hoped as a result of the elections. With respect to the issues of ERA and abortion, Conover and Gray argue that mass constituencies in support of conservative candidates were not mobilized, thereby dening the latter a conservative mandate upon which to act. *Feminism and the New Right,* chaps. 6, 7. It should be noted, however, that both ratificationist and antiratificationist women showed increasing skill in the 1980 elections, and they also spent more heavily. NOW, for example, which had spent only $10,000 in the 1978 elections in North Carolina, spent more than $30,000 in the primary alone, not counting in-kind contributions. See N.C. Financial and PAC files, box 39, NOW Papers. See also Campaign Report Transmittal, Office of Campaign Reporting, State Board of Elections. For the assessment of the impact of STOP ERA PACs in Illinois, see Judson H. Jones, "The Effect of Pro- and Anti-ERA Campaign Contributions on the ERA Voting Behavior of the 80th Illinois House of Representatives," *Women and Politics* 2 (Spring 1982): 71–86, esp. 84. NOW interviews with candidates as well as interviews by the authors indicate heightened activity on the part of fundamentalists, one Senate aspirant indicating that the "fundies" were all over him for a commitment. See NC 1980 Interviews: Senate, box 38, NOW Papers. Reporters for some of North Carolina's leading newspapers had begun zeroing in on the growing political impact of the fundamentalists and their opposition to ERA during the 1979 legislative session as well as in the 1980 elections. See, for example, Raleigh *News and Observer,* February 15, 19, 1979; *Charlotte Observer,* February 11, 13, 1979. Legislators interviewed also noted the increased visibility and skill of fundamentalist lobbyists in the 1979 session, one veteran representative ranking them near the top in effectiveness—a position usually claimed by the bank lobbyists and savings and loan companies along with the insurance and utilities industries.

41. For legislators' appraisals of ERA's chances, see Raleigh *News and Observer,* November 7, 1980, January 2, 17, 21, 1981. For appraisals by ratificationist women leaders, see Raleigh *News and Observer,* December 11, 1980. McAllister's optimism in the wake of the election was also evident in her communications to the coalition. See, for example, NCUERA newsletter, December 1980.

42. For the original quotation (which referred to "pauseless campaigns"), see Carrie Chapman Catt and Nettie Rogers Schuler, *Woman Suffrage and Politics* (New York: Charles Scribner's Sons, 1926), pp. 107–8.

43. Raleigh *News and Observer,* February 6, 10, 11, 13, 14, 18, 19, 1981. Proponent strategy was to try to defer a vote in the hopes of picking up additional support later in the session—an approach that many pro-ERA legislators considered futile. See the comments of Senator Carolyn Mathis, a longtime supporter from Charlotte, in Winston-Salem *Sentinel,* February 18, 1981.

44. Raleigh *News and Observer,* February 27, 28, 1981; *Greensboro Daily News,* February 27, 28, 1987; *Charlotte Observer,* February 27, 28, 1981. See also Winston-Salem *Sentinel,* February 27, 1981; *Durham Morning Herald,* February 28, 1981.

45. See the sources listed in n. 44, above. See also Raleigh *News and Observer,* March 3, 4, 1981; *Durham Morning Herald,* March 3, 4, 1981; Winston-Salem *Sentinel,* March 3, 1981; *Charlotte Observer,* March 3, 4, 1981. As the press noted, one of the ironies of the session was that women legislators garnered a greater number of chairs of important committees in the Senate under Green, as well as in the House, than they had ever had before. *Charlotte Observer,* December 21, 1980; Raleigh *News and Observer,* January 20, May 3, 25, 1981.

46. The March 1981 NCUERA newsletter, sent out immediately after the gentleman's agreement, called on ratificationists not only to march on May 2 but to attend weekly candlelight vigils in front of the Legislative Building, circulate petitions affirming continued support for the amendment—one million signatures was the goal—and become more involved in the Democratic party, ousting Democrats at precinct meetings who were hostile to ERA and other women's issues.

47. The account of the march is based on a detailed report in the Raleigh *News and Observer* May 3, 1981, and on participant observation.

48. McAllister, speech to May 2, 1981 march, reprinted in NCUERA newsletter, August 1981, pp. 5–6.

49. A brief account of the meeting is contained in the background section of the 1982 campaign report prepared for NCAE and NEA. See Jo Ann Norris, Janet Evans, and Pam Ryan, NCAE-NEA ERA Legislative Lobbying Campaign, p. 2 (hereafter Norris Report), authors' files. The authors are grateful to Jo Ann Norris, NCAE political affairs specialist and president of NCUERA in 1982, for providing us with a copy.

50. See report of ERA endorsements in county conventions in NCUERA newsletter, August 1981, p. 3; also in Raleigh *News and Observer,* June 19, 1987.

51. Norris Report, p. 3. The responses of anti-ERA legislators was typified by Cass Ballenger, who pronounced ERA "deader than a door nail"; Lawing and other supporters noted "signs of life." *Greensboro Daily News,* April 6, 1982; *Charlotte Observer,* April 7, 1982. The press, supportive as ever, welcomed the "resurrection" in editorials. See *Greensboro Daily News,* April 10, 1982; *Asheville Citizen-Times,* April 17, 1982; *Charlotte Observer,* April 20, 1982.

52. Norris Report, pp. 3–4. The description of Liston B. Ramsey is based on a feature story in the Raleigh *News and Observer,* January 2, 1981; other appraisals of the Speaker are based on interviews.

53. Norris Report, pp. 3–4.

54. Ibid., pp. 4–5. The meeting in which Martha McKay and Jane Patterson had tried to impress upon Eleanor Smeal and Mollie Yard that NOW would have to be part of a coalition effort and should keep a low profile in the conservative eastern part of the state occurred October 22, 1979. NOW's Alice Cohan recorded the message

quite clearly in her own notes, but it went unheeded. See N.C., Cohan Notes, box 38, NOW Papers. For comments of North Carolina NOW leaders, see N.C. ERA Countdown Campaign, Field Organization and Campaign Chronology, Outreach Section, box 37, NOW Papers. Negative references to NCUERA and especially to MCAllister and McCain, which crop up with some frequency in the NOW Papers, serve to document further the mutual mistrust evident in interviews. See, for example, not only the Field Organization and Campaign Chronology, but also N.C. Legislative and Chapter Analysis, February 13, 1981, box 38, NOW Papers. Similar difficulties were reported in Florida and Oklahoma and, according to Mansbridge, also in Illinois. *Why We Lost ERA,* pp. 136–37. While differences in perspective between national leaders and those in charge of state efforts are virtually inevitable, as both groups readily acknowledged, our interviews indicate that the tension between ERAmerica representatives such as Jean McLean and Jane Campbell and NCUERA's McAllister was negligible by comparison.

55. Norris Report, p. 4. NOW officials were aware of McCain's discomfort, but there was apparently no lasting alienation between her and the organization. In addition to Terry Schooley and NOW, NCUERA was, of course, represented, having defined itself under Norris's leadership as a grass-roots support network for McCain with an ERA PAC headed by Ann Chipley as the special legislative arm of the coalition. (Chipley wore various hats since she was not only an active member of the NCWPC but state president of AAUW.) Other constituent organizations sending representatives included the league, NCAE, and AFL-CIO. Martha McKay was on hand for the NCWPC, while Jane Patterson, now secretary of the Department of Administration, spoke on behalf of the Hunt administration. Wilma Woodard represented the Women's Legislative Caucus. Jane Campbell of ERAmerica was also present, as was a national represenitive of NEA. Representatives of the North Carolina Democratic party frequently popped in to share information and coordinate activities.

56. The Raleigh *News and Observer* and other major newspapers provided extensive coverage of the press conference and poll results on May 20, 1982, as well as NCAERA leaders' criticism summarized in the following paragraph. Advance information on the Harris poll had already circulated in the press along with reports that ERA supporters were lobbying hard. *Charlotte Observer,* May 14, 15, 1982; Winston-Salem *Sentinel,* May 14, 1982; *Asheville Citizen-Times,* May 15, 1982; *Durham Morning Herald,* May 15, 1982.

57. Jim Hunt's task was complicated by the fact that there was little room to maneuver. As Lawing observed, voting against ERA had become for many a "matter of principle"; Ballenger, in a more colloquial fashion, stated "Everybody's got their mind set in concrete." *Charlotte Observer,* May 14, 20, 1982. For other efforts by ratificationists, see Norris Report, pp. 6–7. The press carried reports of additional endorsements by the mainstream religious community, by Hunt's secretary of commerce, and others. *Charlotte Observer,* May 23, 27, 1982.

58. For McCain's instructions and the accompanying packet of ERA material, see McCain to ERA County Coordinators, May 7, 1982, Betty R. McCain Papers, Wilson. McCain's efforts were underscored by Russell Walker and Doris Cromartie of the Democratic party. Russell Walker and Doris Cromartie to ERA County Coordinator, March 5, 1982, McCain Papers. Information on NCUERA activities is based on interviews.

59. Extensive documentation of NOW's impressive organizational effort is contained in N.C. ERA Countdown Campaign, Field Organization and Campaign Chronology, box 37, NOW Papers. Months of careful preparation on the part of the core

NOW staff beginning in August 1980, the excellent materials with which both volunteers and staff worked, and the efforts of these new recruits all contributed to the effectiveness of a superbly orchestrated lobbying effort. The eagerness of women who had never been active in previous ratification campaigns to become involved in 1982 was a development attested to not only by NOW organizers but also by the press. An editorialist for *Charlotte Observer* reported receiving dozens of phone calls from women who wanted to know what they could do to help get ERA passed. *Charlotte Observer,* June 2, 1982.

60. Norris Report, p. 7.

61. Ibid., pp. 7–8; *Durham Morning Herald,* May 27, June 3, 4, 1982; Raleigh *News and Observer,* May 25, 1982.

62. Norris Report, p. 9. Raleigh *News and Observer,* May 27, 1982.

63. Both sides were already engaging in intense constituent lobbying. See *Charlotte Observer,* June 2, 1982; June 3, 1982, issues of *Asheville Citizen-Times, Charlotte Observer, Greensboro Daily News,* Winston-Salem *Sentinel;* June 4, 1982, issues of *Durham Morning Herald, Greensboro Daily News,* Raleigh *News and Observer,* June 4, 1982. The Raleigh *News and Observer,* put the count of antiratificationists at five thousand. See also Norris Report, p. 9.

64. NOW observed, "The NOW Countdown Campaign again supplied all the money, work, people, and expertise, with NCUERA dragging us two steps backward with every step forward." "The June 6th Rally," N.C. Countdown Campaign Notebooks, book 37, p. 3, NOW Papers.

65. For a list of anticipated opponent maneuvers and proposed countermeasures, see Norris Report, exhibit Y. The account of legislative action and the governor's response in this and the following paragraph rely on the June 5, 1982, issues of Raleigh *News and Observer, Asheville Citizen-Times, Charlotte Observer, Durham Morning Herald, Greensboro Daily News,* Winston-Salem *Sentinel.* Quotations are from the *News and Observer*'s account.

66. Raleigh *News and Observer,* June 7, 1982; Norris Report, p. 10.

67. Raleigh *News and Observer,* June 8, 9, 1982; Norris Report, p. 11.

68. See the June 10, 13, 1982, issues of Raleigh *News and Observer, Durham Morning Herald, Asheville Citizen-Times,* and *Greensboro Daily News;* Norris Report, pp. 11–12.

69. Raleigh *News and Observer,* June 5, 10, 11, 15, 1982; N.C. Countdown Campaign Notebooks, book 37.

Chapter 5. "We Are Called and We Must Not Be Found Wanting": ERA and the Women's Movement

1. Arrington and Kyle, "Equal Rights Activists in North Carolina"; Brady and Tedin, "Ladies in Pink"; Mueller and Dimieri, "Structure of Belief Systems"; Tedin, "Religious Preference and Pro/Anti Activism"; Deutschman and Prince-Embury, "Political Ideology of Pro- and Anti-ERA Women"; Sandra Kay Gill, "Supporting Equality for Women: An Analysis of Attitudes toward the Equal Rights Amendment" (Ph.D. diss., University of Oregon, 1982), pp. 93–141. See also survey by Institute for Research in Social Science, University of North Carolina at Chapel Hill, "North Carolina Women," November 1978, in the institute library.

2. The texts upon which much of the following analysis is based were provided by pro-ERA groups to guide public debate, including typed and mass-reproduced materi-

als as well as printed pamphlets, handouts, and folders with brief statements on specific charges in response to opponents (among them homosexuals; the draft; the right to privacy; alimony, support, and child custody Laws; social security; states' rights; biological differences between men and women), usually relying on Brown, Emerson, Falk, and Freedman, "Equal Rights Amendment," pp. 871–985, and published by the ERA coalition (which included the League of Women Voters, BPW, AAUW, North Carolina Women's Political Caucus, the North Carolina Conference of the National Organization for Women, the North Carolina Nurses Association, the Board of Managers of Church Women United of North Carolina, the North Carolina Association of Women Deans and Counselors, Professional Women's Caucus, and others) Examples are "Dollar and Cents Feminism" (Washington, D.C.: National Organization for Women, n.d.); "The Equal Rights Amendment" (Washington, D.C.: Board of Church and Society, [United Methodist Church], n.d.); "The Equal Rights Amendment, Women and the Draft" (Washington, D.C.: Women United, n.d.); "The ERA: What It Means to Men and Women" (Washington, D.C.: League of Women Voters, n.d.); "How and Why To Ratify the Equal Rights Amendment" (Washington, D.C.: National Federation of Business and Professional Women, n.d.); "The Proposed Equal Rights Amendment to the United States Constitution: A Memorandum" (Washington, D.C.: Citizens Advisory Council on the Status of Women, 1970); "YES/ERA" (N. p.: League of Women Voters, n.d.); "What Happens If This Man Leaves the Picture" (N.p.: NCUERA and Common Cause, n.d.); Katherine Settle Wright, "Why North Carolina Should Ratify the Equal Rights Amendment," Whichard Papers; "Women and the Draft: What Does 'Equal Rights' Mean?" (Washington, D.C.: Women United, n.d.); "Common Cause Action on ERA," an eighty-eight page set of instructions. These materials, in the possession of Jane De Hart, Florence Glasser, Tennala Gross, Miriam Slifkin, and Donald G. Mathews, will ultimately be placed in the special collections of the University of North Carolina at Greensboro. Some materials are also in the Whichard Papers and the Ervin Papers.

3. NCWPC newsletter, January 1972, p. 1.

4. Letters to Ervin from Linda Votta-Sullivan, September 28, 1970; Elizabeth Lansing, December 17, 1970; Pamela Oliver, February 26, 1972; Eunice Storey Deerhake, February 28, 1972; Carol Barnes, March 7, 1972; Margaret W. Bennett, March 16, 1972, all Ervin Papers. See also Ruth Easterling to Ervin, August 28, 1970; Barnes to Ervin, February 20, 1972, both Ervin Papers.

5. Letters to Ervin from Mollie Weaver, March 14, 1972; Gayle E. Bynum, October 19, 1971; Mrs. R. D. Le Brecht, October 14, 1971; Deerhake, March 22, 1972, all Ervin Papers.

6. See, for example, "Fact Sheet on the Equal Rights Amendment" and "Cool Facts for the Hot-Headed Opposition," both (N.p.: League of Women Voters, n.d.), authors' files.

7. See esp. Margaret A. Hunt to Ervin, August 25, 1970, Ervin Papers.

8. Mansbridge, *Why We Lost the ERA,* pp. 26–27.

9. See for example, "Women and the Draft: What does 'Equals Rights' Mean?"; "Draft" (N.p.: Church Women United, n.d.); "The Church, Religion, and the Equal Rights Amendment" (Washington, D.C.: Women's Division fo the Board of Global Ministries [United Methodist Church], n.d.); "The Church and Women's Rights: The Case for ERA" (Raleigh: North Carolina Council of Churches, [1982]).

10. For a discussion of the lack of public debate among pro-ERA activists about the likelihood of assigning women to combat and the presumption that congressional supporters agreed with Van Alstyne, see Mansbridge, *Why We Lost ERA,* pp. 76–89.

11. Hunt, speech to North Carolina House of Representatives, April 16, 1975, Hunt Papers.

12. Mansbridge, *Why We Lost the ERA,* pp. 76–89.

13. Jessie Rae Scott, testimony, Constitutional Amendments Committee Public Hearing, February 13, 1979, tape, authors' files. See also minutes of the hearing, General Assembly of North Carolina Papers, Legislative Building, Raleigh.

14. "Ten Reasons Why the Equal Rights Amendment Should Be Ratified" (N.p.: Republican National Committee n.d.), reason 5, authors' files.

15. Ibid., reason 8. For elaboration, see Ruth Bader Ginsburg, "Women as Full Members of the Club: An Evolving American Idea," *Human Rights* 6 (1977); 7; Brown, Emerson, Falk, and Freedman, "Equal Rights Amendment," p. 884: "The adoption of a constitutional Amendment will also have effects that go far beyond the legal system. The demand for equality of rights before the law is only a part of a broader claim by women for the elimination of rigid sex role determinism."

16. *Raleigh Times,* December 29, 1984.

Chapter 6. "We Don't Want To Be Men!": Women against the Women's Movement

1. Lois J. Watkins to Representative Margaret Kessee, February 24, 1973, copy, authors' files. The following discussion is based on interviews with opponents of the Equal Rights Amendment, all of the available broadsides and information sheets disseminated by STOP ERA and other anti-ERA groups in North Carolina (in authors' files or in the papers of all persons mentioned in this note), approximately thirty-six inches of correspondence to Senator Samuel J. Ervin in SHC; and several feet of letters and published documents in the ERA Collection, NCDAH. These papers include those of Representative J. Allen Adams, NCDAH; Representative A. Hartwell Campbell, NCDAH; Representative David H. Diamont, NCDAH; Senator John R. Gamble, NCDAH; Representative Peter W. Hairston, NCDAH; Senator Cecil Hill, NCDAH; Representative Edward S. Holmes, NCDAH; Representative Robert B. Jones, NCDAH; Representative Margaret Kessee, copies in authors' files; Representative Larry E. Leonard, NCDAH; Representative Ernest B. Messer, NCDAH; Representative George W. Miller, in private possession; Representative Robie Nash, NCDAH; Representative Ned R. Smith, NCDAH; Representative Margaret Tennille, NCDAH; Senator Willis P. Whichard, SHC.

2. Charles Taylor, "Interpretation and the Sciences of Man," in *Interpretative Social Science: A Reader,* ed. Paul Rabinow and Williams Sullivan (Berkeley: University of California Press, 1979), pp. 25–71, esp. pp. 48–65. Harlan Gradin told us that we should read this: he was right.

3. Clifford Geertz, "Ideology as a Cultural System," in Geertz, *The Interpretation of Culture: Selected Essays* (New York: Basic Books, 1973), pp. 193–233.

4. *Phyllis Schlafly Report* 5, no. 7 (February 1972); 6, no. 7 (February 1973), sec. 2; 6, no. 12 (July 1973); 7, no. 10 (May 1974), sec. 10; 8, no. 5 (December 1974), 6, no. 12 (July 1975), sec. 2.

5. Geertz. "Ideology as a Cultural System," p. 220.

6. See letters in Claude Curry Papers, SHC. A few letters paralleled that of a woman who "wondered if the Virgin Mary had encountered her slightly awkward situation in the times of today whether even Jesus Christ would have survived to be born." Mrs. Shelby Tyler to Curry, April 4, 1971, Curry Papers. Jan Anderson, an

anti-ERA activist in Raleigh, believed that abortion "is based on one word—selfishness." She thought of her position on ERA and abortion as highly consistent in the mid-1970s. But she did think that there ought to be provision for therapeutic abortions.

7. Note on petition in George Miller Papers and copy of same petition in Whichard Papers, for the 1979 session of the North Carolina General Assembly. This connection was not an innovation of the 1979 session; it was assumed by opponents in 1977 as well. One among many letters was that of Everett C. Merrill to George Miller, February 7, 1977, in which the correspondent listed eight reasons to be opposed to the ERA. Reason 4 was that homemakers would lose out on social security; reason 6 was that the state would begin establishing day care centers for children; reason 5 was: "Abortion (ERA's answer to household duties)." This short sentence is freighted with a broad range of meaning concerning the gender revolution.

8. Steiner, *Constitutional Inequality,* pp. 98–103, esp. p. 102. Kristen Luker, *Abortion and the Politics of Motherhood* (Berkeley, Calif.: University of California Press, 1984), p. 205, points out how prolife activists in California understood abortion as a threat to family life and the values thought to derive from it. Many women were suspicious of ERA for the implied support it gave to abortion. By the mid-1970s anti-ERA people in North Carolina were agreed on the connection—with the possible exception of pro-ERA antichoice Roman Catholics.

9. Beth Hartshook, speech to Northeast Junior High School, Greensboro, n.d., copy, authors' files.

10. Mrs. J. V. Braswell, Grifton, to Ervin, September 15, 1970, Ervin Papers.

11. Mansbridge, *Why we Lost ERA,* pp. 98–100; Ehrenreich, "Defeating ERA."

12. Mrs. Alonzo Mayfield to Ervin, March [n.d.], 1972, Ervin Papers.

13. Constituent [name in possession of authors] to Whichard, February 1977, Whichard Papers.

14. Note accompanying anti-ERA petition from Mr. and Mrs. J. B. Matthews to Whichard, March 1, 1975, Whichard Papers.

15. Edward A. Tiryakian to Whichard, February 9, 1977, Whichard Papers.

16. The questions asked of interview texts, letters, and printed anti-ERA literature were informed by reading Mary Douglas, *Purity and Danger: An Analysis of Concepts of Pollution and Taboo* (London: Routledge and Kegan Paul, 1966).

17. Annie Pearl Puckett to Ervin, February 10, 1977, Ervin Papers.

18. Alta M. Clements to Whichard, February 16, 1977, Whichard Papers; Mrs. Evelyn M. Butchett to Ervin, October 3, 1970; Mrs. Kenneth O. Eddins to Ervin, September 6, 1970, both Ervin Papers.

19. Ehrenreich, "Defeating ERA."

20. Violet S. Devieux to Ervin, March 23, 1972, Ervin Papers.

21. Undated, unsigned noted to Whichard, clipping files, Whichard Papers.

22. Phyllis Schlafly, speech at Dorton Arena, Raleigh, February 13, 1977, tape, authors' files.

23. The decline among mainline churches has been noted since the 1960s. See Dean Kelly, *Why Conservative Churches Are Growing* (New York: Harper and Row, 1972).

24. Peter Skerry, "Christian Schools versus the I.R.S.," *Public Interest,* no. 61 (Fall 1980): 18–41.

25. Richard Hofstadter, "Pseudo-Conservatism Revisited—1965," in Hofstadter, *The Paranoid Style in American Politics and Other Essays* (New York: Alfred Knopf, 1965), esp. pp. 67–88.

26. Abner Cohen, *Two Dimensional Man: A Essay on the Anthropology of Power and Symbolism in Complex Societies* (Berkeley, Calif.: University of California Press, 1976), pp. 4, 55, 56–58.

27. See Arrington and Kyle, "Equal Rights Activists in North Carolina"; Brady and Tedin, "Ladies in Pink." The discussion of fundamentalism is based on Peter Berger, *The Sacred Canopy: Elements of a Sociological Theory of Religion* (Garden City, N.Y.: Doubleday, 1967); James M. Buckley, *The Fundamentals and Their Contrasts* (New York: Methodist Publishing House, 1906); Walter Edmund Ellis, "Social and Religious Factors in the Fundamentalist-Modernist Schisms among Baptists in North America, 1895–1934" (Ph.D. diss., University of Pittsburgh, 1974); Hugh Eugene Mattson, "The Fundamentalist Mind: An Intellectual History of Religious Fundamentalism in the United States" (Ph.D. diss., University of Kansas, 1971); George Marsden, *Fundamentalism and American Culture, 1880–1920* (New York: Oxford University Press, 1979); Alan Peshkin, *God's Choice: The Total World of a Fundamentalist Christian School* (Chicago: University of Chicago Press, 1986); Gary Schwartz, *Sect Ideologies and Social Status* (Chicago: University of Chicago Press, 1970); Sandra Sizer, *Gospel Hymns and Social Religion* (Philadelphia: Temple University Press, 1978), esp. pp. 138–47; Grant Wacker, "Fundamentalism as Ideology," December 1980, unpublished paper, authors' files; other materials in authors' files, both published and unpublished.

28. Douglas, *Purity and Danger,* p. 53; but see also pp. 53–57.

29. Frances Scoledge, "Vote No," *Amesbury* (Massachusetts) *News* September 15, 1976, emphasis added. See also "Our Readers Write," *Melrose* (Massachusetts) *Evening News,* August 2, 1976; "Gripes, Groans and Cheers," *Melrose Evening News,* August 2, 1976; *Amherst* (Massachusetts) *Record,* September 15, 1976; *Lawrence* (Massachusetts) *Eagle-Tribune,* September 14, 1976; Boston *Pilot,* July 9, 1976; Lynn *Sunday Post,* October 17, 1976.

30. Mrs. Carl A. Embler to Senator Cecil Hill, February 17, 1977, Hill Papers, ERA Collection.

Chapter 7. Men Besieged "On Account of Sex": Bargaining in the General Assembly

1. Old-line Democrats could never really reconcile themselves to what the national party had become. On February 25, 1977, a former state chair, James R. Sugg of New Bern, wrote legislators bewailing the U.S. Senate's passage of ERA and suggesting a state referendum on the issue lest the Democratic party be damaged. He chose as his significant constituency an anti-ERA group working with and through the North Carolina Baptist Fellowship, whose spokesman was E. T. Iseley. Iseley had warned Sugg that the Democratic party was in danger of betraying Democrats to the same people who had been responsible for harassing schools and employers on affirmative action. E. T. Iseley to James R. Sugg, February 24, 1977, copy; Sugg to Whichard, February 25, 1977 (among others), both Whichard Papers. The mood of Democratic conservatives can be inferred from the fact that George Wallace addressed a fund raiser for the North Carolina party on February 16, 1974, at Dorton Arena in Raleigh, although liberals opposed the invitation as well as resolutions praising Senator Ervin. At that time the Executive Committee of the state party also backed off from the quota system to recruit blacks and women, blaming "quotas" for the defeat of 1972.

2. Ferrel Guillory, interview with Jack Bass and Walter DeVries, December 11,

1973, Southern Oral History Project Collection, University of North Carolina at Chapel Hill.

3. *Durham Morning Herald,* January 31, February 4, 25, 1984; *Durham Sun,* February 29, 1984; Raleigh *News and Observer,* February 29, March 8, 1984. For a discussion of the modernization of the North Carolina legislature, see Herbert Wiltsee, "State Legislative Modernization: A Perspective," *Popular Government* 40 (Spring 1975): 1–4, 17; the entire issue is on the General Assembly, which was going through an important metamorphosis. Important sources for helping us understand the North Carolina General Assembly were Milton S. Heath, Jr., "Fifty Years of the General Assembly," *Popular Government* 46 (Winter 1981): 20–26; Michael Crowell and Milton S. Heath, Jr., "The 1973 General Assembly," *Popular Government* 39 (May 1973): 2–11, 58; Michael Crowell, "The 1975 Legislature—Introduction," in *North Carolina Legislation, 1975: A Summary of Legislation in the 1975 General Assembly of Interest to North Carolina Officials,* ed. Joan G. Brannon (Chapel Hill: Institute of Government, 1975), pp. 1–20, 39; James C. Dennard, "Overview of the 1977 General Assembly," in *North Carolina Legislation, 1977: A Summary of Legislation in the 1977 General Assembly of Interest to North Carolina Officials,* ed. Joan G. Brannon (Chapel Hill: Institute of Government, 1977), pp. 1–23; Robert L. Farb, "An Overview of the 1979 General Assembly," *North Carolina Legislation, 1979: A Summary of Legislation in the 1979 General Assembly of Interest to North Carolina Officials,* ed. Joan G. Brannon and Ann L. Sawyer (Chapel Hill: Institute of Government, 1979), pp. 1–10; Robert L. Farb, "Overview of the 1981 General Assembly," in *North Carolina Legislation, 1981: A Summary of Legislation in the 1981 General Assembly of Interest to North Carolina Officials,* ed. Ann L. Sawyer (Chapel Hill: Institute of Government, 1981), pp. 1–14. See also the informative and insightful lecture by Willis P. Whichard, "The Legislature and the Legislator in North Carolina," Whichard Papers; Thad L. Beyle and Merle Black, eds., *Politics and Policy in North Carolina* (New York: MSS Information Corporation, 1975); interviews of legislators and other political observers.

4. Ned Cline, in *Greensboro Daily News,* January 30, 1977. See also Bill Gilkeson, in *Durham Morning Herald,* April 24, May 27, 1977. Of the 170-member General Assembly, 161 were men in 1973, 165 in 1975, 161 in 1977, 152 in 1979, and 148 in 1981–82; 120 were Democrats in 1973, 160 in 1975, 159 in 1977, 150 in 1979, and 134 in 1981–82, 167 were whites in 1973, 164 in 1975, 165 in 1977, 166 in 1979, and 166 in 1981–82.

5. Guillory interview, December 11, 1973, Southern Oral History Project Collection.

6. Stephen N. Dennis, "Recent Changes in the Appropriations Process," *Popular Government* 40 (Spring 1975): 11–17; Michael Crowell, "1975 Legislature—Introduction."

7. Responses and copies of responses to the questionnaire are in Petersen Papers. The numbers were gleaned from tallying responses and votes as the latter were anticipated in press releases by NCWPC, also in Petersen Papers, and report of the vote in the North Carolina Senate, February 28, 1973.

8. The following tabulations were developed this way: Using official documents such as *The North Carolina Manual* (issued for each legislature), maps outlining Standard Metropolitan Statistical Areas, Counties and Selected Places, maps of House and Senatorial Districts, and reports of how legislators voted on the Equal Rights Amendment, we developed short biographies of each legislator and placed them in their appropriate constituency together with their position on ratification.

9. Ratificationists lost five votes and gained ten, one from conversion to ratifica-

tion and nine from the election of 1976. Of the five votes they lost, three came from conversion and two from electoral losses for reasons other than support of ratification. The final vote was 61–55.

10. Drum to Howard Twiggs, Gladys Tillett, Julia Miller, Carolyn Williams, and Gerry Hancock, November 12, 1974, Drum Papers.

11. Members of the North Carolina General Assembly and lobbyists on both sides of the ratification issue helped us understand lobbying. So did articles by Rob Christensen in Raleigh *News and Observer,* May 26, August 23, 25, and other unsigned articles in the same paper, August 27, 1985. See also Guillory interview, December 11, 1973, Southern Oral History Project Collection.

12. Rob Christensen, in Raleigh *News and Observer,* March 29, 1985. Opponents argued that if the study found that women had been treated inequitably, it would "provide ammunition for a lawsuit against the state on the question of pay discrimination." They concluded that it would cost the state less if no one knew whether or not there had been a lack of equity in dealing with women employees.

13. Richard Hatch, producer, "The General Assembly Today," WUNC-TV, March 4, 1977, videotape, ERA Collection. Pro-ERA activists were often frightened by the violence they inferred from the language and threats of opponents.

14. In 1970 Ervin argued in the U.S. Senate that ERA was not needed because the Supreme Court was correct in its position in *Hoyt* v. *Florida,* a case upon which proponents had based an important part of their argument. Seven years later, before North Carolina legislators, Ervin argued that ERA was "not needed" because the Supreme Court had, in *Taylor* v. *Florida,* reversed its position in *Hoyt* v. *Florida.* In other words, Ervin argued that ERA was not needed because the Court had overruled a decision which he had once argued had not been discriminatory in the first place.

15. Frederick Brooks, statement, Constitutional Amendments Committee Public Hearing, February 13, 1979, copy, authors' files. See also minutes of the hearing, General Assembly of North Carolina Papers.

16. Material on James Baxter Hunt is gleaned from interviews of persons close to him and articles in *Charlotte Observer,* February 7, 1982, August 13, 1983, December 1, 1985; *Greensboro News and Record,* October 19, 1980, December 28, 1984; Raleigh *News and Observer,* August 22, October 17, 1976, September 28, 1978, October 7, December 28, 1984.

17. The sketch of James Collins Green is based on interviews and Ned Cline, in *Charlotte Observer,* April 3, 1977, February 24, 1983; *Greensboro Daily News,* March 23, 1975, December 12, 1976, July 31, 1977; *Greensboro News and Record,* July 1, 1984; Raleigh *News and Observer,* February 9, 1975, April 8, 1977; Bruce Siceloff, in Raleigh *News and Observer,* July 21, 1978, March 30, 1980; Winston-Salem *Journal,* April 3, 1983.

18. This comment by Jack Rhyne was on his reply to Elisabeth Petersen's polling candidates for state legislative office in 1972. Copy, Petersen Papers.

19. *Durham Morning Herald,* January 17, 1979.

20. *Charlotte Observer,* May 29, 1977. The article was on the "sexism" of the General Assembly.

21. John Ed Davenport to Ervin, Jr., September 19, 1972, Ervin Papers.

22. See Report of the Task Force on the Effect of Ratification of the Equal Rights Amendment on the Law of Virginia to the Virginia House of Delegates, Messer Papers, ERA Collection. Many state legislators' papers had copies of this report, which Davenport had circulated as an anti-ERA statement.

23. Louis Harris 1979 Survey on the Equal Rights Amendment, no. 794022; Louis

Harris 1982 ERA Survey in North Carolina, no. 822029, both Data Library, Institute for Research in Social Science, University of North Carolina at Chapel Hill.

24. *Washington Post,* February 12, 1979.

25. *Charlotte Observer,* January 15, 1979.

26. Peter W. Hairston to Mary Louise Wilkerson, March 18, 1975, Hairston Papers, ERA Collection.

27. See Bernard Schwartz, *Swann's Way: The School Busing Case and the Supreme Court* (New York: Oxford University Press, 1986); Mark Schafer, "The Desegregation of a Public University System" (Ph.D. diss., University of North Carolina at Chapel Hill, 1980).

28. Judson D. De Ramus, Jr., memorandum (statement prepared for delivery, April 16, 1975, but sent to colleagues in the 1977 session), Nash Papers, ERA Collection.

29. Katherine Sebo to ERA Supporters, June 26, 1975, authors' files; Legislative Research Commission, *Report to the 1977 General Assembly of North Carolina: Sex Discrimination.* (Raleigh:

30. The discussion of the domestic violence bill is based on interviews with Miriam Dorsey, Mary Susan Parnell, and I. Beverly Lake, Jr., as well as Minutes of the Judiciary II Committee of the North Carolina Senate, February 8, 13, 15, March 8, 1979, General Assembly of North Carolina Papers; "Criminal Law and Procedure: Domestic Violence," in *North Carolina Legislation, 1979,* pp. 61–62.

31. Raleigh *News and Observer,* July 6, 1985.

Chapter 8. ERA and the Politics of Gender

1. A "debate" is an exchange of ideas and analyses on agreed upon issues in which differences are conceded and explained. There is an implied acknowledgment of issues raised by the opposition in that they are taken into account and specifically refuted. A "litany" is merely a statement-and-response pattern in which the issues are not mutually conceded and in which the analysis of each side is ignored by the other in favor of repeating statements that symbolize but do not explain a position.

2. Joan Wallach Scott, "Deconstructing Equality-Versus-Difference; or, The Uses of Poststructuralist Theory for Feminism," *Feminist Studies* 14 (Spring 1988): 33–50; Michael Walzer, *Spheres of Justice: A Defense of Pluralism and Equality* (New York: Basic Books, 1983), pp. 202–3, cited by Scott, p. 50.

3. Jane J. Mansbridge, "Who's in Charge Here? Decision by Accretion and Gatekeeping in the Struggle for ERA," *Politics and Society* 13 (1984): 343–82, esp. pp. 344, 349, 375–76; Brown, Emerson, Falk, and Freedman, "Equal Rights Amendment," pp. 871–985, esp. pp. 967–78. See also Mansbridge, *Why We Lost the ERA,* pp. 67–89.

4. Brown, Emerson, Falk, and Freedman, "Equal Rights Amendment," pp. 872, 884, 885, 890–92, 905–7, 939.

5. Berry, *Why ERA Failed,* pp. 86–120, esp. 120; Mansbridge, *Why We Lost the ERA,* pp. 216–18.

6. In his *Paranoid Style in American Politics,* Richard Hofstadter discussed a style that was apocalyptic and metaphorically paranoid in that it imagined a threat where none existed—especially the threat as imagined by those who defined it. Hofstadter's insights are appropriate here as well; but since there are clinical connotations to the word "paranoid," we have preferred the word "apocalyptic," in part because it flows from the religious perceptions of many opponents of the Equal Rights Amendment

and feminism and is not as pejorative as the more popular term. In addition, the apocalyptic style is, in the minds of those to whom it comes easily, justified by the moral issues thought to be at stake. A literal statement of issues or of positions on issues may not contain the necessary moral message, thus encouraging exaggeration and distortion so that the real moral character of the issue is understood. The style has been perfected in North Carolina by the pseudoconservative National Congressional Club.

7. The style reflects one of the major themes of Western culture, best exemplified by the Revelation of Saint John in the Christian Bible. There a slave writes to his fellow Christians in highly metaphorical language of his vision of the future. Mingling themes of cosmic judgment and expectation as if things hoped for were things actually seen, the writer reports the future as if it were history. He discusses moral judgment as if it were action and makes observations about "meaning" as if they were statements of fact. Thus the judgment that he sees and the symbols of evil he describes are but representations of a reality that lies behind what we would call history; things are not what they seem to be. To take an example: The powerful may rule now but their rule is illusory; God is really acting behind the scenes to vindicate us. The Evil One may appear to be winning, but he is in "fact" about to be cast down!

8. Hendrik Hertzog, "The Constitution of Aspiration and 'The Rights That Belong to Us All,' " *Journal of American History* 74 (December 1987): 1013–34, esp. p. 1017.

9. Martha Minow, "We, the Family: Constitutional Rights and American Families," *Journal of American History* 74 (December 1987): 959–983; Brown, Emerson, Falk, and Freedman, "Equal Rights Amendment," p. 890.

10. That equality is a "tough" concept, in Barbara Jordan's phrase, is underscored by Joan Hoff, whose critique of liberal legalism incorporates arguments similar to those implied in the statements of anti-ERA women. See Hoff, *Too Little—Too Late: Changing Legal Status of U.S. Women from the American Revolution to the Present* (Bloomington, Ind.: Indiana University Press, forthcoming), chap. 1. Jordan's comment, quoted by Hoff, was made in an address to the First Ladies' Conference on the Constitution, Atlanta, February 11–12, 1988.

11. Carol Gilligan, *In a Different Voice: Psychological Theory and Women's Development* (Cambridge, Mass.: Harvard University Press, 1982).

12. Feminist critics of liberalism who emphasize affiliative values of the private sphere are often referred to as maternal feminists. For various maternalist views, see Jean Bethke Elshtain, "Antigone's Daughters," *Democracy* 2 (April 1982): 46–59; Elshtain, "Feminism, Family and Community," *Dissent* 29 (Fall 1982): 442–49; Elshtain, "Feminist Discourse and Its Discontents: Lanugage, Power, and Meaning," *Signs* 7 (Spring 1982): 603–21; Sara Ruddick, "Maternal Thinking," *Feminist Studies* 6 (Spring 1980): 342–67.

13. The critique of liberal theory is extensive. See, for example, Lorrenne M. G. Clarke, "Women and John Locke: Who Owns the Apples in the Garden of Eden?" in *The Sexism of Social and Political Theory: Women and Reproduction from Plato to Nietzsche,* ed. Lorrenne M. G. Clark and Lynda Lange (Toronto: University of Toronto Press, 1979); Zilliah Eisenstein, *The Radical Future of Liberal Feminism* (New York: Longman, 1981); Jean Bethke Elshtain, *Public Man, Private Woman: Women in Social and Political Thought* (Princeton, N.J.: Princeton University Press, 1981); Susan Moller Okin, *Women in Western Political Thought* (Princeton, N.J.: Princeton University Press, 1979); Carole Pateman, *Sexual Contract* (Stanford, Calif.: Stanford University Press, 1988), among others.

14. Suffragists' language reflected their being rooted in the separate female culture of the late nineteenth century. The threat of gender change loomed larger in the second woman's movement because of the explicit critique of gender roles.

15. See, for example, Elizabeth Fox-Genovese, "Women's Rights, Affirmative Action and the Myth of Individualism," *George Washington Law Review* 54 (January–March 1986): 336–74. Some feminist scholars focus on the need for a feminist jurisprudence. See, for example, Catherine MacKinnon, "Feminism, Marxism, Method, and the State: Towards a Feminist Jurisprudence," *Indiana Law Journal* 56 (1980–81): 375–444. Ratificationists had only the public language of rights available to them in their fight for placing women in the law, thus demonstrating to some the inadequacy of patriarchal language in dealing with women.

16. Linda K. Kerber, "Separate Spheres, Female Worlds, Woman's Place: The Rhetoric of Women's History," *Journal of American History* 75 (June 1988): 9–39; Karen Offen, "Defining Feminism: A Comparative Historical Approach," *Signs* 14 (Autumn 1988): 119–57.

17. Mansbridge, *Why We Lost the ERA,* pp. 149–50, 18486.

18. Ibid., pp. 68, 71, 75–77, 79, 80–85, 88, 261, 274.

19. This is not to identify "imaginary" and "symbolic" as having nothing to do with the way in which people actually live. Symbols are real and lively things in that they direct, shape, and encapsulate meaning; they are the creation of the human imagination as it wrestles collectively with history and future, and to that extent are "imaginary." But the connotation that this agonistic creation is somehow not "real" has to be resisted in order to understand collective action and human creativity.

20. The fact that Protestant fundamentalists, Pentecostals, and conservative Roman Catholics could side so easily against the feminist assault on gender does not mean that fundamentalist culture is embraced by all antifeminists. There is a difference between "fundamentalist culture" and "cultural fundamentalism." The former refers to the culture developed over the past century structured on religious fundamentalism—the core of thought rests on a radical theism that concretizes God. The latter refers to an approach to culture that denies its relativity, claiming for it instead an absolute quality inconsistent with the implications of cultural anthropology—the core of thought rests on this radical absolutism that concretizes culture.

21. This idea is based on Shelby Steele, "I'm Black, You're White, Who's Innocent: Race and Power in the Age of Blame," *Harpers* 276 (June 1988): 45–53.

Appendix 1

North Carolina

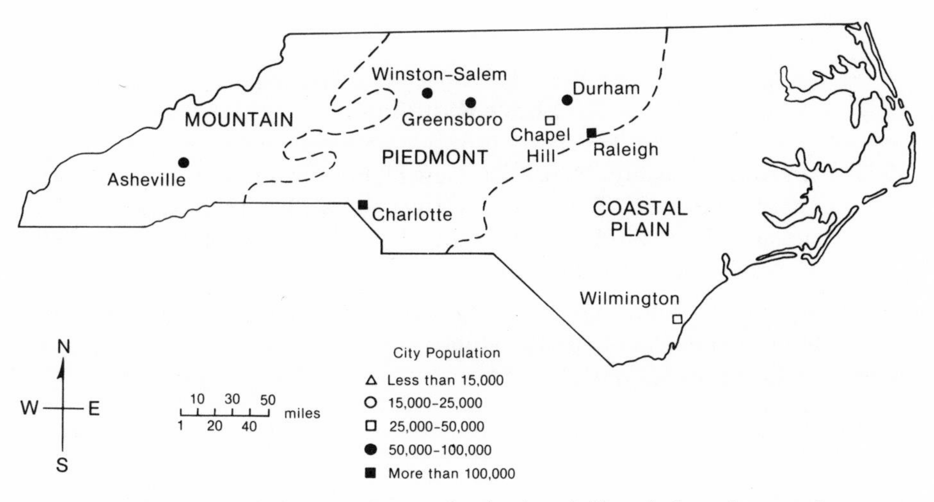

The North Carolina piedmont, more urbanized and liberal than the state's western mountains or agricultural coastal plain, was the major source of support for the Equal Rights Amendment.

Appendix 2

State Actions on the Equal Rights Amendment, 1972–1982

States Ratifying

1972: Hawaii, Delaware, Nebraska, New Hampshire, Idaho, Iowa, Kansas, Texas, Maryland, Tennessee, Alaska, Rhode Island, New Jersey, Wisconsin, West Virginia, Colorado, New York, Michigan, Kentucky, Massachusetts, Pennsylvania, California

1973: Wyoming, South Dakota, Minnesota, Oregon, New Mexico, Vermont, Connecticut, Washington

1974: Main, Montana, Ohio

1975: North Dakota

1976:

1977: Indiana

1978:

1979:

1980:

1981:

1982:

Source: Janet K. Boles, *The Politics of the Equal Rights Amendment* (New York: Longman, 1979), pp. 2–3.

States Rescinding Ratification

1973: Nebraska

1974: Tennessee

1977: Idaho

1978: Kentucky (rescission bill vetoed by acting governor)

States Not Ratifying

Alabama, Arizona, Arkansas, Florida, Georgia, Illinois, Louisiana, Mississippi, Missouri, Nevada, North Carolina, Oklahoma, South Carolina, Utah, Virginia

Appendix 3

Votes on Bills to Ratify the Equal Rights Amendment in the North Carolina General Assembly, 1973–1982

1973: Governor: James H. Holshouser (Republican)
Lt. Governor: James Baxter Hunt (Democrat)
Senate Vote on S 8, February 28
For: 23 (19 Democrats, 4 Republicans)
Against: 27 (16 Democrats, 11 Republicans)

1975: Governor: James H. Holshouser (Republican)
Lt. Governor: James Baxter Hunt (Democrat)
House Vote on H 15, April 16
For: 57 (54 Democrats, 3 Republicans)
Against: 62 (56 Democrats, 6 Republicans)

1977: Governor: James Baxter Hunt (Democrat)
Lt. Governor: James Collins Green (Democrat)
House Vote on H 43, February 9
For: 61 (58 Democrats, 3 Republicans)
Against: 55 (52 Democrats, 3 Republicans)
Abstentions: 3 Democrats
Senate Vote on H 43, March 1
For: 24 (23 Democrats, 1 Republican)
Against: 26 (23 Democrats, 3 Republicans)

1979: Governor: James Baxter Hunt (Democrat)
Lt. Governor: James Collins Green (Democrat)
S 234 was killed in the Senate Constitutional Amendments Committee

on February 15 by a pro-ERA majority because ratificationists lacked the requisite votes for Senate passage.

1981: Governor: James Baxter Hunt (Democrat)
Lt. Governor: James Collins Green (Democrat)
S 173 was allowed to die in the Senate Constitutional Amendments Committee on February 27 by a "gentleman's agreement" in which opponents agreed not to defeat the bill on the floor of the Senate in exchange for supporters' assurances that a ratification bill be neither revived nor voted upon in the remainder of the session.

1982: Governor: James Baxter Hunt (Democrat)
Lt. Governor: James Collins Green (Democrat)
Senate Vote to table S 803, June 4
For: 27 (17 Democrats, 10 Republicans)
Against: 23 (23 Democrats)

Appendix 4

Persons Interviewed

Jan Anderson	January 18, 1977
Sylvia Arnold	July 19, 1977
Mercedith Bacon	August 26, 1979
Anne Bagnal	September 25, 1979
Cammy Bain	June 13, 1979
Beverly Batson	April 15, 1977, June 14, 1979
Birch Bayh	June 19, 1986 (telephone)
Marilyn Bissell	October 9, 1979
Robert Blackburn	September 14, 1977
Maria Bliss	June 30, 1977, July 17, 1979
Teela Bennett	December 26, 1979
Louise Brennan	October 23, 1979
Elizabeth Farrior Buford	December 20, 1986
A. Hartwell Campbell	June 24, 1986
Jane Campbell	April 28, 1987
Wanda Canada	July 7, 1977, August 30, 1979
Mary Cantwell	April 12, 1977, July 13, 1979
Daniel Carr	August 6, 1979, November 17, 1980
Peggy Carter	July 7, 1977
Lela Chesson	March 16, 1977
Ruth Cook	September 9, 1986
Thelma Crawford	June 28, 1979
Wilma Davidson	June 23, 1977

Nancy Dawson-Sauser (formerly Drum)	November 26, 1976, June 20, 1977
Marilyn DeVries	March 10, 1977
Terry Donahue	March 16, 1977
Miriam Dorsey	August 8, 1979
Bobette Eckland	August 13, 1979
Peg Edwards	September 9, 1979
Ruth English	June 28, 1979
Pamela Fender	October 5, 1982
Agnes Fisher	June 25, 1979
Jane Fonvielle	April 14, 1977
Karen Gaddy	March 16, 1977
Alice Wynn Gatsis	March 15, 1977, August 8, 1978, April 28, 1987 (telephone)
Kaye Gattis	October 5, 1982
Florence Glasser	May 6, 1973, July 9, 17, 1977, July 20, 1982
Karen Gottovi	June 13, 1979
Lynne Gottlieb	August 25, 1979
Marse Grant	September 7, 1977
Rachel Gray	August 25, 1979
Linda Grimsley	March 26, 1977
Carolyn Hackney	September 12, 1979
Bett Hargrave	August 19, 1979
Renee Hartman	April 16, 1977
Esmerelda Hawkins	March 15, 1977
John T. Henley	June 23, 1986
Karolyn Kay Hervey	August 8, 16, 1979
Sherry Hicks	July 13, 1977
Carolyn Highsmith	September 3, 12, 1979
Mary Hopper	October 24, 1979
Patricia S. Hunt	August 1, 1979
Sonia Johnson	May 1980
Margaret Keesee	September 2, 1979
Kent Kelly	August 7, 1979
Collins Kilburn	September 7, 1979
Nancy Klein	April 19, 1978
Betty Ann Knudsen	August 22, 25, 1979

I. Beverly Lake, Jr.	November 7, 1979
Lora Lavin	April 12, 1977
Patricia Locke	June 25, 1977
Martha Lowrance	August 25, 1979, October 5, 1982
Beth McAllister	August 7, 1979, September 5, 28, 1982
Betty McCain	September 16, 1982
Martha McKay	May 6, 1973, March 29, 1974
Joan McLean	April 27, 1987
Ann Mackie	July 19, 1977
Helen Mahlum	October 20, 1979
Anneliese Markus-Kennedy	August 25, 1979
Barbara Richie Marsh	August 21, 1979
Helen Rhyne Marvin	October 25, 1979
Alice Elaine Mathews	July 30, 1982
Bobbie Matthews	March 15, 1977
Ruth Mary Meyer	September 14, 1979
George W. Miller	September 30, 1979
Marlyn Miller	October 5, 1982
Lamar Mooneyham	November 3, 1980
Catherine Moore	April 15, 1977, June 14, 1979
Jane Morse	June 26, 1977
Riky Mottershead	March 15, 1977
Marsha Newton-Graham	October 25, 1979
Veronica Nicholas	June 26, 1979
Jo Ann Norris	October 19, 1982
J. Mallory Owen	September 14, 1977
Jan Parker	October 5, 1982
Mary Susan Parnell	July 23, 1979
Jane Patterson	March 18, 1977
Gene Payne	February 21, 1978
Doris Pearce	April 15, 1977
Susan Pfeiffer	June 25, 1977
Mary Pegg	October 13, 1979
Elisabeth Petersen	May 6, 1973, February 19, 26, 1977 July 26, 1977
Charles V. Petty	September 14, 1977

Jane Porter	July 13, 1977
Janice Ramquist	September 12, 1979
Marshall Rauch	October 25, 1979
Nancy Roberts	August 25, 30, 1979
Tibbie Roberts	April 2, 1977, October 20, 1979
Grace Rohrer	September 30, 1979
Benjamin D. Schwartz	June 14, 1979
Jessie Rae Scott	August 28, 1979
Kay Sebian	June 12, 1979
Miriam Slifkin	June 8, 1977, March 28, 1979
Dorothy Slade	May 16, 1977, October 5, 1976 (telephone)
Eleanor Smeal	April 1, 1987
Katherine Stark	July 18, 1977
Jean Stevens	June 25, 1979
Carl Stewart	December 8, 1979
Toni Suiter	March 15, 1977, August 8, 1978
Lib Sykes	June 24, 1977
Margaret Tennille	October 9, 1979
Daisy Thorpe	March 15, 1977
Gladys Tillett	June 26, 1977
Willis P. Whichard	September 17, 1979
Carolyn Williams	June 27, 1977
Marilyn Williams	October 25, 1979
Charles Winberry	September 24, 1979
Betty Wiser	January 18, 1977
Liz Wolfe	April 30, 1977

Index

5.00